A SENSE OF THE AMERICAN WEST

Historians of the Frontier and American West Series
Richard W. Etulain, Series Editor

A SENSE OF THE AMERICAN WEST
An Anthology of Environmental History

edited by
JAMES E. SHEROW

Published in cooperation with the University of New Mexico Center for the American West

University of New Mexico Press
Albuquerque

Prologue and Introduction © 1998 by the University of New Mexico Press
All rights reserved.
First edition

Library of Congress Cataloging-in-Publication Data
A sense of the American West : an anthology of environmental history /
   edited by James E. Sherow. — 1st ed.
      p.   cm. — (Historians of the frontier and American West)
   "Published in cooperation with the University of New Mexico Center for the
American West."
   Includes bibliographical references and index.
   ISBN 0-8263-1913-0. — ISBN 0-8263-1914-9 (pbk.)
   1. West (U.S.)—Environmental conditions—History.   I. Sherow,
James Earl.   II. Series.
GE155.W47S46   1998
333.7'0978—dc21   98-8387
                                                                                    CIP

CONTENTS

PROLOGUE viii

SECTION I

1. Introduction
   An Evening on Konza Prairie 3
2. Spirit of Place and the Value of Nature
   in the American West 31
   Dan Flores

SECTION II

3. Environmental Change in Colonial
   New Mexico 41
   Robert MacCameron
4. Bison Ecology and Bison Diplomacy
   The Southern Plains from 1800 to 1850 65
   Dan Flores
5. Workings of the Geodialectic
   High Plains Indians and their Horses in the Region of the
   Arkansas River Valley, 1800–1870 91
   James E. Sherow

SECTION III

6. Emerging Desert Landscape in Tucson 117
   E. Gregory McPherson and Renee A. Haip
7. Reclaiming the Arid West
   The Role of the Northern Pacific Railway in
   Irrigating Kennewick, Washington 129
   Dorothy Zeisler-Vralsted

8. Wildlife, Science, and the National Parks,
   1920–1940   147
   Thomas R. Dunlap

9. Manipulating Nature's Paradise
   National Park Management Under
   Stephen T. Mather, 1916–1929   161
   Richard West Sellars

SECTION IV

10. Battle for Wilderness
    Echo Park Dam and the Birth of the
    Modern Wilderness Movement   181
    Mark W. T. Harvey

11. "We Can Wait. We Should Wait."
    Eugene's Nuclear Power Controversy, 1968–1970   201
    Daniel Pope

12. The Community Value of Water
    Implications for the Rural Poor
    in the Southwest   221
    F. Lee Brown and Helen M. Ingram

13. Mexican American Women Grassroots
    Community Activists
    "Mothers Of East Los Angeles"   243
    Mary Pardo

14. The Drought of 1988,
    The Global Warming Experiment,
    and Its Challenge to Irrigation
    in the Old Dust Bowl Region   261
    John Opie

PERMISSIONS   291

INDEX   293

CONTRIBUTORS   309

A man said to the universe:
"Sir, I exist!"
"However," replied the universe,
"The fact has not created in me
A sense of obligation."
STEPHEN CRANE,
War is Kind (1899)

PROLOGUE

Naturally, I think environmental history, and its application to the American West, is an exciting and interesting field of study. This work is challenging to do, and its practitioners are some of the most interesting scholars around. The ranks of environmental historians have grown considerably in the last decade and a half, and even an anthology such as this one can only highlight the work of a few when many deserve recognition. Three of the best-known environmental historians of the West, William Cronon, Richard White, and Donald Worster, have been excluded from this anthology. Their ideas are already common intellectual currency and are widely published. Many other scholars have long been at work, and the inclusion of just a few of these suggests the greater dynamism of the field.

This anthology is divided into sections, and each reflects some of the variety in the environmental history of the West. The first section addresses the ways in which environmental history is done. The Introduction explores theories and practices about environmental history, and it also suggests a methodology. Dan Flores, in the following piece, challenges scholars and laypeople to rethink the way in which they relate to a place. In so doing, he asks us all to take the time to sense where we live, or where we write about, and to become native to that place.

The next three sections are organized chronologically. The second covers the period before Anglo-American culture dominated the region. Robert MacCameron illustrates how colonial Hispanics and the subsistence agriculture they practiced in New Mexico modified and shaped the environment. In the next two chapters, which roughly cover the years from 1780 to 1870, Dan Flores and I have purposefully viewed the activity of humans as simply one interrelated aspect of the entire High Plains environment. Flores analyzes the relationship of High Plains Indian peoples and bison, while I examine the connections between Indians and horses.

The third section deals with the crucial period of the late nineteenth and early twentieth centuries, when the conservation movement achieved its greatest strength. Gerald Nash observed years ago that the West was, and is, a society of urban oases. And Martin Melosi has pleaded the case for the study of cities as environments, even suggesting that the environmental history of cities is just as important, maybe more important, than any study of Redwood stands and the forester's ax. For the most part, Melosi's plea for more research in this area has gone largely unheeded. Chapter 6 of this book presents one exceptional study, by E. Gregory McPherson and Renee A. Haip, which charts how Anglo-American culture shaped the urban environment of Tucson. Reclamation was another important part of conservationism, and in chapter 7, Dorothy Zeisler-Vralsted reveals the important role railroad corporations played in developing viable irrigation in the state of Washington.

An important part of the conservation movement was the creation and preservation of national parks. In chapters 8 and 9, Thomas Dunlap and Richard Sellars relate how American science, markets, and social trends shaped the historical environments of national parks. Certain plants and animals thrived and others did not, and conscious, human activity caused these results.

The fourth section covers the post–World War II era, a time marked by the rise of environmentalism and encompassing issues of class, race, ethnicity, pollution, and natural resource use. In Chapter 10, Mark W. T. Harvey depicts the rise of contemporary environmentalism in the West as activists fought to preserve Dinosaur National Monument. The Daniel Pope essay reprinted here conveys how an urban, post–World War II middle class effectively stopped nuclear power development around Eugene, Oregon. Serious issues of "environmental racism" are related in F. Lee Brown and Helen M. Ingram's portrayal of Southwestern Hispanic and Indian peoples' fight to protect their communal water use and community values. The following essay, by Mary Pardo, analyzes the successful campaign of poor, Hispanic women in East Los Angeles to halt the despoilment of their neighborhoods.

Some historians are beginning to view humans as a part of, and not removed from, nature. In this holistic view, humans, guided by their cultural constructs, simply become another force at work in a given environment. In the final essay, John Opie uses recent holistic understandings of physics, and especially chaos theory, to analyze the problems farmers faced in pumping the Ogallala aquifer.

In assembling this work I have been blessed with the critical insights and support of many people. I have always enjoyed the love and acceptance of my two daughters, Brie and Evan, even when I have been terribly cranky as a result of this work. I extend my sincerest thanks and appreciation to Susan Flader, Dan Flores, John Opie, Chuck Rankin, Virginia Scharff, and Donald Worster for reading and commenting on my thoughts; and to Bob Irvine, Wendy Kyle, and Bob Rook, former graduate students here at Kansas State University, who waded through a mass of articles to help me select those that appear in this anthology. Special thanks go to Doug Weiner and Elliott West, whose criticisms of this work I have highly valued. I also thank Durwood Ball, editor at the University of New Mexico Press, and Richard Etulain, for their unwavering interest and support in seeing this anthology published. My heartfelt appreciation goes to my dear soulmate and friend, Bonnie Lynn-Sherow, whose keen criticism and insights always strengthen my thinking.

SECTION I

We are the offspring of history, and must establish our paths in this most diverse and interesting of conceivable universes— one indifferent to our suffering, and therefore offering us maximal freedom to thrive, or to fail, in our own chosen way.

STEPHEN JAY GOULD, WONDERFUL LIFE

The following Introduction focuses on some of the different notions historians have about writing both environmental and western history. It also argues for a particular way of doing environmental history. Although the essay stresses only one approach to the subject, other approaches may be found in the additional works referenced in the notes.

In my introductory essay, I make explicit my discomfort in turning environmental histories into morality plays based upon the science of ecology. Environmental historians often rely upon ecological theory in building their works. Some scholars have tended to purport the existence of a stable natural order, and to show in their writing how people have been the great disrupters of nature. But contemporary theories of physics, biology, and mathematical modeling reveal flux, diversity, and change as the only standards in biosystems. As Stephen Jay Gould contends, people devise their own destinies in the randomness of creation.

One historian growing in stature is Dan Flores, the John Hammond Professor of History at the University of Montana. He believes that an environment, region, or place are all best understood when first sensed. This view, of course, flies in the face of scholars who rely upon empirical evidence, or who proclaim their histories as some form of social science. The past, according to this school of thought, can be proven. The geographer Yi-Fu Tuan, however, has some very keen insights about place, or our environment, and how it shapes culture. "People go through their lives," he believes, "making only very limited demands on their perceptual powers."[1] In short, people do not fully use their senses to perceive their environments, or to discern how their environments shape them.

Professor Flores makes some compelling arguments for immersing oneself into a place before attempting to write about it. He is open to sensing his surroundings. In his piece, he discusses "going native" and writing the "natural history of a spirit of a place." This he did, in his acclaimed *Caprock Canyonlands* (1990), and again in the article reprinted here. Yi-Fu Tuan's monograph, *Topophilia*, has greatly influenced Flores's thinking about how a person acquires a spirit, or sense, of place. What does Flores mean by "spirit of place," and how is a person to realize this? Why, in his view, have individualism and capitalism kept Americans from developing a sense of place? What kind of "paradigm shift" does Flores advocate?

CHAPTER ONE

INTRODUCTION
An Evening on Konza Prairie

WHERE WESTERN AND ENVIRONMENTAL HISTORY MEET

With the approach of dusk, on September 1, 1993, my students and I sat on a high, creamy-white limestone ridge that was capped with thin prairie soils and tallgrasses in the Flint Hills of Kansas. Our vista created an illusion of a world surrounding us in a grand circle. The sky appeared like an inverted blue bowl, the edge of which touched an encircling horizon. The sun had slowly begun its descent to the west, nearly touching the line between earth and sky, and it lighted the northward-drifting dust in the air above in blazing streaks of yellows, oranges, purple, reds, and blues. The southern wind blew gently through tall stands of nearly mature golden Indian grass, and created in them a soft, crackling, sound, like thousands of tiny brushes whisking across snare drums. Tall bluestem swayed and waved its three-fingered seed spikes, as if in some choreographed dance. Nighthawks, with their underside, white-striped wings outstretched, glided lazily on the currents, then suddenly tucked their wings and dove earthward. They swept into their wide-open beaks insects rising to greet the cooling evening air. In the east, a bright, white, full moon began its ascent into a deep, impenetrable blue sky.

My students and I were there for more reasons than just to behold this spectacular setting. Together, we discussed the meanings of such words as *nature*, *environment*, and *place*. We took note of our surroundings and mused over what forces had shaped these varied environments through time. We identified quite readily one environment, the tallgrass prairie. But others began coming into view, too. North of the wide floodplain of the Kansas River, we

BISON AREA AT KONZA PRAIRIE NATURAL RESEARCH AREA, MANHATTAN, KANSAS. (PHOTOGRAPH © JAMES NEDRESKY, LAWRENCE, KANSAS.)

identified the heavily timbered, urban environment of Manhattan, Kansas. On the southern floodplain sprawled an agricultural environment in square-shaped fields of corn and soybeans. To the west, we could see the remnants of a ranching environment, with pastures of imported brome grass near a limestone bunkhouse and barn. We were sitting in a place now called the Konza Prairie Research Natural Area, an 8,600-acre reserve of tallgrass prairie owned by the Nature Conservancy and managed by the Division of Biology at Kansas State University. We identified our immediate surroundings as a scientific environment, where researchers, carefully controlling, counting, and measuring physical phenomena, study the fauna, flora, hydrology, and climate of this place. One feature, a paddock enclosing a herd of around 150 bison, particularly shaped this scientific environment. Public hiking trails cut through the area where we were sitting, helping to create an environment partially shaped by a middle-class, consumer culture that valued wildlife preservation. The tallgrass prairies themselves, these treeless expanses, are remnants formed in large part by the historical burning practices of such Indian peoples as the Kaws, who once inhabited this region.

Altogether, we had identified six historical environments still existing in the greater biome of the tallgrass prairies of the Flint Hills. We observed how each stood distinct, yet also blended into all the others. We imagined and debated the historical forces that had shaped the landscape before us. We considered the interacting influences of human culture, plant and animal communities, physical terrestrial and atmospheric forces, and solar and cosmic powers. We had begun exploration of the environmental history of the Flint Hills, a tiny microcosm of the West.

Quickly, questions arose about the ways in which environmental historians might approach the task of studying the tallgrass prairies, or any other place in the West. Most assuredly, scholars in both disciplines, the history of the American West and environmental history, greatly resemble temperamental chefs, each with his or her own special mix of ingredients. Environmental historians, until recently largely self-trained, have taken a variety of approaches toward their work. Some deal strictly with ideas about the environment or nature, others examine public policy and law, and still others concern themselves with such social movements as environmentalism. Some portray "non-human nature" as a major force and co-actor with human culture. They espouse the classic nature–history dichotomy: human culture working outside of nature, yet affecting, and being affected by, natural forces at work in the environment. Other historians of the environment advocate regional studies, and some emphasize a global approach. Some would include the study of cities as ecosystems, or environments, yet others believe urban systems should be excluded from their stories.²

There are at least as many different "recipes" for doing environmental history among western historians as there are among environmental historians. Next, take a look at the chuckwagons of western history and their cookies. Recently, the differences among western historians have reached a high level of publicity. Supposedly, the vying western cooks who are all hurling prairie pastries at each other are corralled into two opposing outfits. Both the academic and popular press have labeled the "cookoff" as a duel between the "new" and the "traditional" western historians.³

But the debate among western scholars is actually more fractured than this superficial division would indicate. "New" western historians cannot agree on a common approach among themselves, any more than can the traditionalists. Both schools often splinter into many smaller camps: those who see the West as a place marked by geographical similarities; those who separate the West into regions and ever smaller subregions. There are still those who find value in Turner's frontier thesis. Others prefer to stress ethnic and racial diversity. And several simply believe "the West" a fictitious place without any regional identity, but they do very well for themselves by writing about it anyway.⁴

True enough, environmental history shares something of the time-honored methods for studying the American West. The way in which the history of the West has lent itself to regional and frontier analyses probably explains why a large part of American environmental history has risen out of western stories. As Richard White pointed out several years ago, many of the origins of the field can be traced to Frederick Jackson Turner, Walter Prescott Webb, and, particularly, James Malin.⁵

According to Turner, the "frontier" Europeans transformed into a new people, "Americans." Turner, in part, used Herbert Spencer's interpretation of Charles Darwin's theory of evolution to reach this conclusion, and in so doing he argued the power of environment to shape culture. Europeans, by adapting to a new environment and through the forces of natural selection, had evolved into a new social critter.⁶

There is an environmental determinism in Turner's thesis, but it brings about its own subjugation. While European society was initially mastered by the wilderness, a "social evolution" takes place. The land is transformed, from a place of "savagery" into the "manufacturing organization with city and factory system." Out of this comes "a new product that is American," and this new society shows an independence from both its initial European and wilderness antecedents. The evolution from European to American occurred, so Turner argued, through the conquest of the frontier environment.⁷

Walter Prescott Webb gave Turner's ideas a regional flavor in his book, *The Great Plains* (1931). His story also depicts cultural change resulting from environmental adaptation. The American way of life failed when settlers emerged out of the shadows of eastern forests and onto boundless grasslands of the Plains. People adjusted their culture to a lack of trees and water, so Webb believed, by inventing new technologies and laws. With an abundance of timber in the East, they had built split-rail fences, and now on the shortgrass plains, they invented barbed wire for fences. They had regulated the full-flowing eastern rivers with the riparian water doctrine, which protected downstream water quality and quantity from upstream uses. In the West, they devised the prior appropriation doctrine, which protected a person's right to consume limited stream flows in economic pursuits no matter where along the river a person used the water.[8]

In Webb's thinking, environment was paramount to human culture. Americans and nature stood apart, but both felt the influence of the other. In order for American culture to survive, so Webb argued, it had to adapt to new environments. In the East, American culture rested on three legs: water, timber, and land. In the West, aridity, grasslands, and deserts prevailed, leaving only an abundance of land. When Americans encountered this new region, their culture collapsed, until they formed new institutions and invented new technologies to offset the lack of water and timber. Webb also argues for a kind of environmental determinism, but in this case, land shapes culture. If one believes Turner, then Americans conquered the frontier environment; Webb believed they adapted to it.

Advancing his thinking beyond Turner's and Webb's, James Malin, known to his peers as a crusty and ornery fellow, established a more sophisticated methodology for doing environmental history. Malin used the science of ecology to understand how people had shaped, and been shaped by, their environments. Moreover, Malin conducted his own environmental field studies, collecting specimens and noting the seasonal rhythms of grassland plants and animals. He took careful note of how nonhuman forces combined with the power of people, through their ideas and tools, to shape environments.[9]

Malin focused upon finding the ingredients of successful human adaptation to the grasslands. People, Malin always believed, and not any other force at work on the Plains, should set the norms for human occupation. He also believed in unfettered market forces, not government planning, and he believed in the human mind's ability to create new environments rather than letting non-human forces determine what people could, or could not, do with the land. And, like many good, rock-ribbed, Kansas Republicans, he distrusted any form of "New Deal" liberalism.

Malin's way of doing history won few adherents. For one thing, his irascible nature and unflinching political partisanship perturbed his predominately liberal colleagues. At the same time, other scholars who were emphasizing the new social history came to dominate the profession. In contrast, Malin's parochial insistence on the importance of winter wheat and alfalfa was poorly received by those who were inspired by the high drama of the civil rights movement in the 1950s and early 1960s.

In the 1960s, however, environmental historians made a comeback, riding a wave of mounting concern over pollution and newsworthy ecological catastrophes. These scholars catered to the tastes of a powerful middle-class consumer society and its preoccupation with quality-of-life issues. In 1962 the new environmentalists readily embraced Rachel Carson's warnings, in *Silent Spring*, about the dangers of DDT, and launched the contemporary environmental movement. During Lyndon Johnson's administration, Congress passed several important environmental laws such as the Clear Air Act (1963), the Wilderness Act (1964), the Endangered Species Act (1966), and the National Wild and Scenic Rivers Act (1968).

Environmental activists found outlets for their energies in several new groups: the Environmental Defense Fund, founded in 1967; Friends of the Earth, founded by David Brower in 1969; and Greenpeace, established in 1969. The Santa Barbara oil spill of 1969 led many people to realize just how vulnerable biosystems were to serious industrial accidents. On university campuses throughout the nation, rapidly growing concerns over pollution prompted hundreds of thousands of environmentalists to celebrate Earth Day in April 1970. Many of them were already familiar with Samuel Hays's academic monograph, *Conservation and the Gospel of Efficiency* (1959), but Roderick Nash's more palatable *Wilderness and the American Mind* (1967) reached out beyond the ivory tower to catch the attention of a newly receptive public. In the next decade several like-minded historians such as Susan Flader, Samuel Hays, Donald Hughes, Roderick Nash, Wilbur Jacobs, John Opie, Harold Pinkett, John Perkins, Morgan Sherwood, Keir Sterling, and Donald Worster discussed forming a society to promote the study of environmental history.[10] In 1976, at the annual meeting of the American Historical Association in Denver, Colorado, they held the first meeting of the American Society for Environmental History, and, in the same year, John Opie began editing the *Environmental Review*, and Kent Shifferd served as its book review editor.

While many environmental historians were trained as western historians, others reached far beyond the West to consider topics around the globe—from ancient civilizations to the present. American scholars sought out peers in Europe and Asia, some of whom, not surprisingly, were unschooled in the history of the American West, and together they helped form the intellectual framework for the field. Scholars have taken inspiration from the works of the nineteenth-century American diplomat and writer George Perkins Marsh and the Russian geographer Alexander Ivanovich Woeikof. Historical geographers from Great Britain, and members of the Annales school on the Continent, represent two other contingents of the multiplicity of approaches practiced by environmental historians today. Unquestionably, the field is rich and varied in its topics and global reach, although the American West continues to be one of its central areas of study.[11]

THE STORY OF PEOPLE IN ENVIRONMENTS THROUGH TIME

I must once again return to that small patch of earth some people call Konza. The students who sat with me atop that limestone point wanted an explanation of environmental history, and I had a deceptively simple, one-sentence definition for the field: environmental history is the story of people in environments through time. In my view, people simply exist as one force among many at work in a given environment. I related this definition to my students, and we spent the rest of the semester attempting to understand its ramifications.

Like James Malin, I use science to understand how people, and their conscious activities, have fit into their environments. More recent scholars, such as John Opie, a prominent environmental historian, have correctly chastised the profession for not incorporating enough of the writing and thinking of scientists into their works.[12] Clearly, when historians consider the perspectives of ecologists, biologists, and physicists, new views of human life come into historical focus. I ask, how can environmental historians hope to produce viable, believable, stories about places, like the grasslands, without first understanding the bio-, cultural, and physical forces at work?

The life force is the most important power shaping this planet. Environments on Earth are alive; they would be of little concern to us if they were not. Ecologists have long discussed and debated whether the Earth itself is alive. James Lovelock has stood at the forefront in this debate, proclaiming that the Earth *is* alive. According to his Gaia thesis, biological systems on Earth work to maintain a steady global environment in which life can flourish. Some critics, like Daniel Botkin, fret over the teleological implications of Lovelock's thinking (the Earth works in purposeful ways to achieve its own goals), but Botkin's own *Discordant Harmonies* nevertheless discusses the "biosystems" (life systems) of this planet.[13]

The microbiologist Lynn Margulis, a former colleague of Lovelock's, writes about how all life forms are interdependent and work in a vast microcosmos. "It doesn't matter whether you compare kangaroos, bacteria, humans, or salamanders," she has said, "we all have incredible chemical similarities." All life is connected through time and space. "The carbon dioxide we exhale as a waste product becomes the life-giving force for a plant; in turn, the oxygen waste of a plant gives us life." In addition, the connections on this planet are physical, and "this interdependence is an inexorable fact of life; without it, no organism can hope to survive."[14] All life, so these three scientists agree, is inextricably intertwined.

Daniel Botkin writes: "Life and the environment are one thing, not two, and people, as all life, are immersed in the one system." "When we influence nature," he convincingly adds, "we influence ourselves; when we change nature, we change ourselves."[15] Yet, I wish he had substituted "environment" for "nature." Why?

For at least four hundred years, thought in the Western world has isolated people from everything else called Nature. Those steeped in this intellectual tradition, commonly think of people *and* the environment, rather than people *as a part of* environments. In Botkin's compelling view, people inhabit a place as does any other species, and together the actions of all species form unique places called environments. For example, Kansa Indians, farmers, ranchers, ecologists and hikers all had a part, along with other historical terrestrial and cosmic forces, in shaping the environment of Konza.

The word "environment" works for me, whether or not it is an anthropocentric term. Derived from Old French, the term "environment" denotes a surrounding grouping of things affecting the development or life of someone or something. Naturalists write studies about the environments of insects or other animals all the time.[16] Whereas history itself is an anthropocentric endeavor, and so is telling stories of people's past surroundings, the concept of environment embraces more than humans and their works; it includes all life and its interaction with terrestrial and cosmic forces.

Careful use of the word environment—or perhaps better words might be biosystems or biomes, with the emphasis on life-systems or life-communities, or ecosystems, with its embrace of biotic and abiotic forces, casts aside a pervasive and dangerous dualism in western thought: the antitheses "people" and "Nature." So as to avoid this problem, I now choose terms like environment, biomes, biosystems, or ecosystems, whereas before I might have used nature, or Nature. Most writers still speak of "nature" but Bill McKibben and others persuasively argue the need to eliminate its use, or change its connotations completely.[17]

Most likely, nature (with a lower-case *n*) does not really exist, and more likely, Nature (with a capital *N*) is a conception of the human imagination. If nature can be defined as "all that is," then why should humans somehow be separate from it? The answer lies with the ancient Greeks, who either believed that nature did not contain everything, or that everything might be collected and placed into a thing called nature. This act left outside of nature the minds, the minds cradled in the human cranium, who labeled or named nature and its collective pieces. In their view, nature became a container for everything material. Once, only gods, God, or spirits were thought to move the world, but with this thing called nature, a physical entity subject to human description appeared. Thus, a dualism emerged separating people from the rest, which was called nature.[18]

By the end of the sixteenth century many Europeans had begun using the term Nature (with a capital *N*). Philosophers such as Thomas Hobbes and René Descartes, artists like Leonardo da Vinci, and scientists such as Galileo and Francis Bacon, by capitalizing the term, established an even more distinct dualism than the ancient Greeks. As Neil Evernden has written: the use of the term Nature created "a section of 'everything but,' and another for the exception: God or humanity."[19] Sir Isaac Newton also embraced this view, and gave it a set of mechanical rules. The modern age arrived and every field of study fell sway to a clockwork universe analogy, in which the scientist, as an unobtrusive observer, could take apart the pieces of the clock, Earth, and put them back together, and where the whole was never more than the sum of its parts.[20]

With modern inquiries into the functions of the brain and the workings of genetics, and with the general philosophical and theological questioning regarding the existence of God, humans find themselves more and more inextricably returned to the machinery of Nature. Love, infidelity, intelligence, depression, sexual orientation, drug dependency, aging, violence, and many other human characteristics, so many biologists argue, may be in part programmed by the random ordering of the genes found in our DNA. The domain of the soul and of free will diminishes, while the genetic realm swells with its explicit biological explanations of human behavior.[21] Some social thinkers hope to be able to "deprogram" human "imperfections" once an international team of biologists completes mapping the genetic structure of DNA.[22]

Contemporary biologists immersed in the work of genetics emphasize the material existence of human beings. In Richard Dawkins's view, life has only one purpose, the replication of genetic material through evolution. His work—*The Selfish Gene*—discusses the biological commonalities of human beings

12 Introduction

with other living species. Human beings' biological functions are the same as those of any other species: they work solely to replicate their own genetic material through reproduction.²³

What distinguishes humans as a species is consciousness, along with its higher-order property of individual self-awareness. Many scholars, Dawkins included, believe consciousness is possibly an emergent (that is, suddenly springing into existence within the brain) and co-evolutionary force, along with biological evolution, and its emergence gave rise to human culture. Whereas biologically all species reproduce through genes, culture is reproduced through base building blocks called memes (God is a good example of a meme). Other psychologists, for example, Richard Dennett and William Calvin, to name just two, also believe consciousness and culture have evolved through time.²⁴

Human consciousness, especially its property of self-awareness, distinguishes our species from other life forms and terrestrial and cosmic forces. But consciousness does not lift people out of "Nature." In fact, consciousness acts as one force, among many, that shape environments. Human self-awareness and its directed activities work alongside forces such as lightning, wind, asteroids, comets, sunspots, gravity, snow, rain, or any other physical phenomenon, in shaping the environments of which it is a part.

As is apparent on Konza, the conscious efforts, of Kansa Indians, farmers, ranchers, ecologists, and hikers, all of whom differed according to their respective world views, physically affected the shape of their historical environments. All of the people who used Konza moved through that environment as part of a biome, a community of life. They, like the south winds that carried rainstorm-producing, water-rich Gulf air, or the cold north winds that caused violent thunderstorms and grassfire-setting lightning, coursed over the land and formed its history.

Unlike those scholars who cling to a modern philosophical tradition, I believe consciousness works as a part of the material, objective world, not in some subjective reality. Humans are a part of the biosystems in which they live; the difference between them and other species arises only in the degree to which humans are aware of their place in their environments. Or, phrased in another way, *human beings are a conscious part of environments*. In an important way, *environments are becoming aware of themselves*.

In fact, current theories of consciousness do much to explain why people write history in the first place. Here, I would suggest forgoing the postmodernist critiques of literature and history and coming to grips with consciousness theory as means to understanding history.²⁵ As Dennett explains, the mind works in two ways. The physical grey matter simultaneously receives, analyzes, and stores information drawn from each of the senses and reorders all of it in the mind itself, simultaneously mulling over memories, feelings, and ideas. Another mental property processes information sequentially, one thought followed by another. As individuals, this ability to sort our thoughts and feelings allows us to form stories with a beginning and an end about who we are, and to plan our actions hours, days, and years into the future.[26] Historians do for a society what the conscious mind does for the individual: they tell a sequential, chronological story. These accounts allow human societies to make sense of their world.

Physicists, however, constantly remind us of our limitations in making sense of historical environments like Konza. John Gribbin's *In Search of Schrödinger's Cat* explains how scientific observations affect the results of quantum experiments. Moreover, in this difficult-to-understand realm of tiny particles, a multiplicity, perhaps an infinite number, of scenarios is possible in any given experiment, and the outcome becomes only what is observed. In depicting historical events, people, one could sensibly argue, lack any understanding of what has happened because so much of what they and their predecessors have done goes by unobserved. But even more to the point, historians create a past much the same way physicists create the outcome of quantum experiments. Through the power of observation, a historian decides the outcome of the story, when indeed several different versions of the event could have taken place.[27]

Another aspect to the quantum world is that it is "holistic; the parts are in some sense in touch with the whole."[28] The universe is composed thoroughly of the quantum world, and consequently, all things in the universe are connected in ways only dimly perceived and recognized through unaided human perception. Ironically, science, the very instrument by which humans would study and control the objective, material world ("Nature"), has destroyed the premises undergirding a human–Nature dualism, and has dragged humans down into a physical, or material, soul-less or spiritless, Nature. This correlates with biologists' and ecologists' contentions that humans remain entirely within environments, not separate from them.

Evernden correctly clarifies how this dualism, human–Nature, has never really existed, and calls into question the existence of the container itself, that is, nature. Evernden astutely labels these concepts as "social creations."[29] The sooner environmental historians drop this dualism, then the more poignancy their stories will take on. The story will then become one of humans becoming responsible for their own fates without any moralism derived from either a spiritually transcendental nature or from imagining biologically stable ecosystems sans people, or a world of "natural resources and laws" subject to the untethered whelms of people. Rather, environmental history will teach how people have determined their own fates as they have shaped interconnected relationships with the other forms of life in their environments.

Moreover, once people recognize their place in any given environment, perhaps they will drop the notions of "dominating," or "conquering," these biosystems in the pursuit of "progress" and "growth." Many people still see progress and economic growth as synonymous, and they see no limitations to either. Controlling and dominating nature, or Nature, denies constraints, and more alarmingly, extends mastery and dominion over ourselves as well. The tools of science become the means to power, and those who direct technology control those who do not. People with wealth and power control those without and they ignore environmental constraints to their behavior. What is worse, market forces lead some people to define others as economic commodities that, more often than not, have to accommodate the design of machines and economic systems.

Anti-dualistic notions are not new, but have a venerable heritage. Henry David Thoreau, so Thomas Lyon writes, "sought above all a pure and direct experience . . . which would transcend the usual distance between subject and object and grant participation in the wholeness of nature." Echoing Thoreau, Muir proclaimed: "All things flow here in indivisible, measureless currents." Many others, writers and scientists, including Liberty Hyde Bailey, Aldo Leopold, and Rachel Carson, echoed these sentiments while still holding to the use of nature in their writings.[30]

The critics of dualism refute two dubious, although time-honored, assertions made in the fuzzy use of the term nature: either people should command nature, or they should submit themselves to the laws of nature. This tradition, as critics point out, came fully to light in the sixteenth, seventeenth, and eighteenth centuries in the writings of René Descartes, Francis Bacon, Sir Isaac Newton, John Locke, Jean Jacques Rousseau, and Karl Marx, and their ideas still permeate environmental historical writing today. For example, some think contemporary social change, its rapid pace shaped by market forces, threatens the more gradual change at work in the world of nature, which lies apart from society, but is affected by society, and upon which the existence of society is wholly dependent. Nature becomes the standard by which people (who are not a part of nature, but who can still destroy it) should model their lives, and nature also substitutes for God, a god, or the gods, and at the same time teaches human morality, or at least what makes for right or wrong actions if people want to sustain their existence on this planet.[31]

Yet this view seems to deny the sudden, cataclysmic changes wrought by forces other than culture. The planet quickly changes when large asteroids or comets crash to the surface, or even during some extensive periods of volcanic eruptions, or when deadly pathogens destroy their hosts. As a species of life, humans will have to adapt to these sudden changes or perish. Environmental change is variable, and humans must allow themselves flexibility and diversity in their actions and institutions in order to survive in any environment. And not even this will guarantee human life. In January 1993 an asteroid more than four miles wide missed striking the earth by a mere five to six million miles—in galactic terms, a near-collision.[32] Such an impact could have had sudden and devastating repercussions for humans.[33]

A second assertion, more closely related to the domination of nature than its authors might think, reveals culture and environment coming together to create a "second nature" on top of a "first nature." This "first" nature somehow escaped human shaping toward purposeful ends, whereas "second nature" came into existence in response to cultural designs. A simple example might be the High Plains short-grass prairies ("first nature") turned under and into wheat fields ("second nature").[34]

Of course, the soybean fields of Konza differ from its open expanses of tallgrasses. Both, however, are environments shaped by the life in them. Culture, as well as other forces, helped to shape the tallgrass prairies and soybean fields. Kansa Indians burned the region and stimulated the growth of several select species of grass, whereas American farmers curtailed burning to protect their crops. Sunspots may have played a role in determining rainfall patterns that would have directly affected the vigor of either tallgrasses or soybeans. Sunspots and culture both worked simultaneously to shape Konza. Life is the environment, and as human beings are a life form, then they also are simply one additional element comprising and shaping the environments in which they live.

With "first" and "second" nature, however, Evernden's dualism emerges yet again. This dualism places people and their societies outside of, and in potential command of, nature. A dialectic, born of Hegelian philosophy and Marxist ideology, is at work here. Thesis, nature, is modified by material culture, antithesis, and produces second nature, synthesis. In Marx's view, people remained a part of "natural history" until such time as they could master both first nature *and* their own societies, which were seen as a conscious manifestation lying outside nature. In the unfolding of this dialectic, Marx believed, people would finally emerge out of "mere animal conditions of existence into really human ones," a positive social development. But Marx and Engels were both quite clear on one point: "namely that nothing discredits modern bourgeois development so much as the fact that it has not yet succeeded in getting beyond the economic forms of the animal world." Any writer representing the effects of a market culture as "second nature" is distorting the Marxist critiques of capitalism. In Marxist analysis, second nature arises only when people have control over their own society, a condition Marx found utterly impossible in capitalism.[35]

However, those histories written depicting life as part of the environment have some formidable advantages over those which adhere to a human–nature dichotomy. Many critics might view environmental history as a hybrid field of study, all-too-encompassing and lacking definite boundaries, or born of a trendy social movement. While traditional historians have relied exclusively upon written documents from the past, environmental historians openly incorporate sources from several fields of study, thereby blurring the lines formerly separating history from other academic endeavors. The environmental historian lives in an interdisciplinary, holistic world. Human cultures, Marx fully realized, were a part of nature, and as such, their history remained a part of natural history. However, Marx, like Thomas Hobbes before him, believed that while in a state of nature people remained immoral and uncivilized.

What Marx failed to realize was how thoroughly, and inextricably, people remain enmeshed in his notion of nature, or maybe better yet, in my sense of environment. Life, people included, creates environments, and this is one reality, not a social liability, of our existence on this planet. Human continuity is constrained because of life, and human life is dependent upon the continuance of other life forms, if Lynn Margulis and others are correct. Any ideas and ways of living that deny constraints to human existence poise a serious threat to longevity.

Moreover, human voices and stories are also a part of their environment. Really, environmental historians relate the past, not only of human societies, but also of the environments in which they live. It is as if environments have become aware of themselves through history. In this light, historians relate the biorelationships of which humans are only one part. Humans will probably never relate empathically to collar lizards or sideoats grama. However, people can come to understand how collar lizards, sideoats grama, and they themselves have co-evolved through time, and have either relied upon, or competed with, each other for survival.

ENVIRONMENTAL HISTORIANS AND THE WEST

Creating the history of environments, then, has particular usefulness in doing the study of place or region. Divide the West however you will, as a distinct region west of the Mississippi or Missouri Rivers, or as a region composed of several subregions such as the Great Plains, the Southwest, the Northwest, the Midwest, or maybe even Texas, or even into smaller places like the Kansas River Valley or Konza, and you can still use an environmental approach to define a region or place. As ecologists point out, an environment can be contained in the belly of a moose, or it can encompass the entire planet. When historians study the interactions of all living species occupying a given area as they create a place in a state of constant fluidity through time, environmental history comes into view.

Others will come to terms with this view as the chaotic workings of our planet become ever more apparent. These systems defy the power of traditional scientific methods to explain and measure physical and social phenomenon. The realms of chaos and complexity require a new understanding of place. The forces at work on, and around, this planet create patterns delicately balanced between the powers of stability and instability. Any environment, past or present, is sensitive-dependent on initial conditions. Small changes in these conditions can cause extraordinary change, whereas large fluctuations may produce only minor variations.[36]

This helps explain why scientists have such difficulty understanding the effects of CO_2 buildup in the atmosphere, or the rapid alterations fire ants cause to southern environments. More befuddling in the world of chaos is the fact that "randomness with direction can produce surprising complexity[, and how] dissipation is an agent of order." William M. Schaffer, an ecologist, has given up the notion of "natural balance" and now uses chaos theory to explain the movements of ants, bees, and flowers, in the desert mountains outside Tucson, Arizona.[37] Life's evolution, so Stephen Jay Gould argues, reveals no "predictable evolutionary pathways." Rather, the outcome is contingent, or resting upon "an unpredictable sequence of antecedent states, where any major change in any step of the sequence would have altered the final result."[38] Replay the tape of history and a new story appears.

Still, many historians cling to notions of predictability and stability in nature. They often employ the climax-ecosystem theories of Frederick Clements and Eugene Odum, whereas these theories have fallen out of favor with most modern ecologists in the last two decades.[39] Clements and Odum saw environments as biosystems that would, if left alone by people, develop a self-sustaining equilibrium of all their parts. When historians continue to latch onto these theories, then certain types of lessons always appear. Usually, the moral becomes one of humans destroying viable ecosystems, and replacing them with unsustainable ones.

Moreover, the old, pervasive dualism of people–Nature often underlies climax-ecosystem theories. This is often seen in the writings of many western environmental historians, who stress the transcendent morality of an undeveloped nature calling to save us from ourselves, where the fall from grace is usually seen in terms of modern agriculture and the culture of capitalism, not to mention cities, all of which keep us from the Garden. In these works, one can usually find a transcendental moralism and tempered misanthropy: by not paying attention to what nature teaches about "gradual natural" change, humans are constantly messing up their own nests. It is as though people have paid too much attention to their own undertakings and not enough to those of nature. Once again, I contend, one cannot separate nature from the exertions of people so neatly.[40]

The issues dividing western historians are further compounded by other issues that already separate environmental historians who study the American West. The main source of contention revolves around whether the West is a place, a process, or simply a reiteration of American society. These sources of contention largely engage the anti-Turnerians' disagreements with the Turnerians, both neo- and traditional. For example, those historians who rely on a revision of Webb argue that aridity has shaped the history of the American West, whereas others, who embrace an updated Turner, see the American West molded by a process of "invasion, settlement, and community formation [that] followed certain broad, repeating patterns in most, if not all, parts of North America."[41] Webb versus Turner, place versus process, and the argument goes on, and on, and on.

THE WEST AS A PLACE, OR WESTERN ENVIRONMENTAL HISTORY

Western environmental historians have few difficulties in touting the virtues of their respective methodologies and ideologies. So, let me write out my own recipe for preparing an environmental history. But, remember, like many chefs, I season according to taste and I will discard any ingredient messing up the flavor of the stew. And, I will revise my recipe whenever I discover better elements and methods to cook with.

For the moment, my own sense is to treat the West as a place, a geographical entity shaped by aridity. Take a close look at the composite, nighttime photographs of the North American continent taken by NASA satellites. Bright spots punctuate the darkness engulfing the landmass. These luminary patches are city lights outlining clearly many physical features of the United States. Coastlines stand out, as well as the geography of urban settlement.

Look west, beyond the arc of bright splotches that trace the urban settlements of the mid-continent, from Minneapolis to the north, swinging southwest toward Kansas City, then to Wichita, and south revealing the locations of Oklahoma City, Dallas–Fort Worth, Austin, and San Antonio. Beyond this line, huge patches of darkness appear with isolated bright areas scattered around and about. Welcome to the arid and semi-arid expanses of the West with a few oases scattered about.

This West of arid expanses and oasis settlements has always been a dynamic place where processes have occurred to form well-marked environments. From the wide-open spaces of the tallgrass prairies to the Pacific Ocean, the land has inspired westerners to reflect upon their role in shaping their environments, and themselves in their environments. Mary Austin, Roy Bedichek, Bernard DeVoto, Mari Sandoz, and Wallace Stegner all contemplated the broad, arid expanses of the West and set their descriptions in beautiful prose.[42]

Take the Great Plains, for example. What would I want to consider in writing a history of this place beginning in the early 1800s? I would begin by simply exploring the place, and, if possible, living in it to gain a sense of it. I would get out in the place and experience its winters, summers, rainstorms, hailstorms, its hills and rills, and the wind blowing in my face. I would take in the smell of the dry grasses in the fall; the poignant, sweet scents rising from soils pounded by spring rains; and the biting, crisp, clean, clear air of a plains world frozen in the grip of a hard-driven February blizzard. Only then would I gain some sense of what others have experienced, what their words refer to when they speak of crossing a dust-choked trail in a searing July sun.

Next, I would define the general characteristics and extent of the shortgrass prairies, while realizing just how oscillating and impermanent its borders were. I would explore how other environments and forces intersected with the Great Plains and modified it through time. Most people realized that they were in a distinct place when traveling the central plains, even though the region was always in state of flux. Take the eastern portion of the Great Plains: it lacked precise edges because the mixed prairies of tall and short grasses obscured where one began and the other ended. Moreover, this transition zone constantly shifted directions in response to rainfall, the tallgrasses growing thick and lush and heading westward in wet years, and thinning and wilting and receding eastward in dry years. A global environment intersected with the biosystems of the plains, with the jet stream and Pacific Ocean currents affecting wind and cloud patterns, and in so doing, determining when and where the rains fell. Solar forces need to be accounted for, too, if sunspots established the cyclical appearance of dry years.

Rainfall conditions also had a bearing on disease-carrying insects and other pathogens, which affected horse and human populations. Microbes, for example, the bacilli that causes cholera, inhabit humans and other animals, and mortality rates during epidemics would have contributed to the changing appearance of the shortgrass regions over time. In 1849, for example, a particularly wet year on the plains, a cholera epidemic originating in Asia decimated the Southern Cheyennes.

Then, I would follow with an exploration of the intersection of material culture (looking especially at such factors as gender and age) with other forces at work on the plains. Indians and whites kept large herds of horses and mules on the plains. Where and how grasses grew fixed the difficulty or ease by which they were able to keep their animals. Women would strip and cut branches from acres of riparian cottonwoods to supplement horse feed in the dry winter months. Boys would be assigned the responsibility of guarding small grazing groups of horses, which might range miles from the lodges in the spring months when the animals would fatten on the protein-rich young shoots of buffalo grass. In dry years, these routines became nearly impossible, whereas in wet years, they were easily accomplished.

Explaining the environmental history of the plains in the first half of the 1800s also requires understanding the intersections of cosmic, global, and local forces and how these intertwined with peoples' world views. The Southern Cheyennes obligated themselves to reciprocal relationships with other species on the plains, whereas Anglo-Americans took a decidedly utilitarian view of the land as something to be dominated and rendered into profits through farms and ranches. These peoples' cultures shaped their physical actions on the plains, and these actions merged simultaneously with all the other forces at play. Culture, guided by consciousness, melded with other forces in creating a Great Plains environment constantly undergoing change.

Yet not everyone shares my vision of the West as a place shaped foremost by aridity and human cultures, and these scholars, too, have told some compelling stories. The environmental history of the West, not surprisingly, reflects the same differences characterizing the field of environmental history. Several historians have written about environmental topics within ideological, political, economic, natural-resource, technological, or sociological frameworks.[43] Other historians explore the way in which people mix with other elements in nature, or Nature. These interactions produce, according to these historians, environments, landscapes, or cultural landscapes. In these studies ideas, policy, economics, technology, and culture all interact with the physical forces and forms in nature, or Nature, to render ever-changing landscapes as these various components develop and evolve together through time.[44] In a departure from these approaches, a few scholars are dropping the dualism of "people" and "nature." Their stories stress the interactions of all life in shaping an historical environment without regarding people as some exalted species.⁴⁵

When my students and I hiked down from our vantage point on Konza, the moon cast a soft, strong luminescence on the grasses, trees, and trail. The winds had calmed, the grasses had hushed their songs, and the still night air filled with an orchestra of hoot owls, cicadas, crickets, and frogs. Just above the western horizon, Jupiter and Mars hung in close tandem. As we looked around us, took in our environment, we understood its history differently than we had just hours before. We had begun our course toward environmental history. As Konza served as a starting point for my students on their journey toward environmental history, the mountains, deserts, and coastal forests of the West will serve anyone in the same manner. By re-envisioning how people have been a part of place, then we can acquire a new way of seeing history, and in that vision, a route toward a better understanding of ourselves.

NOTES

1. Yi-Fu Tuan, *Topophilia: A Study of Environmental Perception, Attitudes, and Values* (New York: Columbia University Press, 1974, 1990), 245. Also see, Yi-Fu Tuan, *Passing Strange and Wonderful: Aesthetics, Nature and Culture* (Washington, D.C.: Island Press, 1993).

2. Please regard the sources cited for this essay simply as a cursory, but hopefully useful, introduction to the field. An old, but useful guide to the early historiography of the field is Lawrence Rakestraw, "Conservation History: An Assessment," *Pacific Historical Review* 41 (August 1972): 271–88. More recently, Richard White's, "American Environmental History: The Development of a New Historical Field," *Pacific Historical Review* 54 (August 1982): 297–35, still serves as an excellent guide and introduction to the field. An exceptionally useful guide is Thomas Jaehn, comp., *The Environment in the Twentieth-Century American West: A Bibliography* (Albuquerque, NM: Center for the American West, 1990). For a good discussion of some of the differences among environmental historians, see Donald Worster, ed., "A Round Table: Environmental History," *Journal of American History* 76 (March 1990): 1087–1147. Martin Melosi's, "The Place of the City in Environment History," *Environmental History Review* 17 (Spring 1993): 1–23, is a telling commentary on the difficulties involved in developing the environmental histories of cities. For further reading on the urban context, see Gerald D. Nash, *The American West in the Twentieth Century: A Short History of an Urban Oasis* (Albuquerque: University of New Mexico Press, 1977); and John M. Findlay, *Magic Lands: Western Cityscapes and American Culture After 1940* (Berkeley: University of California Press, 1992).

Several journals carry environmental articles, and among them are *Environmental History* (formerly *The Environmental History Review*), *Forest & Conservation History* (formerly *Forest History Newsletter*, then *Forest History,* and later the *Journal of Forest History*), *Interdisciplinary Studies in Literature and Environment,* and the *Pacific Historical Review,* with many other journals often carrying articles relating to environmental history.

3. For some examples of the "New West" history, see: Patricia Limerick, Clyde A. Milner II, and Charles E. Rankin, eds., *Trails: Toward A New Western History* (Lawrence: University Press of Kansas, 1991); William Cronon, George Miles, and Jay Gitlin, eds., *Under an Open Sky: Rethinking America's Western Past* (New York: W. W. Norton, 1992); Richard White, *"It's Your Misfortune and None of My Own": A New History of the American West* (Norman: University of Oklahoma Press, 1991); Patricia Limerick, *The Legacy of Conquest: The Unbroken Past of the American West* (New York: W. W. Norton, 1987); Donald Worster, *An Unsettled Country: Changing Landscapes of the American West* (Albuquerque: University of New Mexico Press, 1994); Donald Worster, *Under Western Skies: Nature and History in the American West* (New York: Oxford University Press, 1992); Clyde Milner II, ed., *Major Problems in the History of the American West* (Lexington, MA: D. C. Heath and Co., 1989); and Sucheng Chan, et al., eds., *Peoples of Color in the American West* (Lexington, MA: D. C. Heath and Co., 1994).

For examples of the reactions to this approach from the "traditional" western historians, see Gene Gressley, ed., *Old West/New West: Que Vadis?* (Worland, WY: High Plains Publishing Company, 1994); Gerald D. Nash, "Point of View: One Hundred Years of Western History," *Journal of the West* 42 (January 1993): 3–4, and his more temperate *Creating the West: Historical Interpretations 1890–1990* (Albuquerque: University of New Mexico Press, 1991), and William W. Savage, Jr. "The New Western History: Youngest Whore on the Block," *Bookman's Weekly* (4 October 1993): 1242–47.

See Dick Kreck, "Showdown in the New West," *Denver Post Magazine,* 21 March 1993, 6–8; Larry McMurtry, "How the West Was Won or Lost: The Revisionists's Failure of Imagination," *The New Republic,* 22 October 1990, 32–38; and Josh Kurtz, "The Second Battle For the West: Historians Duke It Out Over Kit Carson, Manifest Destiny and Even Each Other," *Santa Fe Reporter,* 30 June–6 July 1993, 11–12, 13, for examples of the way in which the press has discussed the feud.

4. A few articles will bear the point. Donald Worster, "New West, True West: Interpreting the Region's History" *Western Historical Quarterly* 18 (April 1987): 141–56, makes an argument for the West as a place shaped by aridity. William Cronon's "Revisiting the Vanishing Frontier: The Legacy of Frederick Jackson Turner," *Western Historical Quarterly* 18 (April 1987): 157–76, and Robert Lang, Deborah Popper, and Frank Popper, "'Progress of the Nation': The Settlement History of the Enduring American Frontier," *Western Historical Quarterly,* 26 (Autumn 1995): 289–308, explore the continuing influence and importance of the Turner thesis. David Gutiérrez's "Significant to Whom? Mexican Americans and the History of the American West," *Western Historical Quarterly* 24 (November 1993): 519–37, carefully considers the importance of Mexican-American culture to the history of the American West. Those who deny the West as a region still rely upon Earl Pomeroy's classic article, "Toward a Reorientation of Western History: Continuity and Environment," *Mississippi Valley Historical Review* 41 (March 1955): 579–600. Other musings on the direction of

the history of the American West can be found in the respective articles by Walter Nugent, Michael P. Malone, and William Robbins in the *Western Historical Quarterly* 20 (November 1989).

5. White, "American Environmental History," 297–98.

6. Frederick Jackson Turner, *The Frontier in American History*, with a foreword by Ray Allen Billington (New York: Holt, Rinehart & Winston, 1962), serves as a good introduction to his theories, while Wilbur R. Jacobs's, *On Turner's Trail: 100 Years of Writing Western History* (Lawrence: University Press of Kansas, 1994), and David M. Wrobel's, *The End of American Exceptionalism: Frontier Anxiety from the Old West to the New Deal* (Lawrence: University Press of Kansas, 1993) both serve as a critical analysis of the times and influence of Turner.

7. Jacobs does a fine job in identifying the influence of Darwin in Turner's work. See especially the first two chapters of his *On Turner's Trail*. The quotes in this paragraph are Turner's and can be found in his "The Significance of the Frontier in American History," in Turner, *The Frontier in American History*, 4, 11.

8. Walter Prescott Webb, *The Great Plains* (New York: Grosset's Universal Library, 1976).

9. James C. Malin, *The Grassland of North America: Prolegomena to Its History with Addenda and Postscript* (Gloucester, MA: Peter Smith, 1967), and for a good introduction to Malin, see James C. Malin, *History and Ecology: Studies of the Grasslands,* ed. Robert P. Swierenga (Lincoln: University of Nebraska Press, 1984).

10. Rachel Carson, *Silent Spring* (Boston: Houghton Mifflin Co., 1962). For good discussions of post–World War II environmentalism see Samuel P. Hays, *Beauty, Health, and Permanence: Environmental Politics in the United States, 1955–1985* (Cambridge: Cambridge University Press, 1987), and Robert Gottlieb, *Forcing the Spring: The Transformation of the American Environmental Movement* (Washington, D.C.: Island Press, 1993). Samuel Hays, *Conservation and the Gospel of Efficiency: The Progressive Conservation Movement, 1890–1920* (Cambridge: Harvard University Press, 1959), and Roderick Nash, *Wilderness and the American Mind* (New Haven: Yale University Press, 1967).

11. George Perkins Marsh, *Man and Nature* (Cambridge: Harvard University Press, 1864) began tying together the influences of culture and physical forces in shaping environments. Henry Thomas Buckle, *History of Civilization in England,* 2 vols. (New York: D. Appleton, 1885), boldly put forth that history was formed by both "mental and physical" laws: consequently "there can be no history without the natural sciences." These approaches toward history were largely ignored until Fernand Braudel's *The Mediterranean and The Mediterranean World in the Age of Philip II,* 2 vols., trans. Sian Reynolds (New York: Harper & Row, 1972), followed by Vernon Gill Carter and Tom Dale, *Topsoil and Civilization* (Norman: University of Oklahoma Press, 1955). In that same year, under the auspices of the Wenner-Gren Foundation for Anthropological Research, William L. Thomas, Jr. spearheaded an important convening of social scientists, biologists, and other scholars who came together to assess the legacy of Marsh and Woeikof. Thomas identified these two men as the "intellectual fonts in the modern period in developing the theme, 'Man's Role in Changing the Face of the Earth.'" Afterward, Thomas collected and edited the papers of this meeting and this still timely anthology was published as *Man's Role in Changing the Face of the Earth* (Chi-

cago: University of Chicago Press, 1956). Later, world histories followed, with Arnold Toynbee, *Mankind and Mother Earth: A Narrative History of the World* (New York: Oxford University Press, 1976), and more sophisticated approaches in Clive Ponting, *A Green History of the World, The Environment and the Collapse of Great Civilizations* (New York: St. Martin's Press, 1992), and Daniel Hillel, *Out of the Earth: Civilization and the Life of the Soil* (Berkeley: University of California Press, 1991).

A few other examples of studies emphasizing a non-U.S. approach include Clarence Glacken, *Traces on the Rhodian Shore: Nature and Culture in Western Thought from Ancient Times to the End of the Eighteenth Century* Berkeley: University of California Press, 1967); Carolyn Merchant, *The Death of Nature: Women, Ecology and the Scientific Revolution* (San Francisco: Harper & Row, 1980); Ziauddin Sardar, ed., *The Touch of Midas: Science, Values and Environment in Islam and the West* (Manchester, U.K.: Manchester University Press, 1984); Alfred Crosby, Jr., *Ecological Imperialism: The Biological Expansion of Europe, 900–1900,* (Cambridge: Cambridge University Press, 1986); André E. Gullerme, *The Age of Water: The Urban Environment in the North of France, A.D. 300–1800* (College Station: Texas A & M University Press, 1988); J. Baird Callicott, *Nature in Asian Traditions of Thought* (Albany: State University Press of New York, 1989); Richard M. Eaton, *The Rise of Islam and the Bengal Frontier, 1204–1706* (Berkeley: University of California Press, 1993); and J. Donald Hughes, *Pan's Travail: Environmental Problems of the Ancient Greeks and Romans* (Baltimore: The Johns Hopkins University Press, 1994).

12. John Opie, *Ogallala: Water for a Dry Land* (Lincoln: University of Nebraska Press, 1993), 313–35. Arthur R. McEvoy's award-winning book, *The Fisherman's Problem: Ecology and Law in California Fisheries, 1850–1980* (Cambridge: Cambridge University Press, 1986), also explores the environmental effects of science.

13. James E. Lovelock, *Gaia: A New Look at Life on Earth* (New York: Oxford University Press, 1987), and his *The Ages of Gaia: A Biography of Our Living Earth* (New York: W. W. Norton, 1988). Daniel B. Botkin, *Discordant Harmonies: A New Ecology for the Twenty-First Century* (New York: Oxford University Press, 1990), 145–51.

14. Lynn Margulis's quote is taken from "Talking on the Water: Wisdom about the Earth, Dispensed from a Floating Podium," *Sierra: The Magazine of the Sierra Club,* May/June 1994, 72; also see Lynn Margulis, *Microcosmos: Four Billion Years of Evolution from Our Microbial Ancestors* (New York: Summit Books, 1986), and Dorion Sagan and Lynn Margulis, *The Garden of Microbial Delights: A Practical Guide to the Subvisible World* (Boston: Harcourt Brace Jovanovich, 1988).

15. Botkin, Discordant Harmonies, 188.

16. For one example, see Scott A. Elias, *Quaternary Insects and Their Environments* (Washington, D.C.: Smithsonian Institution Press, 1994).

17. Bill McKibben, *The End of Nature* (New York: Random House, 1989).

18. Neil Evernden, *The Social Creation of Nature* (Baltimore: The Johns Hopkins University Press, 1992), 18–56.

The Romans incorporated Greek dualism into their understanding of the word nature, which is derived from the Latin *nascere,* to be born. In this sense, nature describes the inborn, defining traits of a thing. All of creation, the cosmos, can have its own "nature," too, and it is a short logical step from there to assume all that is, is nature. As the western world came to consider humans something special, then it also became easy to consider "human nature" as spiritually imbued, and thereby different from, "nature," the whole of material, spiritual-less, existence.

19. Evernden, The Social Creation of Nature, 98.

20. For a short, but excellent introduction to the historical uses of the term nature, see Raymond Williams, "Ideas of Nature," in his *Problems in Materialism and Culture: Selected Essays* (London: Verso, 1980). I agree with Williams, who says, "In this actual world there is then not much point in counterpoising or restating the great abstractions of Man and Nature. We have mixed our labour with the earth, our forces with its forces too deeply to be able to draw back and separate either out" (p. 83).

21. Michael Harrington, in his *The Politics at God's Funeral: The Spiritual Crisis of Western Civilization* (New York: Holt, Rinehart & Winston, 1983), provides a fine discussion of the political ramifications of materialism for contemporary western civilization.

22. Robert Wright, *The Moral Animal: The New Science of Evolutionary Psychology* (New York: Pantheon Books, 1994); Leslie Roberts, "Taking Stock of the Genome Project," *Science* 262 (October 1993): 20–22; and Marie Anna Gillas, "Getting a Picture of Human Diversity," *Bio Science* 44 (January 1994): 8–11.

23. Richard Dawkins, *The Selfish Gene* (New York: Oxford University Press, 1989), 1–11.

24. For a good introduction to consciousness theory, see especially Daniel C. Dennett, *Consciousness Explained* (Boston: Little, Brown, 1991). Other exemplary works include Nicholas Humphrey, *A History of the Mind* (London: Chatto & Windus, 1992); William Calvin, *The Ascent of Mind* (New York: Bantam Books, 1991); Richard M. Restak, *The Modular Brain: How New Discoveries in Neuroscience Are Answering Age-Old Questions About Memory, Free Will, Consciousness, and Personal Identity* (New York: Charles Scribner's Sons, 1994); William Calvin and George A. Ojemann, *Conversations with Neil's Brain: The Neural Nature of Thought and Language* (Reading, MA: Addison-Wesley, 1994); and Antonio R. Damasio, *Descartes' Error* (New York: G. P. Putnam's Sons, 1994).

25. For an introduction to a postmodernist approach to history, see Hans Kellner, "Narrativity in History: Post Structuralism and Since," *History and Theory* 26 (December 1987): 2–29, and William Cronon, "A Place for Stories: Nature, History, and Narrative," *Journal of American History* 78 (March 1992): 1347–76. My sympathies are in accord with a passage written by the late Paul Sheppard.

> My experience tells me that neither the creationists nor the postmodern critics are right, one thinking that the world was made for us and the other that it is made by us. A better vision of the animals is that we are one among them. The only drama in town, to paraphrase G. E. Hutchinson, is the evolutionary play on the ecological stage; what holds the story together is the transformation of energy and substance. Somehow we must find a way into these exchanges in full awareness and discover how to cherish the world of life on its own terms.

This passage was taken from Sheppard's *The Others: How Animals Made Us Human* (Washington, D.C.: Island Press, 1996), 11–12.

26. Dennett, *Consciousness Explained*, 253–82; and Donald Ostrowski, "The Historian and the Virtual Past," *The Historian* 51 (February 1989): 201–20.

27. Granted, the exploration of the quantum world is not easily accomplished by most of us. Still, some fairly clear guides to this realm exist, and these include John Gribbin, *In Search of Schrödinger's Cat: Quantum Physics and Reality* (New York: Bantam Books, 1984), and Fred Alan Wolf, *Taking the Quantum Leap: The New Physics for Nonscientists* (New York: Harper & Row, 1989).

28. Gribbin, In Search of Schrödinger's Cat, 172.

29. Evernden, The Social Creation of Nature, 107–11.

30. Thomas J. Lyon, *This Incomperable Lande: A Book of American Nature Writing* (New York: Penguin Books, 1989), 52–53, 58.

31. For example, see Donald Worster, "Nature and the Disorder of History," *Environmental History Review* 18 (Summer 1994): 1–15.

I agree with much of Worster's thinking, but not with his propensity to separate people and nature, and his seeming reliance upon nature to guide human behavior. The central question of his essay is "Does the history of nature, as we understand it today, still provide us those compelling norms for the history of humankind?" To this he answers, "There still seems to be good evidence and solid reason to support the [reintegration of human life into the more orderly patterns of nature]." These "patterns of nature," as I read Worster, have much in them to instruct humans in how to live their lives.

32. "An Asteroid Buzzes Earth," *Sky & Telescope* (September 1993), 9; Michael Szpir, "Close Encounters of the Bad Kind," *American Scientist* 82 (May/June 1994): 220–21; and *Time Magazine*, 16 February 1993, 56.

33. Stephen Jay Gould's *Wonderful Life: The Burgess Shale and the Nature of History* (New York: W. W. Norton, 1989), is a great study illustrating the effects of mass extinction and its implications for human notions of "evolutionary progress." See, especially, pp. 299–323.

34. A clear and succinct introduction to first and second nature is in Murray Bookchin, *The Philosophy of Social Ecology: Essays on Dialectical Naturalism* (New York: Black Rose Books, 1990), 7–49.

35. Thomas Hobbes, *Leviathan*, Parts 1 and 2, introduction by Herbert W. Schneider (New York: The Bobbs-Merrill Co., 1958), 104–19; Karl Marx and Frederick Engels, *Selected Correspondence, 1846–1895*, with explanatory notes and translated by Dona Torr (New York: Greenwood Press, 1975), 198; Frederick Engels, *Anti-Dühring*, 2nd ed. (Moscow: Foreign Languages Publishing House, 1959), 390–91; Alfred Schmidt, *The Concept of Nature in Marx*, trans. Ben Fowkes (London: NLB, 1971), 42–43; William Cronon, *Nature's Metropolis: Chicago and the Great West* (New York: W. W. Norton, 1991), xvii, 56–57, 266–67.

William Cronon's essay, "The Trouble with Wilderness; or, Getting Back to the Wrong Nature," in his edited volume, *Uncommon Ground: Toward Reinventing Nature* (New York: W. W. Norton, 1995), 69–90, recognizes the troublesome social constructions in the use of "nature," but he still purports its usefulness as a term.

36. Still the best introduction to chaos theory is James Gleick, *Chaos: Making a New*

Science (New York: Viking, 1987). Closely related to the field of chaos in "complexity," and a good introduction to it, is Roger Lewin, *Complexity: Life at the Edge of Chaos* (New York: Macmillan, 1992).

Gould relates this same phenomenon, but he calls it "contingency" (see *Wonderful Life*, 287–92). For a similar view, see Ernst Mayr, *This is Biology: The Science of the Living World* (Cambridge: The Belknap Press/Harvard University Press, 1997).

Historians have been making "Wonderful Life," creative use of these theories. See George A. Reisch, "Chaos, History, and Narrative," *History and Theory: Studies in the Philosophy of History* 30 (No. 1, 1991): 1–20,; David R. Mutimer and Brian T. P. Mutimer, "Chaos, Complexity and the Study of Sports History," *Canadian Journal of History of Sport* 24 (December 1993): 13–29; H. W. Brands, "Fractal History, or Clio and the Chaotics," *Diplomatic History* 16 (fall 1992): 495–510.

37. Gleick, *Chaos*, 315–17.

38. Gould, *Wonderful Life*, 283.

39. The best history of ecological ideas, and a study that includes an extended discussion of Clements and Odum, is Donald Worster, *Nature's Economy: A History of Ecological Ideas*, 2nd ed. (New York: Cambridge University Press, 1994). Also see Frank B. Golley, *History of the Ecosystem Concept in Ecology: More Than the Sum of the Parts* (New Haven: Yale University Press, 1996).

40. I'm in accord with Don Worster in how the market culture has worked against sustainable agriculture, and perhaps even a sustainable American society as we know it today. Our differences lie in how we see culture and nature related, or whether there is something called nature. Whereas Worster sees a set of separate histories playing themselves out at the same time, "the history of a coral reef alongside the history of a coastal city," I see the history of a coastal environment in which there is a coral reef and a coastal city. What Worster's writing suggests, unintentionally or otherwise, is a "misanthropy" discussed by Richard White and Yi-Fu Tuan. See Richard White, "Back to Nature," *Reviews in American History* 22 (March 1994): 3; and Yi-Fu Tuan, *Topophilia*, xiii–xiv.

41. Cronon, Under an Open Sky, 6.

42. This following admittedly short list is somewhat like "great books I have read and really enjoyed." For sure, there are other "telling" works to be studied and enjoyed, but these should begin filling anyone's appetite for a lyrical depiction of the arid West.

Mary Austin, *Land of Little Rain* (Boston: Houghton Mifflin, 1903); Roy Bedichek, *Adventures with a Texas Naturalist* (Austin: University of Texas Press, 1947); Bernard DeVoto, *Across the Wide Missouri* (Boston: Houghton Mifflin, 1947); Mari Sandoz, *Love Song to the Plains* (New York: Harper & Row, 1961); and Wallace Stegner, *The American West as Living Space* (Ann Arbor: University of Michigan Press, 1987).

43. The following works serve only as samplers, not as a wholly inclusive list.

Works dealing with environmental thought and nature writing include Thomas J. Lyon, *This Incomparable Lande* (New York: Penguin Books, 1991); Peter A. Fritzell, *Nature Writing and American: Essays upon a Cultural Type* (Ames: Iowa State University, 1990); Vera Norwood and Janice Monk, eds., *The Desert Is No Lady: Southwestern Landscapes in Women's Writing and Art* (New Haven: Yale University Press, 1987); Max Oelschlaeger, *The Idea of*

Wilderness: From Prehistory to the Age of Ecology (New Haven: Yale University Press, 1991); and Charles Wilkinson, *The Eagle Bird: Mapping a New West* (New York: Pantheon, 1992).

Works on the formation and preservation of parks and monuments give valuable insights to how American thought, economics, and politics interacted. See especially Richard A. Bartlett, *Yellowstone: A Wilderness Besieged* (Tucson: University of Arizona Press, 1985); Hal Rothman, *Preserving Different Pasts: The American National Monuments* (Urbana: University of Illinois Press, 1989); Alfred Runte, *National Parks: The American Experience*, 2nd ed., (Lincoln: University of Nebraska Press, 1987); Susan R. Schrepfer, *The Fight to Save the Redwoods: A History of Environmental Reform* (Madison: University of Wisconsin Press, 1983); Robert Righter, *Crucible for Conservation: The Creation of Grand Teton National Park* (Boulder: Colorado Associated University Press, 1982); John C. Miles, *Guardians of the Parks: A History of the National Parks and Conservation Association* (Lanham, Maryland: Rowman and Littlefield Publishers, Inc., 1996); George M. Lubick, *Petrified Forest National Park: A Wilderness Bound in Time* (Tucson: University of Arizona Press, 1996); Alexander Brummond, *Enos Mills: Citizen of Nature* (Niwot: University Press of Colorado, 1995); and John R. Jameson, *Big Bend of the Rio Grande: Biography of a National Park* (New York: P. Lang, 1987).

Works on natural resources are particularly abundant. Some useful bibliographies include Ronald J. Fahl, *North American Forest and Conservation: A Bibliography* (Santa Barbara: A.B.C.–Clio Press, 1977); Lawrence Lee, *Reclaiming the American West: An Historiography and Guide* (Santa Barbara: A.B.C. Clio Press, 1980); and Donald Pisani, "Deep and Troubled Water: A New Field of Western History?" *New Mexico Historical Review* 63 (October 1988), 311–31. Useful studies include David Clary, *Timber and the Forest Service* (Lawrence: University Press of Kansas, 1986); William Robbins, *American Forestry: A History of National, State & Private Cooperation* (Lincoln: University of Nebraska Press, 1985); Michael Williams, *Americans and Their Forests: A Historical Geography* (Cambridge: Cambridge University Press, 1990); Paul W. Hirt, *A Conspiracy of Optimism: Management of the National Forests since World War Two* (Lincoln: University of Nebraska Press, 1994); Nancy Langston, *Forest Dreams, Forest Nightmares: The Paradox of Old Growth in the Inland West* (Seattle: University of Washington Press, 1995); Donald Worster, *Rivers of Empire: Water, Aridity, and the Growth of the American West* (New York: Pantheon Books, 1985); Marc Reisner, *Cadillac Desert: The American West and Its Disappearing Water* (New York: Penguin Books, 1986); Donald Pisani, *To Reclaim a Divided West: Water, Law, and Public Policy, 1848–1903* (Albuquerque: University of New Mexico Press, 1992); Norris Hundley, Jr., *The Great Thirst: Californians and Water, 1770s–1990s* (Berkeley: University of California Press, 1992); Catherine Miller, *Flooding the Courtrooms* (Lincoln: University of Nebraska Press, 1993); Mark W. T. Harvey, *A Symbol of Wilderness: Echo Park and the American Conservation Movement* (Albuquerque: University of New Mexico Press, 1994); Richard W. Sadler and Richard C. Roberts, *The Weber River Basin: Brass Roots Democracy and Water Development* (Logan: Utah State University Press, 1994); John E. Thorson, *River of Promise, River of Peril: The Politics of Managing the Missouri River* (Lawrence: University Press of Kansas, 1994); Todd Shallat, *Structures in the Stream: Water, Science, and the Rise of the U.S. Army Corps of Engineers* (Austin: University of Texas Press, 1994); Keith C. Peterson, *River of Life, Channel of Death:*

Fish Dams on the Lower Snake (Lewiston, Idaho: Confluence Press, 1995); William D. Rowley, *Reclaiming the Arid West: The Career of Francis G. Newlands* (Bloomington: Indiana University Press, 1996); Donald C. Jackson, *Building the Ultimate Dam: John S. Eastwood and the Control of Water in the West* (Lawrence: University Press of Kansas, 1995); Donald E. Wolf, *Big Dams and Other Dreams: The Six Companies Story* (Norman: University of Oklahoma Press, 1996); Thomas R. Dunlap, *Saving America's Wildlife* (Princeton: Princeton University Press, 1988); Steven Lewis Yaffe, *The Wisdom of the Spotted Owl: Policy Lessons for a New Century* (Washington, D.C.: Island Press, 1994); Eugene D. Fleharty, *Wild Animals and Settlers on the Great Plains* (Norman: University of Oklahoma Press, 1995); Duane A. Smith, *Mining America: The Industry and the Environment* (Lawrence: University of Kansas Press, 1987); Robert Righter, *Wind Energy in America: A History* (Norman: University of Oklahoma Press, 1996); John B. Wright, *Rocky Mountain Divide: Selling and Saving the West* (Austin: University of Texas Press, 1993); and R. McGreggor Cawley, *Federal Land, Western Anger: The Sagebrush Rebellion and Environmental Politics* (Lawrence: University Press of Kansas, 1993); Dan McGovern, *The Campo Indian Landfill War: The Fight for Gold in California's Garbage* (Norman: University of Oklahoma Press, 1995); Donald A. Grinde and Bruce E. Johansen, *Ecocide of Native America: Environmental Destruction of Indian Lands and Peoples* (Santa Fe: Clear Light Publishers, 1995); Laura Pulido, *Environmentalism and Economic Justice: Two Chicano Struggles in the Southwest* (Tucson: University of Arizona Press, 1996); Peter E. Eishstaedt, *If You Poison Us: Uranium and Native Americans* (Santa Fe: Red Crane Books, 1994).

44. Still a first-rate model for doing regional environmental history is Richard White's *Land Use, Environment, and Social Change: The Shaping of Island County, Washington* (Seattle: University of Washington Press, 1980). Similar in methodology is William deBuys's, *Enchantment and Exploitation: The Life and Hard Times of a New Mexico Mountain Range* (Albuquerque: University of New Mexico Press, 1985); Peter G. Boag, *Environment and Experience: Settlement Culture in Nineteenth-Century Oregon* (Berkeley: University of California Press, 1992); and my *Watering the Valley: Development along the High Plains Arkansas River, 1870–1950* (Lawrence: University Press of Kansas, 1990). A remarkable achievement in mixing law as a reflection of culture and its interaction with historical environments is Charles Wilkinson's, *Crossing the Next Meridian: Land, Water, and the Future of the West* (Washington, D.C.: Island Press, 1992).

45. See Opie's *Ogallala*; and especially Elliott West's *The Way to the West: Essays on the Central Plains* (Albuquerque: University of New Mexico Press, 1995), 1.

CHAPTER TWO

SPIRIT OF PLACE AND THE VALUE OF NATURE IN THE AMERICAN WEST

DAN FLORES

> "*The crucial question of the modern world is, 'How are we to become native to this land.'*"
>
> PAUL SHEPARD, *in Max Oelschaeger, ed., The Wilderness Condition*

I moved to Montana six months ago and find myself, for the third time in my life, engaged in the process of being assimilated into a landscape and a place. As a typically transient American, I have of course actually lived in more than three places in my life. And had I been in Laramie or Sheridan, Wyoming, for more than half a year, or Austin or Santa Fe longer than mere summers (perhaps more significant, had I lived outside rather than in those towns), likely I would know *more* country, bones, and soul. As it is, only northwest Louisiana 'round about Shreveport, and the High Plains canyon country south of Amarillo, Texas, have so far claimed me truly and intimately.

In at least one of these places I consciously went native while an adult, so that I was able to observe the process and write a book that was, in effect, the natural history of a spirit of place. As animals whose evolutionary trajectory kept us intimately aware of local and regional landscapes until only the last few symbolic seconds of our species' life, we are still hardwired to experience the sort of landscape energy that undergirds place. This "sprit of place," or "sense of place," as Wallace Stegner would have it, is not an easy phrase to define. And yet I think it is a tangible phenomenon—one that has been noticed and studied for a long time now, and a phenomenon whose investigation may help explain some of the agonizing the modern American West is enduring.

Let me say first what I do *not* mean when I use the phrase "spirit of place." "Spirit," as I am using it here, does not refer to supernatural entities or qualities, but rather to essential and activating possibilities, to the inspiration that tangible phenomena—in this instance landscapes viewed either as real settings or as cultural texts—can impart to us humans. D.H. Lawrence saw this clearly. As he wrote in his famous "Spirit of Place" essay in 1916.

> Every continent has its own great spirit of place. Every people is polarized in some particular locality, which is home, the homeland. Different places on the face of the earth have different vital effluence, different vibration, different chemical exhalation, different polarity with different stars: call it what you like. But the spirit of place is a great reality.

Or as Cormac McCarthy has a group of Mexican travelers put it in his wonderful novel, *All the Pretty Horses,*

> [The vaqueros] said that it was no accident of circumstance that a man be born in a certain country and not some other and they said that the weathers and seasons that form a land form also the inner fortunes of men in their generations and are passed on to their children and are not so easily come by otherwise.

In the meaning that I am attributing to it here, "place" essentially means "space" (landscape) plus people. Spirit of place, in other words, is grounded in the human interaction with local environments. So investigating spirit of place is an approach for puzzling over the wonderfully diverse ways that women and men have both lived in and reacted to spaces in the landscape.

In the social sciences the term "place" is in effect a synonym of "region." I would suggest that in a more economics-oriented way, spirit of place has long been studied under the relatively mechanistic hubris of *regionalism*. This kind of regional study is, in fact, an old topic in history. In the early twentieth century anthropologists like Clark Wissler and Ellen Semple argued that in pre-Columbian America, Indian cultures had arranged themselves across the continent according to general ecological boundaries (what we should call biomes or ecoregions today), so that in pre-contact North America we had a "Desert Southwest Cultural Region," a "Northwestern Woodland Cultural Region," and so forth. Regionalism (or "sectionalism") once dominated the study of American history. After penning his famous Frontier Thesis in 1893, historian Frederick Jackson Turner spent much of his remaining career on a book called *The Significance of Sections in American History*. In it he further refined thinking about the East, the South, and the West, divvying the West, for example, into four distinctive regions: the Plains, the Rockies, the Deserts, and the Coast. With the continuing spread of the Modern Industrial Age, however, most scholars expected regionalism worldwide to become a casualty

Spirit of Place and the Value of Nature in the American West 33

of the homogenizing qualities of the global market economy. But like the prognostications of Vanishing American Indians, the prediction seems to have been premature. The emergence of environmentalist bioregionalism and the current fragmentation in Europe indicate just how powerful a force regionalism still is. Two million years of hunting and gathering immersion in place can't be purged from us easily, it would seem!

In its combination of words the phrase "spirit of place" thus refers to a more modern—one might even say more holistic—technique for examining the interaction between humans and slices of nature. Beyond mere local economies, that ought to include more esoteric interactions like art, literature, religion, and environmental adaptations, and how they are influenced by the activating inspirations of particular landscapes.

Let me say now that whether landscapes do or do not actually exude a spirit or energy that human beings actually do sense and react to is not demonstrable by any science I am aware of. Beyond simple, universal ideas about balance and harmony, aesthetics among humans seem primarily derived from culture. Yet until fairly recently in history, most human cultures fashioned themselves around the premise that nature was locally and uniquely inspired. One example from the American West is that of the Pueblo Indians of northern New Mexico, who built rock "amplifiers" atop the Jemez and Sangre de Cristo mountains toward their villages. The Pueblos, in other words, entertained no doubts about the energy hologram they occupied.

There is also the phenomenon of attachment to home places, which the geographer Yi-Fu Tuan calls "topophilia." In his book *The Territorial Imperative* of a quarter-century ago, Desmond Morris pointed out that territoriality springs from the reptilian brain, the most ancient part of the human mind. And as Diane Ackermann has written more recently in *A Natural History of the Senses*, the evidence is that human evolution is so tied to topography, that our multiple senses contain such rich receptors and our brains such vivid recallers of natural signals, that all humans react emotionally to familiar home places. This emotional attachment occurs on an ascending spectrum of size scales: from hearth to locale, and even to more artificial creations like states and nations.

On the subject of topophilia, I think I would agree with Yi-Fu Tuan on several other scores. Topophilia happens most naturally not in these large creations but in places small enough to be learned well—local landscapes of familiar rocks and soils, the remembered sounds of local birds, the peculiar cycle of local seasons, the cyclically occurring smells of conifer forests sunned to summer fragrance, or the pungent smell (almost a taste) of valley cottonwoods in the fall. But there are at least three other qualities that make up the kind of interaction with place I have become fascinated with, and that drive the rest of this essay. One is the necessity of sinking roots. "Place is pause," Tuan writes in his book *Space and Place*, an idea all of our modern literary Bioregionalists, from Gary Snyder to Wendell Berry, echo. In the American West, many of our roots are yet shallow, a possible explanation for why so much of what passes for "explanation" of the West is actually myth (think of "the Great American Desert" or "Rain Follows the Plow") imposed by those who scarcely knew the region. Transiency is rare among many people around the world, but is a common trait of Americans, engendering a kind of rootlessness that inhibits developing a sense of place. Transiency also inhibits a second element in developing a sense of place, and that is a shared sense of history. Whether it takes the form of mythology, folklore, or historical literature makes no difference. Where history is too short or too diffuse, spirit of place is weak, still much in flux.

But it is the role that value systems play as a kind of human template in spirit of place that is most interesting of all.

When they have studied what I am calling spirit of place, historians have done so primarily by studying modes of production, using theoretical constructs like environmental determinism or possibilism to explain how particular economies emerge in particular places. Or, they have utilized systems theory or central place theory when they have wanted to demonstrate how places are influenced by outside forces. Yet, as I've said, our reactions to landscape have not just been economic. They have also been aesthetic, creative, mythic, and entirely sensual. They have led to indigenous art, such as the painting and folk art traditions of the modern Southwest or the literary renaissance of the Northern Rockies, that express spirit of place symbolically and metaphorically. Yet even in these artistic responses—perhaps more especially in these responses—spirit of place occurs within the context of history and cultural evolution.

Let me say that in another way so that no one gets the impression that I am advocating a kind of environmental determinism. Despite our "hard-wired" biology, human perceptions and the values we place on nature are so culturally driven that we ought not to believe that certain environments exert an irresistible pull of one kind of another. Think of the diverse lifestyles on the Great Plains across the past twelve thousand years. All those cultures were reacting to the same *space* yet they created different *places*. Even in the life of a single culture, spirit of place changes over time. The phenomenon is dynamic, not static.

So in this tour of how spirit of place operates, we have to consider the culturally inherited value systems (the human "software") that have to do with how humans perceive nature. Spirit of place of one kind or another no doubt exists everywhere humans have paused enough to feel a topographic pull, but the shape and feel and taste of it will be very different from one society to the next. The upper stretches of the Ruby River or the Clark Fork in Montana no doubt fostered a kind of emerging spirit of place in the early years of Bannock or Butte—but it was a very different one from the spirit of place the New Mexican Pueblos felt in the valley of the Rio Grande, or the Mountain Crows experienced in the Absaroka homeland, or the four-hundred-year-old Hispanic culture feels about the Southwest.

If the modern American West can be understood to function as a region with its own diverse spirits of place, then it might be useful to try to define what values of human thought have helped create our contemporary places in the West. When we speak today of "nature," "wilderness," even "ecosystems" or "sacred places," are we speaking of abstract realities or of agreed upon fictions springing from particular value systems? And what does it mean for the West as place when one set of values confronts another, perhaps newly emergent set?

Perhaps what distinguishes Euro-American thought about nature most has been the two-thousand-year perception in Western Europe of nature as a storehouse of natural resources, a storehouse that stands independent from humans, and which was specifically created for our use. This anthropocentric idea, as a host of writers have demonstrated—perhaps the most accessibly Clarence Glacken in *Traces on the Rhodian Shore,* Max Oelschlaeger in *The Idea of Wilderness,* and Clive Ponting in his recent *A Green History of the World*—can be traced to a succession of rachet-like changes in the evolution of human societies beginning as far back as the Neolithic Revolution of ten thousand years ago. This abandonment of the green and leisurely world of gather/hunting for the much harder work of agriculture and the settled life of villages and towns set in motion an initial separation of humans from wild animals, wild plants, and an intensive knowledge of local bioregions. In the literature of many modern writers, this constituted the true "fall" of humanity, cast symbolically in the Bible as the expulsion from the Garden.

The continuing development of Western ideas and values proceeded (this is a shorthand version) with the Greek idea of the existence of a human soul that is separate from the worldly realm; the Judaic idea that since humans were the only life form made in the image of God, only they possessed souls; and the Christian exorcism of the pagan *genii locii* (local spirits) from the world, a step that desacralized nature and made its conquest possible. Then came the secularization of the great mysteries of nature via the Scientific Revolution and an empirical method that initiated a general investigation of the natural laws. The Western Age of Reason (seventeenth century) and the Enlightenment (eighteenth century) furthered the machine metaphor for nature with René Descartes's model of animals as mere machines lacking either souls or consciousness, and Descartes's methodology of examining the world scientifically by starting with the known and proceeding from there. Because nature has proven so very complex, Descartes's reductionist method has tended to fragment knowledge, to divide nature into pieces for study, so that the overall tendency of science has been to lose sight of the whole.

The emergence of individualism and capitalism, both linked to what the historian Walter Prescott Webb called "The Great Frontier"—the European conquest of the world and its incorporation into the global market economy—was the next step in the development of the Western conception of nature. In effect, Adam Smith saw to it that individual greed in the interest of the community good became a dominant Western value. Thus, by the nineteenth century, when the American West was being overrun by Euro-Americans and the outlines of our present land-use system for the West were being drawn, the progression of ideas and values peculiar to the dominant Euro-American culture began to interact with the landscape energy emanating from the Great Plains, the Rocky Mountains, the Western deserts. The result was the skeleton of spirit of place in the modern West.

This skeletal framework included, we might have to admit, some pretty peculiar ideas. Euro-Americans did not apprehend, or would not admit, that the continent they wrested from the native peoples was in fact a human-shaped one. We know better today. We know, for example, that at least twelve thousand years of human interaction had profoundly shaped the ecology of North America—from Paleolithic involvement in the massive ecological simplification that saw some thirty-two genera of large animals go extinct around ten thousand years ago (and that left our modern keystone species in place in the West), to the use of fire and other land-use modification strategies by three hundred generations of Indian peoples. And yet, powerful European ideas like "virgin America" and "wilderness" profoundly shaped the way Americans looked at the natural world.

But by the middle of the last century Euro-Americans were being shaped by some new values and ideas. The most important of them was the Romantic backlash against the Age of Reason. The Romantic Movement, begun only a century earlier in Europe and influencing Americans only after about 1825, stressed intuition and emotion, which could be experienced in a transcendent way in nature as nowhere else. Mountains, regarded as chaotic warts on the landscape by Europeans during the Middle Ages, now could be and were appreciated aesthetically.

Americans also were influenced by their sense of cultural inferiority with respect to Europe, and that—coupled with aesthetic appreciation and a growing perception that there were lands in the West that were economically marginal—led to American leadership in the creation of national parks. In creating national parks, what Americans were most interested in, perhaps all the way through the early 1930s, was preserving monumental landscapes that demonstrated to the world that America was a morally superior nation because it occupied a larger-than-life, aesthetically superior continent. The implication was that a superior culture—great music, art, literature, and architecture—would emanate from such surroundings. Since monumentalism had little to do with ecology, we got artificial, linear boundaries for our parks. On the other hand, in a society where nature had become secularized, the great early national parks like Yellowstone and Yosemite began to function—and still do—as the great sacred places of American culture.

Here on the cusp of the twenty-first century we are still hauling all of this human baggage—a very long baggage train extending ten thousand years and farther into the past—that we have internalized both culturally and individually, and that we are towing along with us like a brontosaurus tail. Our current struggles over relatively new ideas like Greater Yellowstone or Northern Continental Divide ecosystem protection, biodiversity, and biocentrism, have much to do with disjunctions between these internalized values and newer ones that represent a further maturing of Western spirit of place. The result has been an increasingly contentious society in the West lately. Think of it this way: contentiousness in the modern West centers around such jarring collisions as managing parks for new values like ecosystem protection or natural fires, while those lands were created and set in motion by older, nineteenth-century values. Or asserting ideas like the innate rights of all life, the advantages of biodiversity, or the spiritual value of wilderness in a cultural milieu still dominated by unquestioning acceptance of traditional anthropocentric or cost-benefit models. Or recognizing how profoundly humans have occupied and shaped this continent, while managing a wilderness system based on the ancient Western ideas that humans exist apart from nature and that America was a virgin land.

I began by speaking of spirit of place because it embodies so clearly human values engaged in a dialectic with the land. The defining essence of spirit of place is not just that it occurs on a bioregional level, or that humans ought to stay put and sink roots, or even that sense of place requires a shared awareness of history to flower fully. The most important thing about spirit of place is that cultural values and human imagination determine it as much as landscape does—and that it exemplifies history's greatest lesson: everything is always changing.

As scientists have been so willing to point out, history's great flaw is that it is explanatory, but not predictive. Yet something, obviously, is happening in many parts of the West as we eye the twenty-first century. What Thomas Kuhn a quarter-century ago referred to as a "paradigm shift"—a base change in values that occurs when sufficient anomalies crack open a long-accepted worldview—seems to have found fertile soil in many areas of the public lands West. The resulting tension between the past and the emerging future, between the old, internalized values and the new environmentalist ones, is manifest not merely in hoarse debate but in what I think may be more predictive venues, art and literature particularly, that have long sensed an impending sea change.

Perhaps what it all means is that after taking a hard look at ourselves and the places we've created in the vast, stunning, sunlit spaces of the American West, we are starting to assimilate, starting at last to go native. Through a shift in Western spirit of place, the most haunting landscapes on the continent may be finding a new voice.

SECTION II

Satanta, in response to watching soldiers and camp followers kill bison indiscriminately, said: "Has the white man become a child, that he should recklessly kill and not eat? When the red men slay game, they do so that they may live and not starve."

SATANTA, A KIOWA, AT MEDICINE LODGE TREATY COUNCIL, 1867

Often, environmental historians have attributed to Indian peoples and Hispanics a love of the land not present in an Anglo-American capitalistic ethos. This love, these same scholars argue, equated to a greater care of, and better ecological adaptation to, the land. Indian peoples and Hispanics, according to this view, never harmed the natural resources supporting their cultures. This section contains articles relating to the interactions of people and the land shortly after the arrival of Europeans to the Southwest and the Great Plains.

Historians such as Donald Worster, and social-agricultural reformers such as Wes Jackson of the Land Institute, in Salina, Kansas, among many others, search for alternative communal orders, farming practices, and economic systems to those of contemporary urban and farm societies. Sometimes these reformers believe the study of subsistence farming practices, with their supposedly benign effects on an environment, may provide the means, and insights, to "sustainable" agriculture.

Robert MacCameron, a historian at SUNY Empire State College, provides a study of the tightly organized communal structure of colonial New Mexicans, and of their subsistence agriculture. While capitalistic agriculture, he concedes, has greatly diminished the bounty of New Mexico, the farming and ranching practices of colonial New Mexicans, he also reveals, reduced the productivity of the land. In what ways did colonial New Mexicans blend their culture to the land around them, and in what ways did their culture disrupt the ecosystems they inhabited? In this case, does subsistence agriculture provide the means, or insights, to remedy current ecological problems in agricultural production?

American Indian peoples believed the earth, and the forces of it, alive, and part of their social order. Scholars have taken heated positions arguing whether or not this made American Indian peoples ecologists. Often Hollywood and television, along with many environmental groups, portray American Indian peoples in this light. But is this an accurate historical depiction?

Dan Flores's piece addresses this issue by discussing the changes in Plains Indian peoples' bison-hunting practices, and how these practices shaped their cultures and diplomatic initiatives with American and New Mexican traders. In what ways does Flores show Plains Indians in tune with their environment, and what forces does he think led them to deplete the natural-resource base upon which their culture rested?

I have taken up the study of how High Plains Indian peoples used horses. People, I maintain, must adapt themselves to, and realize their responsibility for, environmental change in order to reproduce their cultures. Some adaptation strategies will prove viable while others will not. The questions I ask are how did High Plains Indians' horse practices promote their ambitions and goals, and in what ways did the horse impede, and constrain, their life on the Plains. Two larger questions come into view, too. In what ways are people responsible for determining their own fates, and in what ways do other forces, both cultural, physical, and organic determine their fates?

CHAPTER THREE

ENVIRONMENTAL CHANGE IN COLONIAL NEW MEXICO

ROBERT MACCAMERON

In recent years scholars of North American history have paid increasing attention to the interrelationship between human societies and their physical and natural worlds. Their studies analyze the various ways in which people interact with their surrounding environment and how their choices affect not only the human community but the larger ecosystem as well. Just as nature shapes human society, humans, in significant and far-reaching ways, shape nature.

Exemplary works by William Cronon and Richard White have focused, from an ecological perspective, upon the English frontier experience in North America. Cronon, in his seminal book, *Changes in the Land: Indians, Colonists and the Ecology of New England* (1982) demonstrates that the English colonization of New England produced a number of "fundamental reorganizations . . . in the region's plant and animal communities,"[1] a result, fundamentally, of the "colonists' more exclusive sense of property and their involvement in a capitalist economy."[2] Similarly, White, in his study entitled *Land Use, Environment and Social Change: The Shaping of Island County, Washington* (1980), describes how the introduction of European technologies along with the "Columbian Exchange" of plants, animals, and diseases dramatically altered the operation of natural systems, producing in turn what one botanist has described as the "most cataclysmic series of events in the natural history of the area since the Ice Age."[3]

This present study focuses upon environmental change in another area of North America: the upper Rio Grande Valley which came under Spanish rule between 1598 and 1821. This largely semi-arid ecosystem encompasses today the area of north-central New Mexico, from approximately Belen, just below

Albuquerque in the south, to Taos in the north, and from the Sangre de Cristo and Sandia mountain ranges in the east, to the Chama, Jemez, and Rio Puerco River valleys in the west. The study examines just how Spanish culture and society brought about new relationships between human societies, including Pueblo Indians[4] and the land, and how changes in the land occurred as a result of those relationships. The English experience in North America serves as a principal point of reference.

The kind of physical and natural world in which the Spanish settled provides an important context for any discussion of environmental change. Scholars today describe north-central New Mexico as possessing essentially five different life zones: the Upper Sonoran, Transitional, Canadian, Hudsonian, and Arctic-Alpine. The Upper Sonoran zone includes mostly valleys, foothills, and plains, and extends from approximately 4,000 to 7,000 or 8,000 feet in elevation. Average annual rainfall is approximately ten inches. The Transition zone encompasses the middle slopes of the higher mountains and begins at approximately 7,000 to 8,500 feet above sea level on northeast slopes and 8,000 to 9,500 feet on southwest slopes. The Canadian zone covers most of the higher peaks of the Sangre de Cristo range and extends from approximately 8,500 to 12,000 feet depending upon cold or warm slopes. The Hudsonian and Arctic-Alpine zones are found on peaks that are around or above the timberline. Each of these zones is characterized by distinct flora and fauna, as they were at the time of Spanish arrival, although the boundaries between them are sharply marked only on steep-sided slopes. The zones themselves and their individual characteristics result from differences in altitude, temperature, precipitation, and barometric pressure. Significantly, human settlement has occurred almost exclusively in the first two zones. The Upper Sonoran has encompassed, past and present, most of the agricultural and grazing lands of the region, and here farming, except at the highest elevations, requires irrigation or some other water control system. The Transition zone is a bit wetter and therefore allows for some dry farming as well as seasonal grazing. But in fact, both zones constitute a semi-arid environment in which access to water has been crucial for human survival.

Within this environmental context, Pueblo–Spanish contact produced its own particular form of the Columbian Exchange, that is, the reciprocal introduction of Old and New World plants, animals, and disease. More broadly construed, the exchange also included forms of material culture and even distinct values relating to economic, social, political, and religious organization.[5] Students of New Mexico's environmental history differ in their emphasis on the effects that this exchange has had on the land and the human societies occupying it. An anthropological view emphasizes the front end of the exchange, the Spanish introduction of new seeds, wheat, vegetables, and the tools for environmental destruction, including the iron axe and livestock, which changed the ecology of the upper Rio Grande Valley forever.[6] A historical view more often stresses the rather remarkable adjustment achieved by Hispanic settlers in a very rugged frontier environment. A community-based agro-pastoral system of subsistence evolved that successfully sustained, both environmentally and culturally, many small communities for generations.[7] It was only after the arrival of the Anglo-Americans into the upper Rio Grande Valley and the introduction of commercial agriculture that changes in the land occurred in any dramatic fashion.[8]

This second view, in essence, subscribes to a traditional interpretation of the relationship between economics and ecology, as outlined by Donald Worster and others, that subsistence agriculture, such as that employed by the Spanish in New Mexico, tends to preserve much of an environment's diversity and complexity, producing in turn an ecological and social stability. On the other hand, capitalism, involving specialization in production and competition in markets for profit leads to "a radical simplification of the natural ecological order, number of species found in an area, and in the intricacy of their interconnections."[9]

Interpreted in the broadest sense, this essentially linear model for environmental change in colonial New Mexico is basically correct. There is no question that human impact on the land, from Pueblo to Spanish to Anglo-American, was increasingly severe, but this model also contains certain limitations when applied to the case of colonial New Mexico. A closer look at the Spanish period reveals a far more complex process at work. In fact, the evidence indicates that a number of factors or determinants worked toward or mitigated changes in the land, producing in turn, in a largely nonlinear fashion, different kinds and rates of change.

Factors include demographic features of both Pueblo and Spanish, the Spanish introduction of grazing animals into the upper Rio Grande Valley, colonial New Mexico's restive social-economic isolation, Spanish material culture, dimensions of Spanish land institutions, systems of political control, the varying nature of an inclusive frontier, changes in climate, and the vulnerability and resilience of a semi-arid ecosystem. Many of these factors were paradoxical in their effect. They, at once, simplified the ecosystem and thereby accelerated change, and sustained ecosystem diversity and complexity, and produced little or no change.

Demography, especially numbers of people and where they relocated, is an important starting point for understanding the factors affecting environmental change. Throughout the Spanish colonial period approximately 99 percent of the population of the Province of New Mexico, including the general categories of Indians, Spanish, and *castas* (people of mixed blood), occupied about only 1 percent of the land. Access to water, in the form of rivers, creeks, and streams, was one obvious reason. But so also was the intermittent danger posed by the nomadic tribes—Apaches, Comanches, Navajos, and Utes—surrounding the upper Rio Grande Valley. Attempts to expand beyond the Rio Grande itself out to its principal tributaries and elsewhere were often discouraged by the presence of such nomads. Need for water and the presence of unfriendly Indians acted as a centripetal force upon both Pueblo and Spanish settlements. But at the same time, as Spanish population increased, especially during the eighteenth century, overcrowding along the Rio Grande occurred and access to resources, especially water and pasturage, diminished. These factors served as a countervailing or centrifugal force to Spanish settlement.[10]

The Spanish colonial population of the upper Rio Grande Valley was centered principally on three *villas*: Santa Fe, founded in 1610; Santa Cruz de la Cañada, founded in 1695; and Albuquerque, founded in 1706. The first two are located in the area known as the Rio Arriba, or the "upper river," while Albuquerque is located in the Rio Abajo, or "lower river." *La Bajada* mesa, nineteen miles south of Santa Fe, served as the dividing line; there the Santa Fe Plateau drops about 1,500 feet. Over the course of the colonial period these three centers and their environs, consisting of many smaller communities, contained the bulk of the Spanish population. Colonists also tended to radiate out from the three *villas* and establish small farms and ranches (*ranchos*), and hamlets (*plazas*). Yet frequently these more-isolated settlements were abandoned and sometimes returned to again as shifting migration became an important demographic feature of colonial New Mexico.

The Spanish population (including *castas*) in the upper province grew very gradually at first, set back by the Pueblo Revolt of 1680 and the economic unattractiveness of the region, and then increased more rapidly by the end of the eighteenth century. It was less than 1,000 in 1600, approximately 2,900 in 1680, 2,000 in 1700, 3,400 in 1752, 5,800 in 1776, 19,000 in 1800, and 28,000 in 1821.[11] More specifically, figures for the eighteenth and nineteenth centuries indicate shifts in population between and among the three *villas*: enumerations indicate that Spanish concentration was highest in Santa Fe in 1752, in Albuquerque in 1760 and 1776, and in Santa Cruz in 1790, 1800, and 1817. By 1817 Santa Cruz had a Spanish population of 12,903; Albuquerque, 8,160; and Santa Fe, 6,728.[12]

More Spanish and *castas* eventually came to live in the heart of the Rio Arriba, around the Villa of Santa Cruz, than in the vicinities of either Santa Fe or Albuquerque, reflecting both environmental realities and Spanish policies based upon them. In this protective terrain, with perennial supplies of water for irrigation, Spanish governors awarded communal land grants to groups of people, believing that they could protect themselves from attack by nomadic Indians. These grants, with their concentrated agricultural communities, served the purposes of the government by establishing controls over the citizens themselves and by creating an effective defensive frontier. In contrast, the settlement pattern in the Rio Abajo, along the broad, fertile plain of the Rio Grande came to be characterized by a more dispersed population occupying private land grants. Meanwhile, the Pueblo, dramatically reduced in numbers, clung tenaciously to some twenty-one discrete settlements, ranging from Taos in the north to Isleta in the south, and from Pecos in the east to Acoma, the sky Pueblo, in the west. Paradoxically, this reduction in Pueblo population meant that there were more human beings in the upper Rio Grande Valley at the beginning of the Spanish period in 1598 than at its close in 1821.

The Spanish introduction of grazing animals, termed *ganado mayor* (cattle and horses) and *ganado menor* (sheep and goats), into colonial New Mexico altered the face of the land in dramatic ways. In contrast to most other areas of Spanish America, where cattle constituted the principal livestock, sheep came to dominate the landscape of the upper Rio Grande Valley. There were several principal reasons for this development. Immense herds of buffalo on the plains to the east provided a source for hides, jerky, salted tongues, and tallow, products of a quality equal to those from domestic cattle. The hostile Indians surrounding the province, with the exception of the Navajo, usually coveted cattle and horses because they could be driven easily away. Sheep, on the other hand, could be scattered and then recovered after a raid. Throughout the course of the colonial period, the mining settlements of Nueva Vizcaya in Chihuahua and Durango provided a strong market for New Mexico sheep. During the eighteenth and early-nineteenth centuries the sheep trade became the primary export industry.[14]

Unlike the human population of the region, sheep came to be concentrated principally on the broad plains of the Rio Abajo, in and around the Villa of Albuquerque. A livestock enumeration in 1827, six years after Mexican independence, reveals nearly 250,000 sheep and goats in the province, with Albuquerque possessing 155,000, Santa Fe, 62,000, and Santa Cruz de la Cañada, 23,000. At the same time, there were only 5,000 cattle, 2,150 mules, and 850 horses.[15]

Those grazing animals, both sheep and cattle, had an important impact on the ecology of the area. Various descriptions of the land by early visitors provide a benchmark from which to view later changes. Two pre-settlement observers of New Mexico, Hernan Gallegos and Diego Perez de Luxan, representing the Chamuscado–Rodriquez (1581) and Espejo (1582) expeditions, respectively, noted lush grasslands, untouched pastures, highly suitable for both sheep and cattle.[16] After Spanish settlement, evidence indicates that portions of these grasslands were, over time, dramatically overgrazed. While many fewer in number than sheep, cattle also effected changes in the land in several principal ways. Whereas sheep were often grazed on distant pastures and required intensive labor, cattle were turned loose, unattended, on commons close by agricultural plots for safety from hostile Indians. As a result, the commons were frequently overgrazed. Marc Simmons says that Albuquerque's East Mesa, for example, was virtually denuded of grass by 1750, forcing cattle to invade the fenceless crop fields of both the Spanish and the Pueblo.[17]

Over the course of the colonial period, the Spanish archives of New Mexico contain numerous cases of New Mexico governors warning Spanish settlers to control their livestock, particularly from damaging Pueblo fields and irrigation ditches, or face severe penalties. As the number of sheep dramatically increased over the eighteenth and early-nineteenth centuries Spanish settlers frequently petitioned governors for fresh pasturage away from the core settlements of the Rio Abajo and Rio Arriba, looking particularly to the south of Albuquerque, to the Rio Puerco, and to the Chama Valley. Petitioners in the late 1730s, for example, complained about the inadequacy of land and water, especially the shortage of grass, near Albuquerque.[18]

Yet the needs of the settlers and goals of the state were often in conflict. Governors, representing the interests of the defensive function of the frontier, wanted to control and regulate shifts in settlement, especially on frontiers where conflicts with Indians were likely to occur. As a result, settlers' requests for fresh Crown land were often denied, and even in cases where their petitions were approved, they frequently were forced to return to the safe harbor of the three *villas* to escape Indian depredations. These barriers to expansion merely exacerbated livestock pressures on already settled land. By the 1820s, New Mexico sheep growers still sought fresh grass, this time to the east, onto the plains beyond the Sandia and Manzano Mountains.[19]

In contrast, the isolation of colonial New Mexico from the rest of New Spain and English North America clearly moderated the process of environmental change. In 1803 Governor Fernando de Chacon, in a report to officials in New Spain, vividly portrayed the region's general economy by noting that New Mexico's "natural decadence and backwardness is traceable to the lack of development and want of formal knowledge in agriculture, commerce and the manual arts."[20] Chacon thought that Spaniards and *castas* were little dedicated to farming and contented themselves with sowing and cultivating only what was necessary for their sustenance. While remarking on the abundance of sheep in the province, he indicated that no great number of swine existed. The continual raids of the nomadic Indians discouraged raising horses and mules. Deposits of minerals such as lead, tin, and copper were located in various parts of the province but they were virtually untapped, while large-scale smelting or amalgamation operations were nonexistent. Throughout New Mexico, high quality mica or gypsum (*yeso*) covered windows in place of glass panes. Formal apprenticeships, examinations for the office of master, and organized guilds, customary elsewhere in New Spain, did not exist. But out of necessity and the natural ingenuity of the people, according to Chacon, some trades were practiced skillfully, including weaving, shoemaking, carpentry, tailoring, smithing, and masonry. The exports of the colony, transported by the annual mule trains to Sonora, Vizcaya, Coahuila, and points south, consisted of oxen, sheep, woolen textiles, some raw cotton, hides, and piñon nuts. The total value of these products, including wine from the El Paso district, was estimated at only 140,000 pesos annually. The products brought back into the colony included linen goods, chocolate, sugar loaves, soap, rice, iron, leather goods of all sorts, pelts, paper, drugs, and some money. Because hard currency was in chronically short supply throughout the colonial period, barter was the principal mode of exchange.

Trade also existed between the Spanish, the Pueblo, and nomadic tribes. Taos became a principal center for exchange during times of peace, where, for horses, an array of metal tools, corn, and trinkets, the "uncivilized" Indians traded pelts, buffalo skins, and Indians whom they had captured from other tribes.[21] The fact that pelts were obtained either through import or at Taos indicates that little hunting of small or large game went on in the upper Rio Grande Valley itself.

In essence, colonial New Mexico demonstrated a low level of technological development, little or no occupational specialization, generally self-reliant local production for local consumption, and a broad utilization of the entire environment—all characteristics of a subsistence economy. It would be a mistake to assume that such characteristics did not lead to the overuse of resources or environmental degradation, but surely the degree of change was less than what might have occurred in a society distinguished by specialization in production and labor, and by the export and import of products to and from similarly specialized producers.

As an example, if large deposits of silver had been discovered in New Mexico during the Spanish period, as they were in other semi-arid regions of central and northern New Spain, the ecology of the area would have been transformed to a significantly greater degree. Silver mining required the special input of both raw materials and labor: a large supply of wood for fuel, shoring in mines, construction in buildings, and machinery; water for power and washing; and thousands of grazing animals to produce hide, leather, meat, and energy for transport and powering machines. Shifts in population occurred as well, to meet the intensive labor demands in the mines, and the human wastes from such concentrations were merely dumped into the local waterways. The amalgamation of silver also required the use of such products as copper sulfate, common salt, and in large quantities, mercury. Metallic mercury, mercury vapor, and lead, the residues of this process, poisoned both plants and animals as they readily invaded the air, water, and ground.

After only a few decades, the environmental effects of this economic activity were devastating. Entire areas in and around mining communities were denuded of grass, deforested, and eroded by wind and water. Vegetation loss also reduced transpiration, leading in turn to a decline in local rainfall. As food chains were destroyed, animals and fish disappeared, and it can be assumed that the results of mercury and lead poisoning in human beings, mostly Indian labor, were frequently fatal.[22] Nothing on this ecological scale occurred in the Province of New Mexico during the Spanish period.

Yet the Spanish introduction of metal tools into the upper Rio Grande Valley did allow both the Spanish and Pueblo to manipulate that environment in entirely new ways. The introduction of the axe alone enabled bench lands, bounded above and below by steeper slopes, to be cleared of dense vegetation, and woodland along the rivers and in the higher elevations to be cut. Wood became an important source for construction, tools, and fuel. The piñon, a scrub pine, was used to make plowshares and the legs of spinning wheels; the cottonwood to make wine barrels and *carreta* (cart) wheels; the oak to make stirrups and the *carreta* frame; and the Douglas fir to make the shafts of plows and provide large timbers for bridges and *vigas* (beams) in roofs.[23]

The degree to which the upper Rio Grande Valley was deforested as a result can be inferred from several commentaries made at the beginning and end of the Spanish period. In 1582, Luxan wrote that "this Province [New Mexico] boasts of many pine forests..." and that "there are also fine wooded mountains with trees of all kinds."[24] In 1839, in one of the first Anglo accounts of the region, Josiah Gregg, explorer and trader, wrote that "on the water-courses there is little lumber to be found except cottonwoods, scantily scattered along their banks. Those of the Rio del Norte [Rio Grande] are now nearly bare throughout the whole range of the settlements, and the inhabitants are forced to resort to the distant mountains for most of their fuel."[25] Another Anglo account sixteen years later corroborates this description. W.W.H. Davis observed that "wood is exceedingly scarce all over the country. The valleys are generally bare of it, and that found upon the mountains consists of a growth of scrub pine called the piñon. The country is said to have been well wooded when the Spanish first settled it, but in many parts it has been entirely cut off, and in some instances without even leaving a tree for shade."[26]

The slow development of Spanish material culture is most apparent when viewed in relation to that of English North America. As a result of the colony's isolation from European influence, the use of agricultural tools remained largely unchanged over the period of Spanish rule. The scratch plow, essentially as described in the Bible, was equipped with an iron, steel, or wooden share and cut a shallow furrow about six inches deep instead of horning the soil. It was not until the end of the Mexican period in 1846 that two-handle steel plows with a moldboard for horning the soil reached New Mexico.[27]

The ecological implication of this low-level technology is significant. Whereas in English North America the use of deep-cutting plows, in the absence of contour plowing, crop rotation and manuring, caused soil erosion on a massive scale, the scratch plow of New Mexico generated comparatively little soil loss. In fact, the farmlands of the upper Rio Grande Valley were auto-replenishing through the agency of silt-laden irrigation and flood waters and appeared, over time, to a number of colonial victors as particularly abundant and fertile.

Another side of material culture was the Pueblo and Spanish use of soil as a building material. While the Spanish introduced the formed standard adobe brick to the Pueblo, the techniques and materials for construction remained essentially unchanged from what the Indians had used before.[28] Adobe construction tapped several available and replenishing natural resources: loamy soil, sand, water, and straw. Wood was used only for ceiling beams or *vigas*. For the Pueblo, the adobe was one of the most prized Indian crafts as it represented the sacredness of the land itself. In contrast, the English use of wood for both construction and sale, while entailing a sense of craft, was predicated largely on an ethos of function and profit. As a result, Pueblo and Spanish buildings, decrepit when seen through the eyes of some nineteenth-century Anglo-American visitors, arose out of and were an integral part of the physical landscape, and unlike the acute deforestation which occurred in English North America, caused few changes in the land.

After the Pueblo Revolt the Spanish imposed a pattern of land use and settlement on the upper Rio Grande Valley in marked contrast to the widely scattered large estates, worked primarily by Indians through the *encomienda*, of the early seventeenth century. On the basis of individual and communal land grants the Spanish came to live in smaller units strung out along the Rio Grande and its tributaries. In the Rio Arriba, where the colony's Spanish population became most concentrated, the communal land grant dominated.

In this case individuals in a group of settlers would each receive an allotment of land for a house, a plot for irrigation, and the right to use the unallotted land on the grant in common with the other settlers for pastures, watering places, firewood, logs for building, and rock quarrying, among other activities.[29] The intensity of land use under these circumstances certainly differed by degree between and among the various grants. But the very configuration of these settlements, coupled with barriers to out-migration, helped to produce such changes in the land as overgrazing and deforestation, the subsequent loss of topsoil, and the silting of irrigation ditches and streams.

A pattern of land-holding developed in the region, both on and outside of the community grants, of long-lots or narrow strips of land emanating from the waterways. They were a necessary response to the physical environment and represented a practical and equitable method of partitioning irrigable land. Yet over generations, through the institution of partible inheritance, long-lots were divided and redivided lengthwise, leading to more densely populated communities living on increasingly smaller parcels of land.[30] As a result settlers had to move their grazing and cutting operations ever higher up the mountain sides to accommodate the loss of grass and trees on the original commons. At the same time overused irrigation plots for growing wheat and garden vegetables became increasingly less productive.

On the other hand, metes and bounds, a legal system by which property boundaries were determined, was a feature of Spanish land institutions that favored environmental conservation. Boundaries were not established with any prescribed shape in mind, nor were they necessarily contiguous with any others. Instead, they were drawn to follow the natural contours of the land and to include the most valuable resources: soil, woodlands, and access to water. It was a system highly adaptive to the local environment,[31] and predicated, at least in the case of colonial New Mexico, on the needs of subsistence agriculture.

A system of grids, based upon the rectangular survey, came to prevail in many parts of Anglo-America with far more severe environmental consequences. There, little allowance was made for local topography, hydrology, or climate. Fields were arranged according to a rigid north-south, east-west alignment, often resulting in enhanced soil erosion. The distribution of surface water, arable land, grass suitable for grazing, and timber for securing wood was often unequal between and among individual holdings. As a result, property owners tended to make extreme demands upon natural resources that were frequently in short supply. In contrast to the communal restraints evident in parts of colonial New Mexico, in order to maximize profit in a market economy, owners were free to exploit their local environment as they saw fit.[32]

With roots deep in Iberia, Spanish irrigation was at once vital to agricultural productivity and a source of land change and deterioration. The pre-contact Pueblo use of water control devices, the complexity of which is still debated, led to the accumulation of salts and other mineral deposits in the soil and the consequent need to seek new growing areas in the face of an expanding population and shrinking bottomland. The Spanish system, borrowed by the Pueblos, placed a more complex pattern of ditches on the face of the land with even greater natural and physical effects. In both the Rio Arriba and Rio Abajo, the Spanish system of irrigation consisted primarily of a main ditch, the *acequia madre*, receiving water from a river or stream at a higher elevation than the lands being irrigated, and then relying on gravity to carry the flow. Secondary ditches (*sangrias*) branched off the main ditch and directed water to individual fields. Gates were made of earth and boulders to regulate the flow, and flumes (hollowed out logs), provided an elevated channel for water to cross gullies and ravines.[33]

Over the years riparian lines of trees and shrubs developed alongside well-maintained ditches, replicating the biotic environment found along unimpeded streams. In other instances, grazing animals invaded both Spanish and Pueblo fields and trampled the sandy banks along the water courses, filling them or causing breaks which allowed water to escape.[34] Abandoned ditches were often transformed into gullies or small arroyos as the result of runoff and soil erosion.

The problem of salinization that confronted pre-contact Pueblo farmers was intensified under the Spanish. Alkali compounds, consisting of various salts, are characteristic components of arid and semi-arid soils; they are also highly soluble in water. When dissolved by irrigation, the water and alkali enter the soil together, then return to the surface by capillary action, much the same way that oil flows up a lamp wick. The sun then evaporates the water and leaves behind the salts to act as corrosive agents on the stems of plants.[35]

The effects of such phenomena as salinization, deforestation, overgrazing, and population increase reveal that selected areas in both the Rio Arriba and Rio Abajo were exceeding their carrying capacity by the end of the eighteenth century. While remaining relatively small in numbers and employing a subsistence-based economy, the Spanish in the upper Rio Grande Valley had not yet achieved a totally sustainable society in relation to their environment.

While issues of power and control, particularly during the seventeenth century, were often contested at the local level between the Franciscan missionaries and the governor, the latter official influenced the course of environmental change more strongly. Directly responsible to the Viceroy, and after 1776, the Commandante General of the Provincias Internas, the governor appointed or approved lesser officials, enforced royal decrees, and ordered the formation of militia from settlements. More importantly from an environmental perspective, he made land grants which were intended primarily to manage the settlement of the Spanish and *casta* population within the colony. For in theory at least, people were simply not allowed to live where they liked or to move at their own discretion to another location. In the process, Spanish law strictly outlawed such hallmarks of the Anglo-American frontier as land speculation and absentee ownership.

This kind of social control had several ecological effects. In some cases, a governor's decision to award, or even revoke, a land grant led to further deterioration of already occupied land. One such instance occurred in 1735 when Governor Cruzat y Gongora nullified a number of grants made by the acting governor, Juan Paez Hurtado, in the area of the Chama Valley. Facing overcrowded living conditions in the vicinity of the lower valley and Santa Cruz, settlers had sought fresh land for grazing. Gongora's decision to rescind the grants was based in part on his personal need to control and regulate the advancement of settlements, but it was also based on his belief that the upper Chama was a place where colonists and Utes in close contact might precipitate a war.[36] Here, environmental concerns took a back seat to issues of colonial security.

Built into the very land grant process was a step tantamount to the environmental impact study of today. The *alcalde mayor*, a lower-level but important official appointed by the governor, had to determine whether the proposed land grant in his jurisdiction would adversely affect any Pueblo settlement or other third party, and whether there was sufficient water for irrigation and livestock and enough cultivable, grazing, and wood-producing land to support the proposed number of settlers.[37] While this system was hardly perfect, the land grant papers of New Mexico are replete with examples of the *alcalde mayor* addressing these issues in writing to the governor, and sometimes recommending to him that a grant be denied because criteria were not met. While the system was intended to preserve the economic survival of the colony and not the environment, it nonetheless had the long-term effect of conserving the land and water of colonial New Mexico.

The degree to which Pueblo and Spanish society created an inclusive frontier also directly affected issues of land use and change. While Pueblo people embraced Old World plants, animals, and material culture introduced by the Spanish, they did not accept, in any profound sense, Western belief systems, particularly those centering upon religion or property. In anthropological terms, the Pueblo became acculturated, but not assimilated into Spanish society. Although greatly reduced in numbers, they lived under Spanish rule in their own compact, autonomous communities, as they do in the modern era.

Through intermarriage and other forms of contact, the Spanish borrowed culturally from the Pueblo. This occurred more frequently in outlying communities rather than the missions or *villas*. The long-term effect of this two-way process was to produce both a hard and soft impact upon the land. The soft impact emphasizes the influence of Pueblo beliefs and customs upon the Spanish settlers. The hard impact focuses principally upon Pueblo acculturation.

Richard Ford has offered a succinct view of the existence of one Pueblo group, the Tewa, prior to contact and the subsequent effects of the Columbian Exchange upon them and their relationship to the land. The Tewa lived on maize supplemented with squash and beans, and gathered plant products. Rabbits, hares, deer, and other game provided a source for meat. Firewood came from deadwood, and construction timbers were usually recycled from older structures. In essence, according to Ford, this was an ecosystem that could rapidly recover when fields were abandoned or when the human population founded a new village elsewhere.[38]

With the arrival of don Juan de Oñate in 1598, the Tewa economy changed dramatically. Five Spanish contributions had particular impact: the introduction of spring wheat; kitchen gardens grown with irrigation water; orchards of peach, apricot, plum, and cherry trees; grazing animals; and the metal axe. The result for the Pueblo was a strange admixture of environmental degradation to their land and "a beneficial and more secure subsistence base."[39] Deforestation and overgrazing occurred on Pueblo land just as it did on that of the Spanish. But the new food sources only reinforced the adaptive capacity of the Tewa. Wheat, in particular, while not displacing corn, came to serve as "a high yield caloric safety valve" for the Pueblo. It is likely that the Pueblo partially compensated for the local overgrazing and loss of cool-season grasses such as mutton grass, Indian rice grass, and June grass, by raising wheat varieties that matured about the same time.[40] These additions to the Pueblo economy also tempered their need to migrate as they might have in pre-contact times owing to political disputes, scarcity of wood, or unproductive fields. Even as their landscape was simplified, the Tewa expanded their control over the productivity of their land and created a measure of economic security.[41] At least in a material sense, the Pueblo came to more closely resemble their European neighbors.

The inclusive nature of Pueblo–Spanish society entailed accommodative elements as well as more disruptive ones. Whereas students of Pueblo–Spanish relations can determine with some certainty the far-reaching ecological effects of the Spanish donations to the Pueblo in the form of plants, animals, and material culture, it is much more difficult to assess how Pueblo belief systems, and their practice of a particular subsistence agriculture, affected Spanish settlers and the environment of the upper Rio Grande Valley. As Christopher Vecsey has noted, all Native American groups "established a religious association with nature that transcended but did not nullify their effective exploitation of the environment. They achieved an integration of subsistence and religious activities."[42] He includes the Pueblo among those American Indians whose religious core was molded by environmental relations: their fertilization ritual, for example, was about maintaining life in humans, plants, animals, and the world at large, benefiting, in turn, nature and culture alike.

To what degree Spanish settlers adopted these attitudes from the Pueblo is impossible to determine, but more cultural borrowing likely went on between the two groups than in almost any other region of Spanish America. Frances Swadesh has suggested that what evolved in the Province of New Mexico after the seventeenth century was a non-dominant frontier community. Relations between settlers and their Indian neighbors, especially in outlying areas, were far more egalitarian than the colonial model for social relations intended.[43] The traditional Spanish barriers to social mobility, including caste, class, and ethnic identity, were largely absent. Archival records also reveal that a not insignificant number of Spanish settlers moved into Pueblo communities; both men and women married Pueblo Indians and raised their children in the spouse's pueblo. As a result, according to Swadesh, Spanish settlers oftentimes "found themselves more at odds with their own colonial authorities than with their Indian trading partners, compadres, and friends."[44] As settlers acquired knowledge of herbs and wild plant foods from the Pueblo, and as they witnessed or participated in an agricultural cycle both religious and secular in meaning, they may also have internalized a view of the land predicated on principles other than the usual European ones of strict utility. And while this process hardly mitigated the harsher effects of Spanish culture on Pueblo land use, it may have changed or softened the attitude of some Spanish toward exactly how the physical and natural world should be exploited.

Any analysis of land change must also address natural forces, especially variations in climate, and their impact upon biophysical processes. Prolonged drought, heavy rains, and extremes of heat and cold may effect environmental changes entirely independent of human agency. They may also interact with human activities and speed along the process of change. In the case of New Mexico, there are essentially three types of climate: arid, semi-arid and subhumid/humid. Areas of arid climate have scrubby, heat- and drought-resistant desert plant cover; semi-arid areas have a vegetation of short, bunchy grasses; and the subhumid/humid areas, in the hilly or mountainous regions, have woodland or forest cover.[45] Latitude, elevation, and location with regard to moisture-laden winds primarily determine these variations in climate and consequent vegetation. Albuquerque, at the northern point of the arid range, and Santa Fe, in a semi-arid range closely bounded by the mountains, although only sixty miles apart, often exhibit marked differences in local climate. Such differences mean that it is very hard to associate general climate patterns with ecological change in any specific locale.

From a broader perspective, evidence from tree-ring studies (dendroclimatology) indicates that the climate of the upper Rio Grande Valley between 1598 and 1821 fluctuated fairly regularly between wet and cold and warm and dry periods. For western North America, periods of widespread drought occurred between 1626 and 1635, and 1776 and 1785, while periods of above-average moisture occurred from 1611 to 1625, 1641 to 1650, and 1741 to 1755. Chronologies from the upper Rio Grande Valley (lat. 35–43; long. 106–06; based on rings from Douglas fir, ponderosa pine, and piñon pine) reveal dryer than normal periods between 1661 and 1675, 1726 and 1765, and 1796 and 1830, and wetter than normal periods between 1601 and 1645, 1706 and 1730, and 1781 and 1800.[46] These changes in climate clearly had the potential to effect changes in the land, both on their own and in concert with human activity. Drought or extended periods of deficient precipitation may produce, among other effects, a marked decline in plant cover, even in the absence of grazing animals. Particularly vulnerable to drought are perennial grasses like the gramas, which may die off and then be replaced by invaders such as sagebrush, snakewood, and rabbit brush.

But several variables bear strongly on the amount of plant life lost during any particular drought. They include the time of year in which the drought occurs, winter and spring having more adverse effects than summer and fall, and the texture of the soil. More heavily textured soils may retain more moisture, release it to plant use more slowly, and thus allow perennial grasses to survive. Likewise, periods that are colder and wetter than normal can produce environmental change. An excess of rain and snow, and their runoff, have the potential to cause soil erosion, arroyo formation, changes in stream flow, and flooding. Such effects are particularly acute when the climate changes quickly and dramatically from an excessively dry to wet period.[47]

Further inferences can be drawn from the relationship between fluctuations in climate, human activity—including that of Pueblo and Spanish—and ecological change. While both climate and humans, acting independently of one another, may simplify an ecosystem, together their effects may be far greater and more long-lasting. The formation of arroyos, valley bottom gullies characterized by steeply sloping or vertical walls, offers a case in point. Both human land use and climatic changes are among the complex causes leading to their creation. On the human side, logging, fire, grazing, cultivation, and the existence of roads, trails, and irrigation ditches can lead to the local removal of vegetation on valley floors and alter valley bottom soils. This increases the erodibility of valley floor material. Both an increase and decrease in humidity can contribute to arroyo formation. Increase in precipitation and decrease in temperature may increase runoff from slopes, while a decrease in precipitation and increase in temperature may reduce the vegetation cover over drainage basins. Both phenomena increase the velocity of flows through valley bottoms. Individually, but especially together, human and climatic factors lead to localized erosion and arroyo initiation.[48]

Finally, this discussion of environmental factors must be viewed in the context of the inherent vulnerability or resilience of a semi-arid ecosystem. Ecologists no longer assume that the degree of diversity and complexity of an ecosystem determines stability or change. Other factors such as elasticity (ability to recover from damage), and inertia (ability to resist displacement), are now considered more important. Thus desert or semidesert regions may be more resilient and less vulnerable to change than other, more complex ecosystems such as woodlands or rain forests. The flora and fauna of arid regions evolved in an environment where the normal pattern is more or less random alteration of short favorable periods and long stress periods. Ecologists posit that plants and animals have preadapted resilience. Applying this idea to the upper Rio Grande Valley leads to several conclusions. On the one hand, grassland and woodland, for example, might well have survived in some instances the invasion of sheep, cattle, and axe. In fact, some argue that limited or optimal grazing may even enhance the growth of certain grasslands.[49] In contrast to the total absence of grazing or excessive grazing, light grazing favors production of some grasses by restoring nutrients through feces and urine. Even if damaged, grasslands have the potential often to undergo a regeneration of growth. The carrying capacity of any specific ecosystem might be sustained on account of plant and animal resilience, or even through the "benefits" of limited human economic activities upon them. Yet it is clear that human agency, both Pueblo and Spanish, in this area contributed to fundamental changes in the land: the carrying capacity of selected areas was exceeded as the land proved to be more fragile than resilient. It can even be argued that a process of desertification began in selected areas where grazing was particularly heavy. That Anglo-Americans, upon their arrival after 1821, exacerbated those changes manifoldly does not negate this fact.

It is tempting to assess the long-run effects of the Spanish colonial presence in the upper Rio Grande Valley on environmental change as constituting a middle ground between those changes which occurred first under the Pueblo and then later under the Anglo-Americans. The interaction of various factors, in effect, produced a rate and kind of change certainly exceeding that experienced under the Pueblo but falling far short of that produced by Anglos. The acceleration in deforestation and overgrazing in New Mexico after the Anglo-American occupation in 1846, and especially after the introduction of the railroad in 1880, demonstrates that this assessment is partially correct. But the kind of change that occurred between 1598 and 1821 is more complex than merely a linear one. Change was also cyclical and layered or superimposed in nature. The rate of change was at once fast, slow, intermittent, and inexorable, and only rarely constant.

Several primary factors account for these variations in change. People did not settle evenly or randomly on the land. Instead they concentrated themselves in some areas, settled sparsely in others, on some occasions moved for environmental or other reasons, and on other occasions abandoned their land and moved back to where they had come from. Nor were these same people engaged in uniform economic activities having equal impact upon the land. Natural forces acted both alone and in concert with human activities on the land in an unpredictable and capricious manner. For instance, the frequently changing course of the Rio Grande, resulting from its own natural meander, precipitated the cyclic destruction and regeneration of the woodland or bosque habitat of the river.[50] Overgrazing and drought together, followed by above-normal precipitation, enhanced the opportunity for arroyo formation. Finally, the resilience or fragility of the land itself depended upon variations in the adaptability of local flora and fauna, and the degree to which both humans and natural forces could disturb or modify them. How changes occurred in the land was integrally tied to all of these factors.

Change was most linear in the heavier population centers of Santa Fe, Santa Cruz de la Cañada, and Albuquerque. Descriptions of these areas from the sixteenth century *entradas,* when juxtaposed to Anglo-American accounts of the early- to middle-nineteenth century, indicate the sort of dramatic shift akin to descriptions of environmental change in colonial New England and the Pacific Northwest. The archival record for Santa Fe over the colonial period notes steady loss of natural resources in the form of water, grass, and wood to the point that a governor in the late-eighteenth century recommended that the capital be moved to the confluence of the Santa Fe and Rio Grande Rivers.[51] Similarly, the Albuquerque area underwent a gradual yet inexorable process of desertification that resulted from overgrazing by sheep.

In contrast, cyclical change took place most frequently when settlers, usually on account of land pressure in and around the three *villas,* occupied land at the edge of the New Mexico frontier. Settlements in the north around Taos, in the northwest on the Chama River Valley, in the West out to the Puerco River, in the south beyond Isleta, and in the east to the Pecos River, throughout most of the colonial period, were often in flux, sometime abandoned outright on account of Indian hostilities, and sometimes even resettled after a several-year hiatus. As a result, changes in the land that would have occurred from permanent settlement might have commenced only to be interrupted by abandonment, and then recommenced. Deforestation and overgrazing in these areas rarely took place in a sustained way.

In this context, the year 1790 is notable in colonial New Mexico history in that it marked the beginning of relatively peaceful times between the Spanish, the Pueblo, and the nomadic tribes, especially the Apache and Comanche. The outward expansion that followed came to resemble more closely the Anglo-American frontier of an ever-receding line. It also meant that the cyclical nature of ecological change came largely to an end, as places such as the Puerco River Valley began to fill with settlers so steadily that by the middle- to late-nineteenth century become as severely overgrazed and eroded as any place in New Mexico.

Environmental change might also be characterized as layered or superimposed, as in a kind of palimpsest. This occurred most often where Pueblo communities adopted the plants, animals and material culture of the Spanish, moving from a system of extensive to intensive agriculture. In effect, Pueblo land use changed by degree rather than by radical transformation. The growing of wheat required the expansion of Spanish-style irrigation, with concomitant environmental effects: and as the Pueblo came to rely on new domesticated plants and animals, their hunting and gathering of wild plants and animals declined. So the pre-contact Pueblo practices of land use and their effects on land change intensified as Spanish agricultural practices became superimposed on Pueblo practices through the process of acculturation. The dramatic decline in Pueblo population prevented their communities from exceeding carrying capacity as, in selected cases, they were in danger of on the eve of contact.

Environmental change in colonial New Mexico was *sui generis*, predicated on circumstances and conditions reflective of the particular climate, geography, and cultures of the region. The dialectical relationship between different kinds of factors produced varied kinds and rates of change unlike that experienced on any other North American frontier. This study shows the limitation of relying on mode of production or economic systems as a single or even primary explanation for change. They are clearly more useful as a means of analysis from a macro perspective such as representing, in broad strokes, the environmental change which occurred under preindustrial Pueblo and Spanish society through Anglo occupation. A look at the micro level reveals the difficulty in assigning any such clear causality for change. Perhaps the challenge before us is to mediate between the two approaches.

ACKNOWLEDGMENTS

Research for this article was partially funded by a grant from the National Endowment for the Humanities. I would especially like to thank Dan Scurlock, Rose Diaz, Bill Tydeman, Don Dreesen. and Bob Delaney, in Albuquerque, and Phyllis Monday MacCameron, Tom Rocks, Peter Murphy, Anne Bertholf, and Candy Carroll, in Buffalo, for their encouragement and assistance. I would ago like to thank Hal Rothman of *Environmental History Review* and two anonymous readers for helpful comments on the manuscript.

NOTES

1. William Cronon, Changes in the Land: Indians, Colonists and the Ecology of New England (New York: Hill and Wang, 1983), vii.
2. Cronon, Changes in the Land, vii.
3. Richard White, *Land Use, Environment, and Social Change: The Shaping of Island County, Washington* (Seattle: University of Washington Press, 1980), 36. Neither Cronon nor White subscribes to the notion that Native Americans lived in any sort of perfect harmony or static relationship with nature prior to contact with Europeans. Cronon notes that the Indian practice in New England of burning forests to clear land for agriculture and to improve hunting "could sometimes go so far as to remove the forest altogether, with deleterious effects for trees and Indians alike." And White points out that the natives of Island County did not hesitate to alter natural systems when it was to their advantage to do so. For example, through the manipulation of the environment the Salish increased such desired plant species as bracken, nettles, and camas for use as food crops. Rather, most remarkable was the degree of environmental change that occurred in these far corners of English North America after contact. For both Cronon and White such change was both rapid and essentially linear.
4. Scholars generally agree, on the basis of archaeology and Spanish accounts, that pre-contact Pueblo settlements used some form of irrigation farming to grow maize, beans, squash, cotton, and tobacco. While access to surface water removed certain restrictions and risks to agriculture, and, indeed, created surpluses allowing for greater social elaboration, studies now also indicate that environmental factors, some the result of using water-control devices, constrained Pueblo farming and in the process effected changes in the land.

Irrigation farming, even in its crudest forms, likely sets in motion ecological chain reactions. Plants are introduced into habitats where they could not have sustained themselves previously; terracing changes the natural flow of streams; and man-made water diversions modify the natural vegetation, change the organic matter in the soil, and perhaps alter the migrating pattern of birds and animals. See Michael C. Meyer, *Water in the Hispanic Southwest: A Social and Legal History, 1550–1850* (Tucson: University of Arizona Press, 1984), 19.

In addition, poor drainage and high evaporation can lead irrigation waters to deposit salts and other minerals that inhibit crop production. And as crop irrigation brings both fields and plants closer together the risk of crop loss due to diseases and insect pests is increased. See Linda S. Cordell, *Prehistory of the Southwest* (Orlando: Academic Press, Inc., 1984), 203–4.

Due to these factors, Cordell believes that good bottomland in the upper Rio Grande Valley was very likely in increasingly short supply over the two centuries prior to the arrival of the Spanish. Studies of Pueblo land use and evidence of expansion, away from the river onto the plains, indicate that intensive efforts were being made to support a large population increase as pueblos were abandoned and new communities founded. See Cordell, "Prehistory: Eastern Anasazi," in *Handbook of North American Indians*, vol. 9, ed. Alfonso Ortiz (Washington: Smithsonian Institution, 1979), 151. If this interpretation is correct, the carrying capacity (the maximum population that a particular environment can support indefinitely without leading to environmental degradation) along the Rio Grande itself may well have been reaching its limit at the time of Spanish contact.

5. A useful point of departure for understanding similarities and differences in frontier societies, and their relationship to the land, is the notion of an inclusive versus exclusive frontier. Spanish colonists frequently settled in areas of sedentary Indians, seeking Indian labor at the same time that they strove for Indian souls and mated with their women. Where there were no Indians, there were no Spanish. Without a sharply defined racial barrier, this was a frontier of inclusion. In contrast, English colonists, while integrating economically with certain native populations to a degree, did not intermarry. Nor on any significant scale did they attempt to convert Indians to Protestantism. This then was a frontier of exclusion. See Alistair Hennessy, *The Frontier in Latin American History* (Albuquerque: University of New Mexico Press, 1978), 146–47; and Alfred Crosby, Jr., *The Columbian Exchange: The Biological and Cultural Consequences of 1492* (Westport, CT: Greenwood Press, 1972), and *Ecological Imperialism: The Biological Expansion of Europe 900–1900* (Cambridge: Cambridge University Press, 1986).

Add to this broad characterization the fact that the Province of New Mexico served as a defensive and missionary outpost over most of its colonial history. It never was an economic frontier on the order of New England, for example, a region which quickly became integrated into an international commercial trade network. In fact, Spain's loss of portions of its North American empire can be attributed to the advantages of England's expanding economic frontier, over Spain's defensive frontier, operating at long distance from centers of resources and population.

6. Richard I. Ford, "The New Pueblo Economy," in *When Cultures Meet: Remembering San Gabriel Del Yuge Oweenge* (Santa Fe: Sunstone Press, 1987), 73.

7. John R. Van Ness, "Hispanic Land Grants: Ecology and Subsistence in the Uplands of Northern New Mexico and Southern Colorado," in *Land, Water and Culture: New Perspec-

tives on Hispanic Land Grants, ed. Charles L. Briggs and John R. Van Ness (Albuquerque: University of New Mexico Press, 1987), 204.

8. Hal Rothman, "Cultural and Environmental Change on the Pajarito Plateau," *New Mexico Historical Review* 64 (April 1989): 186. Rothman agrees fundamentally with Van Ness. He states that "American influence telescoped into a few years much more environmental and cultural change than Spanish practices had produced in nearly three hundred years." There were two reasons for these varying rates of change, according to Rothman. First, a marginal area such as New Mexico did not attract sufficient numbers of Europeans to effect dramatic environmental change; and, second, the "un-European," semi-arid climate of New Mexico protected it from the "full brunt of the portmanteau biota of the Spanish" (188). Because many Old World plants, such as fruit trees, melons, and wheat, could exist only in proximity to water, more was needed than merely the presence of the Spanish and their descendants to "Europeanize" the plants and animals of New Mexico. What was required was the transformation of New Mexico into an economic frontier, creating opportunities to produce and transport commodities to market on a large scale.

9. Donald Worster, "Toward an Agroecological Perspective in History," *Journal of American History* 76 (March 1990): 1101.

10. Over the period of colonial rule, the Spanish conducted regular, and, by most accounts, fairly accurate censuses of the Province of New Mexico. (Because this study focuses upon the upper Rio Grande Valley, the following population figures excluded the areas of El Paso and Zuni.) Only the estimates of Pueblo population at the time of the first Spanish settlement in 1598 and the Benavides counts of 1630 and 1635 have been the subject of any real debate. The figure for 1598 is generally agreed to be around 38,000, while Benavides estimated some 26,000 Pueblo Indians in 1630. The sharp decline was attributable to disease, starvation owing to Spanish tribute and labor institutions, and flight to western Pueblos. Throughout the remainder of the seventeenth century and for most of the eighteenth century, Pueblo population continued to decline. It was approximately 23,600 in 1680, the year of the Pueblo Revolt, 7,200 in 1706, 5,200 in 1752, 6,500 in 1776, 6,400 in 1805 and 1821. See Marc Simmons, "History of Pueblo-Spanish Relations to 1821," in *Handbook of North American Indians*, vol. 19, ed. Alfonso Ortiz (Washington: Smithsonian Institution, 1979), 185.

11. Oakah L. Jones, Jr., *Los Paisanos: Spanish Settlers on the Northern Frontier of New Spain* (Norman: University of Oklahoma Press, 1979), 117–29.

12. Jones, *Los Paisanos*, 117–29.

13. Alvar W. Carlson, The Spanish-American Homeland: Four Centuries in New Mexico's Rio Arriba (Baltimore: The Johns Hopkins Press, 1990), 9.

14. Marc Simmons, "The Rise of New Mexico Cattle Ranching," *El Palacio* 93 (Spring 1988): 7.

15. John O. Baxter, *Las Carneradas: Sheep Trade in New Mexico, 1700–1860* (Albuquerque: University of New Mexico Press, 1987), 90.

16. George P. Hammond and Agapito Rey, eds., *The Rediscovery of New Mexico, 1580–1594* (Albuquerque: University of New Mexico Press, 1966), 89, 230.

17. Simmons, "New Mexico Cattle Ranching," 7.

18. Baxter, *Las Carneradas*, 24.

19. Ibid., 92.

20. Marc Simmons, "The Chacon Economic Report of 1803," *New Mexico Historical Review* 60 (January 1985): 87.

21. David J. Weber, *The Taos Trappers: The Fur Trade in the Far Southwest, 1540–1846* (Norman: University of Oklahoma Press, 1971), 10.

22. Peter Bakewell, "Ecological Effects of Silver Mining in Colonial Spanish America" (paper presented at the annual meeting of the American Historical Association, December 1985.

23. Hester Hones, "Uses of Wood by the Spanish Colonists in New Mexico," *New Mexico Historical Review* 7 (July 1932).

24. Hammond and Rey, *The Rediscovery of New Mexico*, 221, 230.

25. Josiah Gregg, *Commerce of the Prairies* (Norman, University of Oklahoma Press, 1954), 113.

26. W.W.H. Davis, *El Gringo: New Mexico and Her People* (Lincoln: University of Nebraska Press, 1982), 356.

27. Simmons, "New Mexico's Colonial Agriculture," *El Palacio* 89 (Spring 1983): 9.

28. Edward W. Smith, *Adobe Bricks in New Mexico* (Socorro: New Mexico Bureau of Mines and Mineral Resources, 1982), 1113.

29. Malcolm Ebright, "New Mexican Land Grants: The Legal Background," in *Land, Water, and Culture: New Perspectives on Hispanic Land Grants*, 23.

30. Carlson, *The Spanish-American Homeland*, 31, 69–70.

31. D.W. Meinig, *The Shaping of America: A Geographical Perspective on 500 Years of History* (New Haven: Yale University Press, 1986), 1:240.

32. John Van Ness, "Hispanic Land Grants," 193–94.

33. Nancy Hunter Warren, "The Irrigation Ditch (Photo Essay)," *El Palacio* 86 (Spring 1980): 28. See also Daniel Tyler, "Dating the Caño Ditch: Detective Work in the Pojoaque Valley," *New Mexico Historical Review* 61 (January 1986), and Stanley Crawford, *Mayordomo: Chronicle of an Acequia in Northern New Mexico* (Albuquerque: University of New Mexico Press, 1988).

34. Marc Simmons, "Spanish Irrigation Practices in New Mexico," *New Mexico Historical Review* 47 (April 1972): 145.

35. Arthur Goss, "The Value of Rio Grande Water for the Purpose of Irrigation" (New Mexico College of Agriculture and the Mechanic Arts, Agricultural Experiment Station, Bulletin, November 1893), 34.

36. Frank E. Wozniak, "Irrigation in the Rio Grande Valley, New Mexico: A Study of the Development of Irrigation Systems Before 1945" (The New Mexico Historic Preservation Division, Santa Fe, 1987, photocopy), 38.

37. Wozniak, "Irrigation in the Rio Grande Valley, New Mexico," 25.

38. Ford, "The New Pueblo Economy," 74.

39. Ibid., 86.

40. Vorsila L. Bohrer, "The Prehistoric and Historic Role of the Cool-Season Grasses in the Southwest," *Economic Botany* 29 (July–September 1975): 203.

41. Ford, "The New Pueblo Economy," 87.

42. Christopher Vecsey, "American Indian Environmental Religions," in *American Indian Environments: Ecological Issues in Native American History*, ed. Christopher Vecsey and Robert W. Venables (Syracuse: Syracuse University Press, 1980), 10.

43. Frances Leon Swadesh, "Structure of Spanish-Indian Relations in New Mexico," in *The Survival of Spanish American Villages*, ed. Paul Kutsche (Colorado Springs: Research Committee, Colorado College, 1979), 53–61.

44. Swadesh, "Structure of Spanish-Indian Relations in New Mexico," 61.

45. Yi-Fu Tuan, Cyril E. Everard, and Jerold G. Widdison, *The Climate of New Mexico* (Santa Fe: State Planning Office, 1969), 158.

46. Harold C. Fritts, "Tree-Ring Evidence for Climatic Changes in Western North America," *Monthly Weather Review* 93 (July 1965): 421, 430–31.

47. Fluctuations in the climate of colonial New Mexico can be viewed as well in the broader context of the climatic phenomenon known as the Little Ice Age. Between 1430 and 1850, the Northern Hemisphere's climate was allegedly cooler than periods either before or after. Therefore, the upper Rio Grande, over this time, may have experienced, on average, larger snow cover, enhanced freezing, and, therefore, shorter growing seasons. On the other hand, cooler and more moist summers may have produced excellent harvests. More importantly, however, climate conditions during the Little Ice Age were far from stable, and there were complex spatial patterns of warming and cooling throughout the period. To draw causal relationships between the LIA and long-term environmental change in the upper Rio Grande Valley is therefore risky at best. See T.M.L. Wrigley et al., eds., *Climate and History: Studies in Past Climates and Their Impact on Man* (Cambridge: Cambridge University Press, 1981), 17.

48. Ronald U. Cooke and Richard W. Reeves, *Arroyos and Environmental Change in the American South West* (Oxford: Clarendon Press, 1976), 16.

49. John R. Kummel and Melvin I. Dyer, "Consumers in Agroecosystems: A Landscape Perspective," in *Agricultural Ecosystems: Unifying Concepts*, ed. Richard Lowrance, Bejamin R. Stinner, and Garfield J. Hause (New York: John Wiley & Sons, 1984), 65.

50. Dan Scurlock, "The Rio Grande Bosque: Ever Changing," *New Mexico Historical Review* 63 (April 1988): 135.

51. , Spanish Archives of New Mexico I, 1118, New Mexico State Archives and Records Center, Santa Fe, NM.

CHAPTER FOUR

BISON ECOLOGY AND BISON DIPLOMACY
The Southern Plains from 1800 to 1850

DAN FLORES

In bright spring light on the Great Plains of two centuries ago, governor Juan Bautista de Anza failed in the last of the three crucial tasks that his superiors had set him as part of their effort to reform New Mexico's Comanche policy. Over half a decade, Anza had followed one success with another. He had brilliantly defeated the formidable Comanche *nomnekaht* (war leader) Cuerno Verde in 1779, and as a consequence in 1786, he had personally fashioned the long-sought peace between New Mexico and the swelling Comanche population of the Southern Plains. His third task was to persuade the Comanches to settle in permanent villages and to farm.[1]

But the New Mexico governor found the third undertaking impossible. Observers of Plains Indian life for 250 years and committed to encouraging agriculture over hunting, the Spaniards were certain that the culture of the horse Indians was ephemeral, that the bison on which they depended were an exhaustible resource. Thus Anza pleaded with the tribes to give up the chase. The Comanches thought him unconvincing. Recently liberated by horse culture and by the teeming wildlife of the High Plains, their bands found the Arkansas River pueblo the governor built for them unendurable. They returned to the hunt with the evident expectation that their life as buffalo hunters was an endless cycle. And yet Anza proved to be a prophet. Within little more than half a century, the Comanches and other tribes of the Southern Plains were routinely suffering from starvation and complaining of shortages of bison. What had happened?[2]

Environmental historians and ethnohistorians whose interests have been environmental topics have in the two past decades been responsible for many of our most valuable recent insights into the history of Native Americans since their contact with Euro-Americans.[3] Thus far, however, modern scholarship has not reevaluated the most visible historic interaction, the set piece if you will, of Native American environmental history.[4] On the Great Plains of the American West during the two centuries from 1680 to 1880, almost three dozen Native American groups adopted horse-propelled, bison-hunting cultures that defined "Indianness" for white Americans and most of the world. It is the end of this process that has most captured the popular imagination: the military campaigns against and the brutal incarceration of the horse Indians, accompanied by the astonishingly rapid elimination of bison, and of an old ecology that dated back ten thousand years, at the hands of commercial hide hunters. That dramatic end, which occurred in less than fifteen years following the end of the Civil War, has by now entered American mythology. Yet our focus on the finale has obscured an examination of earlier phases that might shed new light on the historical and environmental interaction of the horse Indians and bison herds on the Plains.

In the nineteenth-century history of the Central and Southern Plains, there have long been perplexing questions that environmental history seems well suited to answer. Why were the Comanches able to replace the Apaches on the bison-rich Southern Plains? Why did the Kiowas, Cheyennes, and Arapahos gradually shift southward into the Southern Plains between 1800 and 1825? And why, after fighting each other for two decades, did these Southern Plains peoples effect a rapprochement and alliance in the 1840s? What factors brought on such an escalation of Indian raids into Mexico and Texas in the late-1840s that the subject assumed critical importance in the Treaty of Guadalupe Hidalgo? If the bison herds were so vast in the years before the commercial hide hunters, why were there so many reports of starving Indians on the Plains by 1850? And finally, given our standard estimates of bison numbers, why is it that the hide hunters are credited with bringing to market only some 10 million hides, including no more than 3.5 million from the Southern Plains, in the 1870s?

Apposite to all of these questions is a central issue: How successful were the horse Indians in creating a dynamic ecological equilibrium between themselves and the vast bison herds that grazed the Plains? That is, had they developed sustainable hunting practices that would maintain the herds and so permit future generations of hunters to follow the same way of life? This is not to pose the "anachronistic question" (the term is Richard White's) of whether Indians were ecologists.[5] But how a society or a group of peoples with a shared culture makes adjustments to live within the carrying capacity of its habitat is not only a valid historical question, it may be one of the most salient questions to ask about any culture. Historians of the Plains have differed about the long-term ecological sustainability of the Indians' use of bison, particularly after the Euro-American fur trade reached the West and the tribes began hunting bison under the influence of the market economy. The standard work, Frank Roe's *The North American Buffalo*, has generally carried the debate with the argument that there is "not a shred of evidence" to indicate that the horse Indians were out of balance with the bison herds.[6] Using the new insights and methods of environmental history, it now appears possible systematically to analyze and revise our understanding of nineteenth-century history on the Great Plains. Such an approach promises to resolve some of the major questions. It can advance our understanding of when bison declined in numbers and of the intertwining roles that Indian policies—migrations, diplomacy, trade, and use of natural resources—and the growing pressures of external stimuli played in that decline. The answers are complex and offer a revision of both Plains history and western Indian ecological history.

Working our way through to them requires some digression into the large historical forces that shaped the Southern Plains over the last hundred centuries. The perspective of the *longue durée* is essential to environmental history. What transpired on the Great Plains from 1800 to 1850 is not comprehensible without taking into account the effect of the Pleistocene extinctions of ten thousand years ago, or the cycle of droughts that determined the carrying capacity for animals on the grass lands. Shallower in time than these forces but just as important to the problem are factors that stemmed from the arrival of Europeans in the New World. Trade was an ancient part of the cultural landscape of America, but the Europeans altered the patterns, the goods, and the intensity of trade. And the introduction of horses and horse culture accomplished a technological revolution for the Great Plains. The horse was the chief catalyst of an ongoing remaking of the tribal map of western America, as Native American groups moved onto the Plains and incessantly shifted their ranges and alliances in response to a world where accelerating change seemed almost the only constant.

At the beginning of the nineteenth century, the dominant groups on the Southern Plains were the two major divisions of the Comanches: the Texas Comanches, primarily Kotsotekas, and the great New Mexico division, spread across the country from the Llano Estacado escarpment west to the foothills of the Sangre de Cristo Mountains, and composed of Yamparika and Jupe bands that only recently had replaced the Apaches on the High Plains. The Comanches' drive to the south from their original homelands in what is now southwestern Wyoming and northwestern Colorado was a part of the original tribal adjustments to the coming of horse technology to the Great Plains. There is reason to believe that the Eastern Shoshones, from whom the Comanches were derived before achieving a different identity on the Southern Plains, were one of the first intermountain tribes of historic times to push onto the Plains. Perhaps as early as 1500 the proto-Comanches were hunting bison and using dog power to haul their mountain-adapted, four-pole tipis east of the Laramie Mountains. Evidently they moved in response to a wetter time on the Central Plains and the larger bison concentrations there.

These early Shoshonean hunters may not have spent more than three or four generations among the thronging Plains bison herds, for by the late-seventeenth century they had been pushed back into the mountains and the sagebrush deserts by tribes newly armed with European guns moving westward from the region around the Great Lakes. If so, they were among a complex of tribes southwest of the lakes that over the next two centuries would be displaced by a massive Siouan drive to the west, an imperial expansion for domination of the prize buffalo range of the Northern Plains, and a wedge that sent ripples of tribal displacement across the Plains.[8]

Among the historic tribes, the people who became Comanches thus may have shared with the Apaches and, if linguistic arguments are correct, probably with the Kiowas the longest familiarity with a bison-hunting lifestyle. Pressed back toward the mountains as Shoshones, they thus turned in a different direction and emerged from the passes through the Front Range as the same people but bearing a new name given them by the Utes: Komantcia. They still lacked guns but now began their intimate association with the one animal, aside from the bison, inextricably linked with Plains life. The Comanches began acquiring horses from the Utes within a decade or so after the Pueblo Revolt of 1680 sent horses and horse culture diffusing in all directions from New Mexico. Thus were born the "hyper-Indians," as William Brandon has called the Plains people.[9]

The Comanches became, along with the Sioux, the most populous and widespread of all the peoples who now began to ride onto the vast sweep of grassland to participate in the hunter's life. They began to take possession of the Southern Plains by the early 1700s. By 1800 they were in full control of all the country east of the Southern Rocky Mountains and south of the Arkansas River clear to the Texas Hill Country. Their new culture, long regarded as an ethnographic anomaly on the Plains because of its western and archaic origins, may not be unique, as older scholars had supposed it to be—at least if we believe the new Comanche revisionists. Irrespective of their degree of tribal unity, however, when they began to move onto the Southern Plains with their new horse herds, their culture was adapting in interesting ways to the wealth of resources now available to them.[10]

To the Comanches, the Southern Plains must have seemed an earthly paradise. The Pleistocene extinctions ten thousand years earlier had left dozens of grazing niches vacant on the American Great Plains. A dwarf species of bison with a higher reproductive capability than any of its ancestors evolved to flood most of those vacant niches with an enormous biomass of one grazer. In an ecological sense, bison were a weed species that had proliferated as a result of a major disturbance.[11] That disturbance still reverberated, making it easy for Spanish horses, for example, to reoccupy their old niche and rapidly spread across the Plains. Those reverberations made the horse Indians thrive on an environmental situation that has had few parallels in world history.

The dimensions of the wild bison resource on the Southern Plains, and the Great Plains in general, have been much overstated in popular literature. For one thing, pollen analysis and archaeological data indicate that for the Southern Plains there were intervals, some spanning centuries, others decades, when bison must have been almost absent. Two major times of absence occurred between 5000 and 2500 B.C. and between A.D. 500 and 1300. The archaeological levels that lack bison bones correspond to pollen data indicating droughts. The severe southwestern drought that ended early in the fourteenth century was replaced by a five-hundred-year cycle of wetter and cooler conditions, and a return of bison in large numbers to the Southern Plains from their drought refugia to the east and west. This long-term pattern in the archaeological record seems to have prevailed on a smaller scale within historic times. During the nineteenth century, for example, droughts of more than five years' duration struck the Great Plains four times at roughly twenty-year intervals, in a long-term dendrochronological pattern that seems to show a drying cycle (shorter drought-free intervals) beginning in the 1850s.[12]

More important, our popular perception of bison numbers—based on the estimates of awed nineteenth-century observers—probably sets them too high. There very likely were never one hundred million or even sixty million bison on the Plains during the present climate regime because the carrying capacity of the grasslands was not so high. The best technique for determining bison carrying capacity on the Southern Plains is to extrapolate from United States census data for livestock, and the best census for the extrapolation is that of 1910, after the beef industry crashes of the 1880s had reduced animal numbers, but before the breakup of ranches and the Enlarged Homestead Act of 1909 resulted in considerable sections of the Southern Plains being broken out by farmers. Additionally, dendrochronological data seem to show that at the turn of the century rainfall on the Southern Plains was at median, between-droughts levels, rendering the census of 1910 particularly suitable as a baseline for carrying capacity and animal populations.[13]

The 1910 agricultural census indicates that in the 201 counties on the Southern Plains (which covered 240,000 square miles), the nineteenth-century carrying capacity during periods of median rainfall was about 7,000,000 cattle-equivalent grazers—specifically for 1910, about 5,150,000 cattle and 1,890,000 horses and mules.[14] The bison population was almost certainly larger, since migratory grazing patterns and coevolution with the native grasses made bison as a wild species about 18 percent more efficient on the Great Plains than domestic cattle. And varying climate conditions during the nineteenth century, as I will demonstrate, noticeably affected grassland carrying capacity. The ecological reality was a dynamic cycle in which carrying capacity could swing considerably from decade to decade.[15] But if the Great Plains bovine carrying capacity of 1910 expresses a median reality, then during prehorse times the Southern Plains might have supported an average of about 8.2 million bison, the entire Great Plains perhaps 28–30 million.[16]

Although eight million bison on the Southern Plains may not be so many as historians used to believe, to the Comanches the herds probably seemed limitless. Bison availability through horse culture caused a specialization that resulted in the loss of two-thirds of the Comanches' former plant lore and in a consequent loss of status for their women, an intriguing development that seems to have occurred to some extent among all the tribes that moved onto the Plains during the horse period.[17] As full-time bison hunters the Comanches appear to have abandoned all the old Shoshonean mechanisms, such as infanticide and polyandry, that had kept their population in line with available resources. These were replaced with such cultural mechanisms as widespread adoption of captured children and polygyny, adaptations to the Plains that were designed to keep Comanche numbers high and growing.[18] That these changes seem to have been conscious and deliberate argues, perhaps, both Comanche environmental insight and some centralized leadership and planning.

Comanche success at seizing the Southern Plains from the native groups that had held it for several hundred years was likewise the result of a conscious choice: their decision to shape their lives around bison and horses. Unlike the Comanches, many of the Apache bands had heeded the Spaniards' advice and had begun to build streamside gardening villages that became deathtraps once the Comanches located them. The Apaches' vulnerability, then, ironically stemmed from their willingness to diversify their economy. Given the overwhelming dominance of grasslands as opposed to cultivable river lands on the Plains, the specialized horse and bison culture of the Comanches exploited a greater volume of the thermodynamic energy streaming from sunlight into plants than the economies of any of their competitors—until they encountered Cheyennes and Arapahos with a similar culture.[19] The horse-mounted Plains Indians, in other words, made very efficient use of the available energy on the

Great Plains, something they seem instinctively to have recognized and exulted in. From the frequency with which the Comanches applied some version of the name "wolf" to their leaders, I suspect that they may have recognized their role as human predators and their ecological kinship with the wolf packs that like them lived off the bison herds.[20]

The Comanches were not the only people on the Southern Plains during the horse period. The New Mexicans, both Pueblo and Hispanic, continued to hunt on the wide-open *llanos*, as did the prairie Caddoans, although the numbers of the latter were dwindling rapidly by 1825. The New Mexican peoples and the Caddoans of the Middle Red and Brazos Rivers played major trade roles for hunters on the Southern Plains, and the Comanches in particular. Although the Comanches engaged in the archetypal Plains exchange of bison products for horticultural produce and European trade goods and traded horses and mules with Anglo-American traders from Missouri, Arkansas, and Louisiana, they were not a high-volume trading people until relatively late in their history. Early experiences with American traders and disease led them to distrust trade with Euro-Americans, and only once or twice did they allow short-lived posts to be established in their country. Instead, peace with the prairie Caddoans by the 1730s and with New Mexico in 1786 sent Comanche trade both east and west, but often through Indian middlemen.[21]

In the classic, paradigmatic period between 1800 and 1850, the most interesting Southern Plains development was the cultural interaction between the Comanches and surrounding Plains Indians to the north. The Kiowas were the one of those groups most closely identified with the Comanches.

The Kiowas are and have long been an enigma. Scholars are interested in their origins because Kiowa oral tradition is at odds with the scientific evidence. The Kiowas believe that they started their journey to Rainy Mountain on the Oklahoma Plains from the north. And indeed, in the eighteenth century we find them on the Northern Plains, near the Black Hills, as one of the groups being displaced southwestward by the Siouan drive toward the buffalo range. Linguistically, however, the Kiowas are southern Indians. Their language belongs to the Tanoan group of Pueblo languages in New Mexico, and some scholars believe that the Kiowas of later history are the same people as the Plains Jumanos of early New Mexico history, whose *rancherias* were associated during the 1600s and early 1700s with the headwaters of the Colorado and Concho Rivers of Texas. How the Kiowas got so far north is not certainly known, but in historical times they were consummate traders, especially of horses, and since the Black Hills region was a major trade citadel they may have begun to frequent the region as traders and teachers of horse lore.[22]

Displaced by the wars for the buffalo ranges in the north, the Kiowas began to drift southward again—or perhaps, since the supply of horses was in the Southwest, simply began to stay longer on the Southern Plains. Between 1790 and 1806, they developed a rapprochement with the Comanches. Thereafter they were so closely associated with the northern Comanches that they were regarded by some as merely a Comanche band, although in many cultural details the two groups were dissimilar. Spanish and American traders and explorers of the 1820s found them camped along the two forks of the Canadian River and on the various headwater streams of the Red River.[23]

The other groups that increasingly began to interact with the Comanches during the 1820s and thereafter had also originated on the Northern Plains. These were the Arapahos and the Cheyennes, who by 1825 were beginning to establish themselves on the Colorado buffalo plains from the North Platte River all the way down to the Arkansas River.

The Algonkian-speaking Arapahos and Cheyennes had once been farmers living in earth lodges on the upper Mississippi. By the early 1700s both groups were in present North Dakota, occupying villages along the Red and Cheyenne Rivers, where they first began to acquire horses, possibly from the Kiowas. Fur wars instigated by the Europeans drove them farther southwest and more and more into a Plains, bison-hunting culture, one that the women of these farming tribes probably resisted as long as possible.[24] But by the second decade of the nineteenth century the Teton Sioux wedge had made nomads and hunters of the Arapahos and Cheyennes.

Their search for prime buffalo grounds and for ever-larger horse herds, critical since both tribes had emerged as middlemen traders between the villagers of the Missouri and the horse reservoir to the south, hustled the Cheyennes and Arapahos west of the Black Hills, into Crow lands, and then increasingly southward along the mountain front. By 1815 the Arapahos were becoming fixed in the minds of American traders as their own analogue on the Southern Plains; the famous trading expedition of August Pierre Chouteau and Jules De Mun that decade was designed to exploit the horse and robe trade of the Arapahos on the Arkansas. By the 1820s, when Stephen Long's expedition and the trading party including Jacob Fowler penetrated the Southern Plains, the Arapahos and Cheyennes were camping with the Kiowas and Comanches on the Arkansas. The Hairy Rope band of the Cheyennes, renowned for their ability to catch wild horses, was then known to be mustanging along the Cimarron River.[25]

Three factors seem to have drawn the Arapahos and Cheyennes so far south. Unquestionably, one factor was the vast horse herds of the Comanches and Kiowas, an unending supply of horses for the trade, which by 1825 the Colorado tribes were seizing in daring raids. Another was the milder winters south of the Arkansas, which made horse pastoralism much easier. The third factor was the abnormally bountiful game of the early-nineteenth-century Southern Plains, evidently the direct result of an extraordinary series of years between 1815 and 1846 when, with the exception of a minor drought in the late–1820s, rainfall south of the Arkansas was considerably above average. So lucrative was the hunting and raiding that in 1833 Charles Bent located the first of his adobe trading posts along the Arkansas, expressly to control the winter robe and summer horse trade of the Arapahos and Cheyennes. Bent's marketing contacts were in St. Louis. Horses that Bent's traders drove to St. Louis commonly started as stock in the New Mexican Spanish settlements (and sometimes those were California horses stolen by Indians who traded them to the New Mexicans) that were stolen by the Comanches, then stolen again by Cheyenne raiders, and finally traded at Bent's or Ceran St. Vrain's posts, whence they were driven to Westport, Missouri, and sold to outfit American emigrants going to the West Coast! Unless you saw it from the wrong end, as the New Mexicans (or the horses) seem to have, it was both a profitable and a culturally stimulating economy.[26]

Thus, around 1825, the Comanches and Kiowas found themselves at war with Cheyennes, Arapahos, and other tribes on the north. Meanwhile, the Colorado tribes opened another front in a naked effort to seize the rich buffalo range of the upper Kansas and Republican rivers from the Pawnees. These wars produced an interesting type of ecological development that appeared repeatedly across most of the continent. At the boundaries where warring tribes met, they left buffer zones occupied by neither side and only lightly hunted. One such buffer zone on the Southern Plains was along the region's northern perimeter, between the Arkansas and North Canadian rivers. Another was in present-day western Kansas, between the Pawnees and the main range of the Colorado tribes, and a third seems to have stretched from the forks of the Platte to the mountains. The buffer zones were important because game within them was left relatively undisturbed; they allowed the buildup of herds that might later be exploited when tribal boundaries or agreements changed.[27]

The appearance of American traders such as Bent and St. Vrain marked the Southern Plains tribes' growing immersion in a market economy increasingly tied to worldwide trade networks dominated by Euro-Americans. Like all humans, Indians had always altered their environments. But as most modern historians of Plains Indians and the western fur trade have realized, during the nineteenth century not only had the western tribes become technologically capable of pressuring their resources, but year by year they were becoming less "ecosystem people," dependent on the products of their local regions for subsistence, and increasingly tied to biospheric trade networks. Despite some speculation that the Plains tribes were experiencing ecological problems, previous scholars have not ascertained what role market hunting played in this dilemma, what combination of other factors was involved, or what the tribes attempted to do about it.[28]

The crux of the problem in studying Southern Plains Indian ecology and bison is to determine whether the Plains tribes had established a society in ecological equilibrium, one whose population did not exceed the carrying capacity of its habitat and so maintained a healthy, functioning ecology that could be sustained over the long term.[29] Answering that question involves an effort to come to grips with the factors affecting bison populations, the factors affecting Indian populations, and the cultural aspects of Plains Indians' utilization of bison. Each of the three aspects of the question presents puzzles difficult to resolve.

In modern, protected herds on the Plains, bison are a prolific species whose numbers increase by an average of 18 percent a year, assuming a normal sex ratio (fifty-one males to forty-nine females) with breeding cows amounting to 35 percent of the total.[30] In other words, if the Southern Plains supported 8.2 million bison in years of median rainfall, the herds would have produced about 1.4 million calves a year. To maintain an ecological equilibrium with the grasses, the Plains bison's natural mortality rate also had to approach 18 percent.

Today the several protected bison herds in the western United States have a natural mortality rate, without predation, ranging between 3 and 9 percent. The Wichita Mountains herd, the only large herd left on the Southern Plains, falls midway with a 6 percent mortality rate. Despite a search for it, no inherent naturally regulating mechanism has yet been found in bison populations; thus active culling programs are needed at all the Plains bison refuges. The starvation-induced population crashes that affect ungulates such as deer were seemingly mitigated on the wild, unfenced Plains by the bison's tendency—barring any major impediments—to shift their range great distances to better pasture.[31]

Determining precisely how the remaining annual mortality in the wild herds was affected is not easy, because the wolf/bison relationship on the Plains has never been studied. Judging from dozens of historical documents attesting to wolf predation of bison calves, including accounts by the Indians, wolves apparently played a critical role in Plains bison population dynamics, and not just as culling agents of diseased and old animals.[32]

Human hunters were the other source of mortality. For nine thousand years Native Americans had hunted bison without exterminating them, perhaps building into their gene pool an adjustment to human predation (dwarfed size, earlier sexual maturity, and shorter gestation times, all serving to keep populations up). But there is archaeological evidence that beginning about A.D. 1450, with the advent of "mutualistic" trade between Puebloan communities recently forced by drought to relocate on the Rio Grande and a new wave of Plains hunters (probably the Athabaskan-speaking Apacheans), human pressures on the southern bison herd accelerated, evidently dramatically if the archaeological record in New Mexico is an accurate indication. That pressure would have been a function of both the size of the Indian population and the use of bison in Indian cultures. Because Plains Indians traded bison-derived goods for the produce of the horticultural villages fringing the Plains, bison would be affected by changes in human population peripheral to the Great Plains as well as on them.[33]

One attempt to estimate maximum human population size on the Southern Plains, that of Jerold Levy, fixed the upper limit at about 10,500 people. Levy argued that water would have been a more critical resource than bison in fixing a limit for Indian populations. Levy's population figures are demonstrably too low, and he lacked familiarity with the aquifer-derived drought-resistant sources of water on the Southern Plains. But his argument that water was the more critical limiting resource introduces an important element into the Plains equation.[34]

The cultural utilization of bison by horse Indians has been studied by Bill Brown. Adapting a sophisticated formula worked out first for caribou hunters in the Yukon, Brown has estimated Indian subsistence (caloric requirements plus the number of robes and hides required for domestic use) at about forty-seven animals per lodge per year. At an average of eight people per lodge, that works out to almost six bison per person over a year's time. Brown's article is not only highly useful in getting us closer to a historic Plains equation than ever before; it is also borne out by at least one historic account. In 1821 the trader Jacob Fowler camped for several weeks with seven hundred lodges of Southern Plains tribes on the Arkansas River. Fowler was no ecologist; in fact, he could hardly spell. But he was a careful observer, and he wrote that the big camp was using up one hundred bison a day. In other words, seven hundred lodges were using bison at a rate of about fifty-two per lodge per year, or 6.5 animals per person. These are important figures. Not only do they give us some idea of the mortality percentage that can be assigned to human hunters; by extension they help us fix a quadruped predation percentage as well.[35]

Estimates of the number of Indians on the Southern Plains during historic times are not difficult to find, but they tend to vary widely, and for good reason, as will be seen when we look closely at the historical events of the first half of the nineteenth century. Although observers' population estimates for the Comanches go as high as 30,000, six of the seven population figures for the Comanches estimated between 1786 and 1854 fall into a narrow range between

19,200 and 21,600.³⁶ Taken together, the number of Kiowas, Cheyennes, Arapahos, Plains Apaches, Kiowa-Apaches, and Wichitas probably did not exceed 12,000 during that same period. Contemporaries estimated the combined number of Cheyennes and Arapahos, for example, as 4,400 in 1838, 5,000 in 1843, and 5,200 in 1846.³⁷ If the historic Southern Plains hunting population reached 30,000, then human hunters would have accounted for only 195,000 bison per year if we use the estimate of 6.5 animals per person.

But another factor must have played a significant role. While quadruped predators concentrated on calves and injured or feeble animals, human hunters had different criteria. Historical documents attest to the horse Indians' preference for and success in killing two- to five-year-old bison cows, which were preferred for their meat and for their thinner, more easily processed hides and the luxurious robes made from their pelts. Studies done on other large American ungulates indicate that removal of breeding females at a level that exceeds 7 percent of the total herd will initiate population decline. With 8.2 million bison on the Southern Plains, the critical upper figure for cow selectivity would have been about 574,000 animals. Reduce the total bison number to 6 million and the yearly calf crop to 1.08 million, probably more realistic median figures for the first half of the nineteenth century, and the critical mortality for breeding cows would still have been 420,000 animals. As mentioned, a horse-mounted, bison-hunting population of 30,000 would have harvested bison at a yearly rate of less than 200,000. Hence I would argue that, theoretically, on the Southern Plains the huge biomass of bison left from the Pleistocene extinctions would have supported the subsistence needs of more than 60,000 Plains hunters.³⁸

All of this raises some serious questions when we look at the historical evidence from the first half of the nineteenth century. By the end of that period, despite an effort at population growth by many Plains tribes, the population estimates for most of the Southern Plains tribes were down. And many of the bands seemed to be starving. Thomas Fitzpatrick, the Cheyennes' and Arapahos' first agent, reported in 1853 that the tribes in his district spent half the year in a state of starvation. The Comanches were reported to be eating their horses in great numbers by 1850, and their raids into Mexico increased all through the 1840s, as if a resource depletion in their home range was driving them to compensate with stolen stock.³⁹ In the painted robe calendars of the Kiowas, the notation for "few or no bison" appears for four years in a row between 1849 and 1852.⁴⁰ Bison were becoming less reliable, and the evolution toward an economy based on raiding and true horse pastoralism was well under way. Clearly, by 1850 something had altered the situation on the Southern Plains.

The "something" was, in fact, a whole host of ecological alterations that historians with a wide range of data at their disposal are only now, more than a century later, beginning to understand.

As early as 1850 the bison herds had been weakened in a number of ways. The effect of the horse on Indian culture has been much studied, but in working out a Southern Plains ecological model, it is important to note that horses also had a direct effect on bison numbers. By the second quarter of the nineteenth century the domesticated horse herds of the Southern Plains tribes must have ranged between one-quarter and one-half million animals (at an average of ten to fifteen horses per person).[41] In addition, an estimated two million wild mustangs overspread the country between south Texas and the Arkansas River. That many animals of a species with an 80 percent dietary overlap with bovines and, perhaps more critically, with similar water requirements, must have had an adverse impact on bison carrying capacity, especially since Indian horse herds concentrated the tribes in the moist canyons and river valleys that bison also used for watering.[42] Judging from the 1910 agricultural census, two million or more horses would have reduced the median grassland carrying capacity for the southern bison herd to under six million animals.

Another factor that may have started to diminish overall bison numbers was the effect of exotic bovine diseases. Anthrax, introduced into the herds from Louisiana around 1800, tuberculosis, and brucellosis, the latter brought to the Plains by feral and stolen Texas cattle and by stock on the overland trails, probably had considerable impact on the bison herds. All the bison that were saved in the late-nineteenth century had high rates of infection with these diseases. Brucellosis plays havoc with reproduction in domestic cattle, causing cows to abort; it may have done so in wild bison, and butchering them probably infected Indian women with the disease.[43]

Earlier I mentioned modern natural mortality figures for bison of 3 percent to 9 percent of herd totals. On the wilderness Plains, fires, floods, drownings, droughts, and strange die-offs may have upped this percentage considerably. But if we hold to the higher figure, then mortality might have taken an average of 50 percent of the annual bison increase of 18 percent. Thirty thousand subsistence hunters would have killed off only 18 percent of the bison's yearly increase (if the herd was six million). The long-wondered-at wolf predation was perhaps the most important of all the factors regulating bison populations, with a predation percentage of around 32 percent of the annual bison increase. (Interestingly, this dovetails closely with the Pawnee estimate that wolves got three to four of every ten calves born.) Wolves and other canids are able to adjust their litter sizes to factors like mortality and resource abundance. Thus, mountain men and traders who poisoned wolves for their pelts may not have significantly reduced wolf populations. They may have inadvertently killed thousands of bison, however, for poisoned wolves drooled and vomited strychnine over the grass in their convulsions. Many Indians lost horses that ate such poisoned grass.[44]

The climate cycle, strongly correlated with bison populations in the archaeological data for earlier periods, must have interacted with these other factors to produce a decline in bison numbers between 1840 and 1850. Except for a dry period in the mid- to late–1820s, the first four decades of the nineteenth century had been a time of above-normal rainfall on the Southern Plains. With the carrying capacity for bison and horses high, the country south of the Arkansas sucked tribes to it as into a vortex. But beginning in 1846, rainfall plunged as much as 30 percent below the median for nine of the next ten years. On the Central Plains, six years of that decade were dry.[45] The growth of human populations and settlements in Texas, New Mexico, and the Indian Territory blocked the bison herds from migrating to their traditional drought refugia on the periphery of their range. Thus, a normal climate swing combined with unprecedented external pressures to produce an effect unusual in bison history—a core population, significantly reduced by competition with horses and by drought, that was quite susceptible to human hunting pressure.

Finally, alterations in the historical circumstances of the Southern Plains tribes from 1825 to 1850 had serious repercussions for Plains ecology. Some of those circumstances were indirect and beyond the tribes' ability to influence. Traders along the Santa Fe Trail shot into, chased, and disturbed the southern herds. New Mexican *ciboleros* (bison hunters) continued to take fifteen to twenty-five thousand bison a year from the Llano Estacado. And the United States government's removal of almost fifty thousand eastern Indians into Oklahoma increased the pressure on the bison herds to a level impossible to estimate. The Southern Plains tribes evidently considered it a threat and refused to abide by the Treaty of Fort Holmes (1835) when they discovered it gave the eastern tribes hunting rights on the prairies.[46]

Insofar as the Southern Plains tribes had an environmental policy, then, it was to protect the bison herds from being hunted by outsiders. The Comanches could not afford to emulate their Shoshonean ancestors and limit their own population. Beset by enemies and disease, they had to try to keep their numbers high, even as their resource base diminished. For the historic Plains tribes, warfare and stock raids addressed ecological needs created by diminishing resources as well as the cultural impulse to enhance men's status, and they must have seemed far more logical solutions than consciously reducing their own populations as the bison herds became less reliable.

For those very reasons, after more than a decade of warfare among the buffalo tribes, in 1840 the Comanches and Kiowas adopted a strategy of seeking peace and an alliance with the Cheyennes, Arapahos, and Kiowa-Apaches. From the Comanches' point of view, it brought them allies against Texans and eastern Indians who were trespassing on the Plains. The Cheyennes and Arapahos got what they most wanted: the chance to hunt the grass- and bison-rich Southern Plains, horses and mules for trading, and access to the Spanish settlements via Comanche lands. But the peace meant something else in ecological terms. Now all the tribes could freely exploit the Arkansas Valley bison herds. This new exploitation of a large, prime bison habitat that had been a boundary zone skirted by Indian hunters may have been critical. In the Kiowa Calendar the notation for "many bison" appears in 1841, the year following the peace. The notation appears only once more during the next thirty-five years.[47]

One other advantage the Comanches and Kiowas derived from the peace of 1840 was freedom to trade at Bent's Fort. Although the data to prove it are fragmentary, this conversion of the largest body of Indians on the Southern Plains from subsistence/ecosystem hunters to a people intertwined in the European market system probably added critical stress to a bison herd already being eaten away. How serious the market incentive could be is indicated by John Whitfield, agent at William Bent's second Arkansas River fort in 1855, who wrote that 3,150 Cheyennes were killing forty thousand bison a year.[48] That is about twice the number the Cheyennes would have harvested through subsistence hunting alone. (It also means that on the average every Cheyenne warrior was killing forty-four bison a year and every Cheyenne woman was processing robes at the rate of almost one a week.) With the core bison population seriously affected by the drought of the late–1840s, the additional, growing robe trade of the Comanches probably brought the Southern Plains tribes to a critical level in their utilization of bison. Drought, Indian market hunting, and cow selectivity must stand as the critical elements—albeit augmented by minor factors such as white disturbance, new bovine diseases, and increasing grazing competition from horses—that brought on the bison crisis of the midcentury Southern Plains. That explanation may also illuminate the experience of the Canadian Plains, where bison disappeared without the advent of white hide hunting.[49]

Perhaps that would have happened on the American Plains if the tribes had held or continued to augment their populations. But the Comanches and other tribes fought a losing battle against their own attrition. While new institutions such as male polygamy and adoption of captured children worked to build up the Comanches' numbers, the disease epidemics of the nineteenth century repeatedly decimated them. In the 1820s, the Comanches were rebuilding their population after the smallpox epidemic of 1816 had carried away a fourth of them. But smallpox ran like a brush fire through the Plains villages again in 1837–38, wiping whole peoples off the continent. And the forty-niners brought cholera, which so devastated the Arkansas Valley Indians that William Bent burned his fort and temporarily left the trade that year. John C. Ewers, in fact, has estimated that the nineteenth-century Comanches lost 75 percent of their population to disease.[50]

Did the Southern Plains Indians successfully work out a dynamic, ecological equilibrium with the bison herds? I would argue that the answer remains ultimately elusive because the relationship was never allowed to play itself out. The trends, however, suggest that a satisfactory solution was improbable. One factor that worked against the horse tribes was their short tenure. It may be that two centuries provided too brief a time for them to create a workable system around horses, the swelling demand for bison robes generated by the Euro-American market, and the expansion of their own populations to hold their territories. Some of those forces, such as the tribes' need to expand their numbers and the advantages of participating in the robe trade, worked against their need to conserve the bison herds. Too, many of the forces that shaped their world were beyond the power of the Plains tribes to influence. And it is very clear that the ecology of the Southern Plains had become so complicated by the mid-nineteenth century that neither the Indians nor the Euro-Americans of those years could have grasped how it all worked.

Finally and ironically, it seems that the Indian religions, so effective at calling forth awe and reverence for the natural world, may have inhibited the Plains Indians' understanding of bison ecology and their role in it. True, native leaders such as Yellow Wolf, the Cheyenne whom James W. Abert interviewed and sketched at Bent's Fort in 1845–46, surmised the implications of market hunting. As he watched the bison disappearing from the Arkansas Valley, Yellow Wolf asked the whites to teach the Cheyenne hunters how to farm, never realizing that he was reprising a Plains Indian/Euro-American conversation that had taken place sixty years earlier in that same country.[51] But Yellow Wolf was marching to his own drummer, for it remained a widespread tenet of faith among most Plains Indians through the 1880s that bison were supernatural in origin. A firsthand observer and close student of the nineteenth-century Plains reported,

> Every Plains Indian firmly believed that the buffalo were produced in countless numbers in a country under the ground, that every spring the surplus swarmed like bees from a hive, out of great cave-like openings to this country, which were situated somewhere in the great 'Llano Estacado' or Staked Plain of Texas.

This religious conception of the infinity of nature's abundance was poetic. On one level it was also empirical: Bison overwintered in large numbers in the protected canyons scored into the eastern escarpment of the Llano Estacado, and Indians had no doubt many times witnessed the herds emerging to overspread the High Plains in springtime. But such a conception did not aid the tribes in their efforts to work out an ecological balance amid the complexities of the nineteenth-century Plains.[52]

In a real sense, then, the more familiar events of the 1870s only delivered the *coup de grace* to the free Indian life on the Great Plains. The slaughterhouse effects of European diseases and wars with the encroaching whites caused Indian numbers to dwindle after 1850 (no more than fourteen hundred Comanches were enrolled to receive federal benefits at Fort Sill, in present-day Oklahoma, in the 1880s). This combined with bison resiliency to preserve a good core of animals until the arrival of the white hide hunters, who nonetheless can be documented as taking only about 3.5 million animals from the Southern Plains.[53]

But the great days of the Plains Indians, the primal poetry of humans and horses, bison and grass, sunlight and blue skies, and the sensuous satisfactions of a hunting life on the sweeping grasslands defined a meteoric time indeed. And the meteor was already fading in the sky a quarter century before the Big Fifties began to boom.

ACKNOWLEDGMENTS

The author would like to thank Richard White, Bill Cronon, Donald Hughes, Peter Iverson, Donald Berthrong, Eric Bolen, Jim Sherow, and Kenneth Owens for reading and critiquing all or parts of this article.

NOTES

1. See Jacobo Loyola y Ugarte to Juan Bautista de Anza, 5 October 1786, roll 10, series 11, Spanish Archives of New Mexico [microfilm], New Mexico State Archives and Records Center, Santa Fe.

2. Ibid. One line reads: "use all your sagacity and efficiency, making evident to the [Comanche] Captains . . . that the animals they hunt with such effort at sustenance are not at base inexhaustible." See also Alfred B. Thomas, ed., *Forgotten Frontiers: A Study of the Spanish Indian Policy of Don Juan Bautista de Anza, Governor of New Mexico, 1777–1787* (Norman: University of Oklahoma Press, 1932), 69–72, 82.

3. See Alfred W. Crosby, Jr., *The Columbian Exchange: Biological Consequences of 1492* (Westport: Greenwood Press, 1972); Alfred W. Crosby, Jr., *Ecological Imperialism: The Biological Expansion of Europe, 900–1900* (Cambridge: Cambridge University Press, 1986); Henry Dobyns, *Native American Historical Demography* (Bloomington: Indiana University Press, 1976); Henry Dobyns, *Their Number Become Thinned: Native American Population Dynamics in Eastern North America* (Knoxville: University of Tennessee Press, 1983); Calvin Martin, *Keepers of the Game: Indian-Animal Relations and the Fur Trade* (Berkeley: University of California Press, 1979); Richard White, *Roots of Dependency* (Lincoln: University of Nebraska Press, 1983); William Cronon, *Changes in the Land: Indians, Colonists, and the Ecology of New England* (New York: Hill and Wang, 1983); Paul Martin and Henry Wright, Jr., eds., *Pleistocene Extinctions: The Search for a Cause* (New Haven: Yale University Press, 1967); and Paul Martin and Richard Klein, eds., *Quarternary Extinctions* (Tucson: University of Arizona Press, 1985).

4. See Richard White, "American Indians and the Environment," *Environmental Review* 9 (Summer 1985): 101–3; Richard White, "Native Americans and the Environment," in *Scholars and the Indian Experience: Critical Reviews of Recent Writings in the Social Sciences*, ed. W.R. Swagerty (Bloomington: Indiana University Press, 1984), 179–204; Christopher T. Vecsey and Robert W. Venables, eds., *American Indian Environments: Ecological Issues in Native American History* (Syracuse: Syracuse University Press, 1980); Richard White and William Cronon, "Ecological Change and Indian-White Relations," in *Handbook of North American Indians*, ed. William C. Sturtevant (Washington: Smithsonian Institution, 1982), 4:417–29; and Donald J. Hughes, *American Indian Ecology* (El Paso: Texas Western Press, University of Texas at El Paso, 1983).

5. See White, "American Indians and the Environment," 101; and White and Cronon, "Ecological Change and Indian-White Relations."

6. Several earlier scholars have addressed this question. For an early argument that the horse Indians overhunted bison, see William T. Hornaday, "The Extermination of the American Bison with a Sketch of its Discovery and Life History," *Smithsonian Report* (1887), 480–90, 506. For statements of this position that offer nothing beyond anecdotal evidence, see James Malin, *History and Ecology*, ed. Robert Swierenga (Lincoln: University of Nebraska Press, 1984), 9, 31–54; George Hyde, *Spotted Tail's Folk* (Norman: University of Nebraska Press, 1961), 24; and Preston Holder, *The Hoe and the Horse on the Plains* (Lincoln: University of Nebraska Press, 1970), 111, 118. For the contrary view, see Frank Roe, *The North American Buffalo: A Critical Study of the Species in its Wild State* (Toronto: University of Toronto Press, 1951), 500–5, 655–70; and Frank Roe, *The Indian and the Horse* (Norman: University of Oklahoma Press, 1955), 190–91. The breadth and authority of Roe's books have given him priority in the field.

7. Since it utilizes long-ignored Spanish documents, on the Comanche's migration I follow Thomas Kavanagh, "Political Power and Political Organization: Comanche Politics, 1786–1875". (Ph.D. diss., University of New Mexico, 1986), rather than Ernest Wallace and E. Adamson Hoebel, *The Comanches: Lords of the South Plains* (Norman: University of Oklahoma Press, 1951). Demitri Boris Shimkin, *Wind River Shoshone Ethnography* (Berkeley: University of California Press, 1947); James A. Goss, "Basin-Plateau Shoshonean Ecological

Model," in *Great Basin Cultural Ecology: A Symposium*, ed. Don D. Fowler (Reno: University of Nevada, Desert Research Institute Publications in the Social Sciences, no. 12, 1972), 123–27.

8. Richard White, "The Winning of the West: The Expansion of the Western Sioux in the Eighteenth and Nineteenth Centuries," *Journal of American History* 65 (September 1978): 319–43.

9. Kiowa origin myths set on the Northern Plains are at variance with the linguistic evidence, which ties them to the Tanoan speakers of the Rio Grande pueblos. Scholars are coming to believe that there is a connection between the mysterious Jumano peoples of seventeenth- and eighteenth-century New Mexico documents and the later Kiowas. See Nancy Hickerson's 1989 study, "The Jumano and Trade in the Arid Southwest, 1580–1700" (in author's possession). I am indebted to Ms. Hickerson for allowing me to examine her work. William Brandon, *The American Heritage Book of Indians* (New York: Random House, 1988), 340.

10. Ernest Wallace, "The Habitat and Range of the Kiowa, Comanche and Apache Indians Before 1867," prepared for the United States Department of Justice for Case No. 257 before the Indian Claims Commission, 1959 (Southwest Collection, Texas Tech University, Lubbock). See Thomas Kavanagh, "The Comanche: Paradigmatic Anomaly or Ethnographic Fiction," *Haliksa* 4 (1985): 109–28; Melburn D. Thurman, "A New Interpretation of Comanche Social Organization," *Current Anthropology* 23 (October 1982): 578–79; Daniel J. Gelo, "On a New Interpretation of Comanche Social Organization," *Current Anthropology* 28 (August-October 1987): 551–52; Melburn D. Thurman, "Reply," *Current Anthropology* 28 (August–October 1987): 552–55. For the earlier position that the Comanches are atypical on the Plains, see Symmes C. Oliver, *Ecology and Cultural Continuity as Contributing Factors in the Social Organization of the Plain Indians* (Berkeley: University of California Press, 1962), 69–80.

11. Jerry McDonald, *North American Bison: Their Classification and Evolution* (Berkeley: University of California Press, 1981), 250–63.

12. Tom Dillehay, "Late Quarternary Bison Population Changes on the Southern Plains," *Plains Anthropologist* 19 (August 1974): 180–96; Darrell Creel et al., "A Faunal Record from West-Central Texas and Its Bearing on Late Holocene Bison Population Changes in the Southern Plains," *Plains Anthropologist* 35 (April 1990): 55–69. For Great Plains dendrochronology, see Harry Weakly, "A Tree-Ring Record of Precipitation in Western Nebraska," *Journal of Forestry* 41 (November 1943): 816–19; and Edmund Schulman, *Dendroclimatic Data from Arid America* (Tucson: University of Arizona Press, 1956), 86–88. For use of meteorological data to argue that climate variability was exponentially greater on the Southern Plains than farther north, see Douglas Bamforth, *Ecology and Human Organization on the Great Plains* (New York: Plenum Press, 1988), 74.

13. Weakly, "Tree-Ring Record of Precipitation in Western Nebraska," 819.

14. U.S. Department of Commerce, Bureau of the Census, *Thirteenth Census of the United States, Taken in the Year 1910*, vols. 6 and 7, *Agriculture, 1909 and 1910* (Washington: Government Printing Office, 1914). My method has been to compile 1910 cattle, horse, and mule figures for the then-existing Plains counties of Texas (119), western Oklahoma (45), New

Mexico (10), counties below the Arkansas River in Colorado (8), and counties in southwestern Kansas (19). The carrying capacity for a biome such as the Great Plains ought to be measured by the use of the county figures. The principal problem with this technique in the past has been overgeneralization of stock numbers through reliance on state totals. It was first used by Ernest Thompson Seton, *Life Histories of Northern Animals* (New York: C. Scribner's Sons, 1909), 1:259–63; and more recently by Bill Brown, "Comancheria Demography, 1805–1830," *Panhandle Plains Historical Review* 59 (1986): 8–12. Range management commonly assigns cows a grazing quotient of 1.0, bulls a quotient of 1.30, and horses and mules 1.25.

15. Joseph Chapman and George Feldhamer, eds., *Wild Mammals of North America: Biology, Management and Economics* (Baltimore: Johns Hopkins University Press, 1982), 978, 986, 1001–2. Modern bison ranchers claim that bison achieve greater land use efficiency and larger herd size on native grass compared to cattle. The editors of the above work call for more research into this question. See also Charles Rehr, "Buffalo Population and Other Deterministic Factors in a Model of Adaptive Process on the Shortgrass Plains," *Plains Anthropologist* 23 (November 1978): 25–27.

16. For the use of a different formula (mean potential bovine carrying capacity per acre) to arrive at similar totals, see Tom McHugh, *The Time of the Buffalo* (New York: Alfred A. Knopf, 1972), 16–17. For the reasonably convincing argument that because of climate variability and less nutritious grasses, population densities of Great Plains bison were lowest on the Southern Plains, see Bamforth, *Ecology and Human Organization on the Great Plains*, 74, 78.

17. The figure for loss of plant lore is based on a comparison of remembered ethnobotanies for the Shoshones (172 species) and the Comanches (67 species). See Brian Spykerman, "Shoshoni Conceptualizations of Plant Relationships" (M.S. thesis, Utah State University, 1977); and Gustav Carlson and Volney Jones, "Some Notes on the Uses of Plants by the Comanche Indians," *Papers of the Michigan Academy of Science, Arts, and Letters* 25 (1939), 517–42. On women's loss of status, see Holder, *Hoe and the Horse on the Plains*.

18. Wallace and Hoebel, *Comanches*, 142. On the loss of Shoshone birth control mechanisms among Comanche women, see Abram Kardiner, "Analysis of Comanche Culture," in *The Psychological Frontiers of Society*, ed. Abram Kardiner et al. (New York: Columbia University Press, 1945). For a report of five hundred adopted captives in a decade, see Jean Louise Berlandier, *The Indians of Texas in 1830*, ed. John C. Ewers (Washington: Smithsonian Institution Press, 1969), 119. For estimates on the Euro-American constituency Comanche adoption and captives trade, see Carl C. Rister, *Border Captives: The Traffic in Prisoners by Southern Plains Indians, 1835–1875* (Norman: University of Oklahoma Press, 1940); and Russell Magnaghi, "The Indian Slave Trader: The Comanche, a Case Study" (Ph.D. diss., St. Louis University, 1970). Indian Agent Thomas Fitzpatrick was adamant that Comanche raids were for children, "to keep up the numbers of the tribe." See Kardiner, "Analysis of Comanche Culture," 89. On hunter-gatherer carrying capacity, see Marvin Harris and Eric Ross, *Death, Sex, and Fertility: Population Regulation in Preindustrial and Developing Societies* (New York: Columbia University Press, 1987), 23–26; and Ezra Zubrow, *Prehistoric Carrying Capacity: A Model* (Menlo Park: Cummings Publishing Co., 1975).

19. David Kaplan, "The Law of Cultural Dominance," in *Evolution and Culture*, ed. Marshall Sahlins and Elman Service (Ann Arbor: University of Michigan Press, 1960), 75–82. On Apache vulnerability, see George E. Hyde, *Indians of the High Plains: From the Prehistoric Period to the Coming of Europeans* (Norman: University of Oklahoma Press, 1975), 65, 70, 91. Other explanations include the Spanish refusal to trade guns to the Apaches and Comanche superiority at horse care. For less monocausal interpretations, see Frederick W. Rathjen, *The Texas Panhandle Frontier* (Austin: University of Texas Press, 1973), 47–48.

20. For references to Comanche names containing the word "wolf" (rendered by Euro-Americans as *isa, ysa, esa,* or sometimes with an *sh* second syllable), see George Catlin, *Letters and Notes on the Manners, Customs, and Conditions of the North American Indians* (New York: Dover Publications, 1973), 2:67–69; Noel Loomis and Abraham Nasatir, eds., *Pedro Vial and the Roads to Santa Fe* (Norman: University of Oklahoma Press, 1967), 488 n. 22a; and Thomas, *Forgotten Frontiers*, 325–27.

21. On Comanche trade with Anglo-Americans, see Dan L. Flores, ed., *Journal of an Indian Trader: Anthony Glass and the Texas Trading Frontier, 1790–1810* (College Station: Texas A & M University Press, 1985), esp. 3–33; and Thomas James, *Three Years among the Indians and Mexicans*, ed. Walter Douglas (St. Louis: Missouri Historical Society, 1916), 191–235. For the argument that the Comanches were among the earliest Plains traders, and that Comanche leadership evolved in a trade/market situation, see Kavanagh, "Political Power and Political Organization." See Charles Kenner, *A History of New Mexican Plains Indian Relations* (Norman: University of Oklahoma Press, 1969); and William Swagerty, "Indian Trade in the Trans-Mississippi West to 1870," in Sturtevant, *Handbook of North American Indians*, 4:351–74.

22. See Hickerson, "Jumano and Trade in the Arid Southwest." On Kiowa origins, see also Maurice Boyd, *Kiowa Voices: Ceremonial Dance, Ritual, and Songs*, 2 vols. (Fort Worth: Texas Christian University Press, 1981); and John R. Wunder, *The Kiowa* (New York: Chelsea House, 1989).

23. Elizabeth A.H. John, "An Earlier Chapter of Kiowa History," *New Mexico Historical Review* 60 (1985): 379–97. On traders' contacts with the Kiowas, see, particularly, Maxine Benson, ed., *From Pittsburgh to the Rocky Mountains: Major Stephen Long's Expedition, 1819–1820* (Golden, CO: Fulcrum, 1988), 327–36.

24. Holder, *Hoe and the Horse on the Plains*. See also Bea Medicine and Pat Albers, *The Hidden Half: Studies of Plains Indian Women* (Lanham: Washington, D.C.: University Press of America, 1983); and Katherine Weist, "Plains Indian Women: An Assessment in Anthropology on the Great Plains," ed. Raymond Wood and Margot Liberty (Lincoln: University of Nebraska Press, 1980) 255–71.

25. Joseph Jablow, *The Cheyenne in Plains Indian Trade Relations, 1795–1840* (Seattle: University of Washington Press, 1950); Donald Berthrong, *The Southern Cheyennes* (Norman: University of Oklahoma Press, 1963), 4–21; Loomis and Nasatir, *Pedro Vial and the Roads to Santa Fe*, 256–58.

26. Alan Osburn, "Ecological Aspects of Equestrian Adaptations in Aboriginal North America," *American Anthropologist* 85 (September 1983), 563–91; Berthrong, *Southern Cheyennes*, 25–26; Jablow, *Cheyenne in Plains Indian Trade Relations*, 67; David Lavender, *Bent's*

Fort (Lincoln: University of Nebraska Press, 1954), 141–54; George Phillips, *Chiefs and Challengers: Indian Resistance and Cooperations in Southern California* (Berkeley: University of California Press, 1975), 42–43; Eleanor Lawrence, "The Old Spanish Trail from Santa Fe to California" (M.A. thesis, University of California, Berkeley, 1930).

27. On the intertribal buffer zones, see Berthrong, *Southern Cheyenne*, 76, 93. On their function in preserving wildlife in other ecosystems, see Harold Hickerson, "The Virginia Deer and Intertribal Buffer Zones in the Upper Mississippi Valley," in *Man, Culture, and Animals*, ed. Anthony Leeds and Andrew Vayda (Washington: American Association for the Advancement of Science, 1965), 43–66.

28. Raymond Dasmann, "Future Primitive," *Co-Evolution Quarterly* 11 (1976): 26–31; David Wishart, *The Fur Trade of the American Revolt, 1807–1840* (Lincoln: University of Nebraska Press, 1979); Arthur Ray, *The Fur Trade and the Indian* (Toronto: University of Toronto Press, 1974); White, *Roots of Dependency*, 147–211.

29. Zubrow, Prehistoric Carrying Capacity, 8–9.

30. Chapman and Feldhamer, eds., *Wild Mammals of North America*, 980–83; Arthur Halloran, "Bison (Bovidae) Productivity on the Wichita Mountains Wildlife Refuge, Oklahoma," *Southwestern Naturalist* 13 (May 1968): 23–26; Alisa Shull and Alan Tipton, "Effective Population Size of Bison on the Wichita Mountains Wildlife Refuge," *Conservation Biology* 1 (May 1987): 35–41.

31. Data on the modern bison herds on the Great Plains are from the refuge managers and superintendents of the Wichita Mountains Wildlife Refuge, Theodore Roosevelt National Memorial Park, and Wind Cave National Park. The National Bison Refuge in Montana did not respond to my inquiries. Robert Karges to Dan Flores, 18 March 1988 (in author's possession); Robert Powell to Flores, 10 February 1988, (in author's possession); Ernest Ortega to Flores, 11 February 1988 (in author's possession). See Graeme Caughley, "Eruption of Ungulate Populations, with Emphasis on Himalayan Thar in New Zealand," *Ecology* 51 (Winter 1970): 53–72. This study has been widely cited in wildlife ecology as evidence that starvation rather than predation is often the key to regulating natural population eruptions. The only documentary evidence I have seen for starvation of bison on the Southern Plains is Charles Goodnight's account of seeing "millions" of starved bison along a front 25 by 100 miles between the Concho and Brazos Rivers in 1867, after bison migration patterns had been disrupted by settlements. J. Evetts Haley, *Charles Goodnight, Cowman and Plainsman* (Norman: University of Oklahoma Press, 1949), 161.

32. The nineteenth-century documentary evidence assigns wolves roles as scavengers of bison killed by other agents; as cullers of weak, sick, and old animals; and as predators of bison calves. The last, I believe, best expresses the regulatory effect of wolves on Plains bison population dynamics. See Gary E. Moulton, ed., *The Journal of the Lewis and Clark Expedition* (Lincoln: University of Nebraska Press, 1987), 4:62–63; Donald Jackson and Mary Lee Spence, eds., *The Expeditions of John Charles Fremont* (Urbana: University of Illinois Press, 1970), Vol. 1, 190–91; Henry Boller, *Among the Indians: Four Years on the Upper Missouri; 1858–1862*, ed. Milo Milton Quaife (Lincoln: University of Nebraska Press, 1972), 270–71; Maria Audubon and Elliott Coues, eds., *Audubon and His Journals* (New York: Scribner's Sons, 1986), Vol. 1, 49; and W. Eugene Hollon, ed., *William Bollaert's 'Texas'* (Norman:

University of Oklahoma Press, 1989), 255. For other descriptions, see Stanley P. Young and Edward A. Goldman, *The Wolves of North America* (New York: Dover Publications, 1944), Vol. 1, 50, 218, 224–31.

33. Katherine Spielman, "Late Prehistoric Exchange between the Southwest and Southern Plains," *Plains Anthropologist* 28 (November 1983): 257–79. For the argument that essential plant resources from this trade ended a nutrition "bottleneck" and therefore allowed the buildup of much larger human populations on the Southern Plains, see Bamforth, *Ecology and Human Organization on the Great Plains*, 8.

34. Jerold Levy, "Ecology of the South Plains," in *Symposium: Patterns of Land Utilization and Other Papers*, ed. Viola Garfield (Seattle: American Ethnological Society, 1961), 18–25.

35. Brown, "Comancheria Demography," 10–11; H. Paul Thompson, "A Technique Using Anthropological and Biological Data," *Current Anthropology* 7 (October 1966): 417–24; Jacob Fowler, *The Journal of Jacob Fowler: Narrating an Adventure from Arkansas through the Indian Territory, Oklahoma, Kansas, Colorado, and New Mexico to the Sources of the Rio Grande del Norte*, ed. Elliott Coues (New York: F.P. Harper, 1898), 61, 63.

36. Wallace and Hoebel, *Comanches*, 31–32. The anthropological literature tends to set Comanche population much more conservatively, often at no more than seven thousand. See, for example, Bamforth, *Ecology and Human Organization on the Great Plains*, 104–14. Such low figures ignore eyewitness accounts of localized Comanche aggregations of several thousand. I have a historian's bias in favor of documentary evidence for estimating human populations; Plains observers computed village sizes relatively easily by counting the number of tents.

37. Berthrong, *The Southern Cheyennes*, 78, 92, 107. The Kiowas and Kiowa-Apaches seem to have averaged about 2,500 to 3,000 from 1825 to 1850, and the Prairie Caddoans perhaps 2,000, shrinking to 1,000 by midcentury. *Report, Commissioner of Indian Affairs* (1842), cited in Josiah Gregg, *Commerce of the Prairies*, ed. Max Moorhead (Norman: University of Oklahoma Press, 1963), 431–32 n. 3.

38. Because of the tough meat and the thick hides that made soft tanning difficult, Indians (and whites hunting for meat) rarely killed bison bulls. See Roe, *North American Buffalo*, 650–70; and Larry Barsness, *Heads, Hides, and Horns: The Compleat Buffalo Book* (Fort Worth: Texas Christian University, 1974), 69–72, 96–98. Dean E. Medin and Allen E. Anderson, *Modeling the Dynamics of a Colorado Mule Deer Population* (Fort Collins: Washington, D.C.: Wildlife Society, 1979). Whether sixty thousand hunters ever worked the Southern Plains in precontact times is now unknowable, but Coronado's chronicler, Castañeda, wrote that there were more people on the Plains in 1542 than in the Rio Grande pueblos. Pedro de Castañeda, "The Narrative of the Expedition of Coronado," in *Spanish Explorers in the Southern United States, 1528–1543*, ed. Frederick W. Hodge and Theodore H. Lewis (Austin: Texas State Historical Association, 1984), 362. For a seventeenth-century population estimate in the Rio Grande pueblos of about 30,500, see Marc Simmons, "History of Pueblo-Spanish Relations to 1821," in Sturtevant, *Handbook of North American Indians*, 9:185 (table 1).

39. For Thomas Fitzpatrick's report, see Berthrong, *Southern Cheyennes*, 124. The English traveler William Bollaert mentions that the Texas Comanches supposedly ate twenty thou-

sand mustangs in the late–1840s. See Hollon, *William Bollaert's Texas*, 361. On the escalating stock raids and trade to New Mexico beginning in the 1840s, see J. Evetts Haley, "The Comanchero Trade," *Southwestern Historical Quarterly* 38 (January 1935): 157–76. Haley generally ascribes the situation to Comanche barbarity and Hispanic lack of respect for Lockean private property rights. See also Kenner, *History of New Mexican Plains Indian Relations*, 78–97, 155–200.

40. See James Mooney, *Calendar History of the Kiowa Indians* (Washington: Smithsonian Institution Press, 1979), 287–95; Levy, "Ecology of the South Plains," 19. The decline in the number of bison was becoming noticeable as early as 1844, two years before the 1846–57 drought. See Solomon Sublette to William Sublette, 2 February, 5 May 1844, William Sublette Papers, Archives, Missouri Historical Society, St. Louis, Mo. In 1845 the trader James Webb and his party traveled from Bent's Fort to Missouri without killing a bison. "Memoirs of James J. Webb, Merchant in Santa Fe, N.M., 1844," typescript, p. 69, James Webb Papers, Archives, Missouri Historical Society.

41. John C. Ewers, *The Horse in Blackfoot Indian Culture* (Washington: U.S. Government Printing Office, 1955). For systematic assessment of the effects of horses on seasonal band size, camps, and resources, see John Moore, *The Cheyenne Nation: A Social and Demographic History* (Lincoln: University of Nebraska Press, 1987), 127–75. For a dynamic rather than a static horse ecology, see James Sherow, "Pieces to a Puzzle: High Plains Indians and Their Horses in the Region of the Arkansas River Valley, 1800–1860," paper presented at the Ethnohistory Conference, Chicago, 1989 (in author's possession). Clyde Wilson, "An Inquiry into the Nature of Plains Indian Cultural Development," *American Anthropologist* 65 (April 1963): 355–69.

42. J. Frank Dobie, *The Mustangs* (New York, 1934), 108–9. Dobie's estimate, as he pointed out, was a guess, but my work in the agricultural censuses indicates that it was a good guess. On horse/bovine dietary overlap see L.J. Krysl et al., "Horses and Cattle Grazing in the Wyoming Red Desert, I. Food Habits and Dietary Overlap," *Journal of Range Management* 37 (January 1984): 72–76. On the drier climate on the Plains between 1848 and 1874, see Weakly, "Tree-Ring Record of Precipitation in Western Nebraska," 817, 819; and Levy, "Ecology of the South Plains."

43. I follow Chapman :and Feldhamer, eds., *Wild Mammals of North America*.

44. McHugh, *Time of the Buffalo*, 226–27. For scientific discussion of predation by wolves on large ungulates, see David Mech, *The Wolf* (New York: Natural History Press, 1970); see also Chapman and Feldhamer, eds., *Wild Mammals of North America*, 994–96. For estimates that the reintroduction of wolves to Yellowstone would reduce the bison herd there between 5and 20 percent see Barbara Koth, David Lime, and Jonathan Vlaming, "Effects of Restoring Wolves on Yellowstone Area Big Game and Grizzly Bears: Opinions of Fifteen North American Experts" in *Yellowstone National Park, Wolves for Yellowstone?* (n.p., 1990), Pt. 4, 71–72, and the computer simulation, Pt. 3, 31. Young and Goldman, *Wolves of North America*, II, 327–33.

45. Schulman, *Dendroclimatic Data from Arid America*, fig. 22: Weakly, "Tree-Ring Record of Precipitation in Western Nebraska," 817, 819.

46. For emphasis on disruption by whites, see Douglas Bamforth, "Historical Documents and Bison Ecology on the Great Plains," *Plains Anthropologist* 32 (February 1987): 1–16. On the *ciboleros*, see Kenner, *History of New Mexican-Plains Indian Relations*, 115–17. Jablow, *Cheyenne and Plains Indian Trade Relations*, 72.

47. My interpretation of the great 1840 alliance of the Southern Plains tribes has been much influenced by Jablow, *Cheyenne in Plains Indian Trade Relations*, 72–73; Levy, "Ecology of the South Plains," 19; Mooney, *Calendar History of the Kiowa Indian*, 276–346.

48. John Whitfield, "Census of the Cheyenne, Comanche, Arapaho, Plains Apache, and the Kiowa of the Upper Arkansas Agency," 15 August 1855, Letters Received, Records of the Office of Indian Affairs, RG 75 [microfilm], M234, reel 878, National Archives, Washington, D.C. Letters between the principals at Bent's Fort make it clear that the Comanche trade in robes was Bent and St. Vrain's chief hope for economic solvency in the early 1840s. See W.D. Hodgkiss to Andrew Drips, 25 March 1843, Andrew Drips Papers, Archives, Missouri Historical Society.

49. Records on the robe trade are fragmentary and frequently at odds with one another; see T. Lindsay Baker, "The Buffalo Robe Trade in the Nineteenth-Century West," paper presented at the Center of the American Indian, Oklahoma City, April 1989 (in T. Lindsay Baker's possession). John Jacob Astor's American Fur Company was taking in 25,000–30,000 robes a year from the Missouri River from 1828 to 1830, and St. Louis was receiving 85,000 to 100,000 cow robes a year by the end of the 1840s. Baker cannot yet estimate how many of those came from the Southern Plains, but the trend toward a larger harvest seems apparent. On the Canadian experience, see Ray, *Fur Trade and the Indian*, 228.

50. Berlandier, *Indians of Texas in 1830*, 84–85. Thirteen epidemics and pandemics would have affected the Comanches between 1750 and 1864; see Dobyns, *Their Number Became Thinned*, 15–20. On the abandonment of Bent's first Arkansas River post, see Lavender, *Bents Fort*, 338–39. John C. Ewers, "The Influence of Epidemics on the Indian Populations and Cultures of Texas," *Plains Anthropologist* 18 (May 1973): 106. Ewers bases his decline on an estimated early-nineteenth-century Comanche population of only seven thousand. If my larger estimate is accepted, the Comanche population decline was more than 90 percent. A fall-off in the birthrate as Indian women contracted Bang's disease from brucellosis-infected bison may have contributed importantly to Indian population decline.

51. James W. Abert, The Journal of Lieutenant J.W. Abert from Bent's Forts to St. Louisiana 1845, ed. H. Bailey Carroll (Canyon, 1941), 15–16; Senate, Report of the Secretary of War, Communicating in Answer to a Resolution of the Senate, a Report and Map of the Examination of New Mexico, Made by Lieutenant J.W. Abert, of the Topographical Corps, 30th Cong., 1st sess., 1848.

52. Richard I. Dodge, *Our Wild Indians* (Hartford: A.D. Worthington and Company, 1882), 286. The idea has lingered in the preserved mythologies of the Southern Plains tribes. In 1881 representatives of many of those tribes assembled on the North Fork of the Red for the Kiowa Sun Dance, where a Kiowa shaman, Buffalo Coming Out, vowed to call on the herds to reemerge from the ground. The Kiowas believed the bison had gone into hiding in the earth, and they still call a peak in the Wichita Mountains Hiding Mountain. Alice

Mariott and Carol Rechlin, *Plains Indian Mythology* (New York: Crowell, 1975), 140; Peter Powell, *Sweet Medicine* (Norman: University of Oklahoma Press, 1969), 1:281–82. There is no mention of this idea in the major work on Comanche mythology, but it is far from a complete compilation: Elliott Canonge, *Comanche Texts* (Norman: University of Oklahoma Press, 1958). On the bison's wintering in protected canyons, see Randolph Marcy, *A Report on the Exploration of the Red River, in Louisiana* (Washington, D.C.: 1854), 125–31.

53. Wallace and Hoebel, *Comanches*, 32; Richard I. Dodge in *The Plains of North America and Their Inhabitants*, ed. Wayne Kime (Newark: University of Delaware Press, 1989), 155–57. On these figures for bison, see Roe, *North American Buffalo*, 440–41.

CHAPTER FIVE

WORKINGS OF THE GEODIALECTIC
High Plains Indians and their Horses in the Region of the Arkansas River Valley, 1800–1870

JAMES E. SHEROW

On a cold, clear November morning, the first rays of sun met a Cheyenne woman running across a brownish-grass plain where the wind of a recent storm had filled the low-lying areas with snow, streaking the landscape with white. She raced to her camp breathlessly, calling her people to prepare for a hunt. The year 1864 had been particularly hard on her people, the *Tsis tsis tas*, and the *Hinono'eino'*, or as called by Americans, the Southern Cheyennes and Arapahos. This unknown woman and her people were hungry and exhausted, but the anticipation of fresh meat aroused excitement and activity in the camp.[1]

In the previous months the tribes had been migrating south from the upper watersheds of the Republican and Smoky Hill Rivers toward their traditional campgrounds on the Arkansas and Purgatoire Rivers. The tribes craved rest not only for themselves, but also for their over six hundred horses and mules, many of which were "lame" and desperately needed care and feed. Her people's leaders, Black Kettle, One-Eye, Yellow Wolf, and Left Hand, had pleaded for help from the soldiers at Fort Lyon, who before had promised aid in worthless treaties. Major Scott J. Anthony, the post commander, instructed the tribal chiefs not to go down the Purgatoire River (with its better winter habitat but also where difficulties might arise with newly arrived white ranchers and farmers). Anthony told Indian leaders to remain camped on the banks of Sand Creek where bison, ever more scare in recent years, were supposedly grazing.[2] Perhaps on Sand Creek the woman who was so excitedly announcing the arrival of the bison, and her people, could find rest, peace, and renewal. The black forms approaching from the southwest momentarily confirmed the decision to keep the lodges pitched along Sand Creek.

The encampment had most of what the Southern Cheyennes and Arapahos needed. Water flowed in Sand Creek at this spot and provided a vital supply for the ponies and the tribe. The sharply cut western bank worked to shield the lodges pitched on the open, flat, eastern side of the creek from the cold-blowing westerly and northerly winds. Between the western side of the creek and the bluff grew a scattering of willows and cottonwoods, which provided heating fuel for the tribe and supplementary forage for the herd. The pony herd grazed on the plains above the creek about three miles or so to the west of the encampment on the brownish winter-dormant buffalo grass now drained of its high spring and summer protein contents. This put the herd in a rather difficult place to defend, but the Indians expected no problems, as the Fort Lyon commander had promised the tribes United States Army protection. Regardless, the large numbers of "lame horses" in the tribes' herds needed this grass or else would soon turn seriously ill or die with the fast approach of winter. For the moment, hunters intended to interrupt their grazing animals and mount the nearest, freshest steeds to scout the bison herd.

What the hunters saw as they climbed over the western bank was not a herd of bison, but rather the approach of Colonel John M. Chivington's small army toward their lodges. The colonel had marched his men all night from Fort Lyon to attack these anxious hunters and their village. His men had one distinct advantage over the unsuspecting Indians: they rode well-fed, strong horses. Chivington did not rely upon dormant buffalo grass to feed his animals; rather, a train of wagons trailing behind the attacking units carried hay and grains. He knew the tribes' camping and herding practices made them susceptible to attack in early winter. Cautiously, and mercilessly, Chivington began a day-long assault on a peaceful, desperate people. First, he successfully struck to separate the vulnerable horses from the tribes. Several warriors managed to escape with some ponies kept nearer the lodges, but the women, whose job it was to gather the main herd in times of danger, found themselves unable to position the animals within the warriors' protection. Chivington's men managed to capture most of the herd, and later the colonel would take a great number of these weakened animals and recondition them for sale on a ranch near Denver.

The Cheyennes and Arapahos reacted in confusion. Some tried to defend themselves, others offered themselves to the Coloradans as peaceful folk only to be shot dead, and still others fled to the north, east, and west however they could. Chivington did not give immediate chase as he kept close to his supply trains. He could ill afford to weaken his horses by tracking small straggling groupings of Indians heading off in several directions. Anglo cavalry tactics worked best in attacking fixed camps or large armed groups, but not in chasing small groups fleeing across the plains, and then especially so in winter. When Chivington made his pursuit, he proceeded en masse, which gave the escaping Indians a distinct advantage, but not due to the supposed swiftness of their horses. When he returned triumphant to Denver, he did so before his wagon trains ran short of supplies for his horses.

Two different ways of living on the plains had come into conflict. The Southern Cheyennes and Arapahos had experienced debacle on Sand Creek, but Chivington and his men could not capture the entire camp. Historians have explained this people's predicaments in various ways, but never in terms of incomplete environmental adaptation to the High Plains. One of the factors, among many others, making the Arapahos and Cheyennes so vulnerable was the way in which they kept their horses. What made pursuit difficult for Chivington was the way in which Anglos tended *their* horses. What happened at Sand Creek was more than the murderous cruelty and cultural chauvinism of Colonel Chivington and his soldiers; in part, the massacre was also the result of the Indians' difficulties in adapting their horse-herding practices to the Great Plains.

Many High Plains Indians failed to understand their place in the *geodialectic*. This one-word term combines James Lovelock's notion of Gaia, a living earth, with the idea of dialectical relationships among living and nonliving forces; this process constantly remakes terrestrial environments. In a dialectical sense, Indians and horses interacting with organic forces made sweeping changes to the Great Plains. Nonliving climatic, tectonic, and cosmic forces (like drought, wind, volcanism, earthquakes, meteoroids, comets, sunspots, asteroids) could also add to the complexity of the geodialectic in the region.

The term geodialectic implies ever-changing environments on the planet, and, in order to survive, humans and other living things must adapt to constant environmental flux. Nature, however, is uninterested in human survival. As Lovelock elegantly put it: "When the activity of an organism favors the environment as well as the organism itself, then its spread will be assisted. The reverse is also true," he continued, "and any species that adversely affects the environment is doomed; but life goes on."[3] The earth has always set limits on any organism's environmental adaptation. Many scholars lose sight of this simple axiom, yet as Donald Worster notes, the constraints to human adaptation are the most important lessons environmental historians impart.[4] The ability of any species, including humans, to persist and flourish rests upon how well it adapts to the geodialectic.

Unlike other living things, people are "aware" of taking actions in nature; in many cases they are "aware" of selecting among possible actions those most likely to sustain their own survival.[5] For example, the High Plains Indians, very aware of what they were doing, chose to use horses in developing their material culture. These choices make what John Bennett has called adaptation strategies, or the "patterns formed by the many separate adjustments that people devise in order to obtain and use resources and to solve the immediate problems confronting them."[6] It is wise to take Bennett's thoughts one step farther—the "adjustments" people make must keep pace with the geodialectic. In the case of High Plains Indians, their ideologies or religions may or may not have been in harmony with nature.[7] This debate, however, is beside the question of historical High Plains Indian horse practices.[8] By focusing on material culture it is possible to show how these people's horse-tending problems were of their own making and how other difficulties were entirely beyond their ability to control.[9]

From around 1800 to 1870, the historical dynamics of the geodialectic had made it difficult for the northern Comanches, Kiowas, Plains Apaches, Southern Cheyennes, and Southern Arapahos to maintain and use their horse herds in the region of the upper Arkansas River Valley. These people faced difficulties grooming their animals during winter and summer, coping with parasites, establishing herd security, hunting buffalo, and adapting to a rapidly changing environment. Most scholars have recognized only how Indian horses liberated the tribes in using the resources of the High Plains, not how horses constrained their actions in certain ways and led to environmental degradation. Consequently, the historical antecedents of these Indians' adaptation strategies were to make them particularly vulnerable at Sand Creek by 1864.

Shortly after 1700 the Comanches began adapting to the variable climate of the Arkansas Valley, which supported a wide array of indigenous animal and plant life. Bison, deer, antelope, bears, wolves, cougars, elk, coyotes, and numerous other mammals, as well as hundreds of species of birds, and reptiles, and thousands of species of insects flourished on the plains. Cottonwoods, willows, and berry bushes skirted water courses in many places. One important introduced exotic animal, the horse, also inhabited the High Plains. By the early 1700s Comanches had obtained horses and rapidly learned to use them in developing a hunting culture dominated by the pursuit of bison. By the late 1700s these people had adapted to the southern High Plains by participating in a flourishing market economy dominated by a fur and meat trade with the Spanish in the Taos and Santa Fe region, and in return receiving agricultural and cloth items from New Mexicans and Pueblo Indians. The Kiowas and Plains Apaches joined the Comanches to dominate the High Plains south of the Arkansas River by the early 1800s.[10]

Cheyennes and Arapahos arrived in the Arkansas Valley sometime between 1810 and 1820. Early accounts of them in the valley come from the journals of

Long's expedition in 1820, and Jacob Fowler's expedition in 1821–22. Zebulon Pike made no mention of the Cheyenne in the region during his trek through it in 1806, so it is reasonable to fix their arrival in the area sometime between 1806 and 1820. Prior to 1820, they had been migrating westward from around the upper Mississippi Valley drainage basin. They began their adaptation to the short-grass plains by casting aside agriculture for a buffalo-hunting culture similar to other High Plains Indians. Soon afterward the Cheyennes and Arapahos controlled the area between the South Platte and the Arkansas River.[11]

Wild horses drew both the Cheyennes and Arapahos to the upper portions of the southern High Plains. In the 1820s this region marked the northernmost range of mustangs.[12] The Cheyennes and Arapahos rushed to the area seeking these horse herds, which would give additional support to their trading operations and to their hunting culture. These feral horses, the Indians realized, could potentially enrich and strengthen their positions in the High Plains economic matrix. These tribes could realize their aspirations if they could mount a successful adaptation of their developing hunting culture to their newly occupied environment.

Between 1820 and 1860, these tribes' economic adaptation to the region depended upon trade with New Mexicans in the Spanish borderlands to the west and south and with Anglo-Americans to the east. The horse trade, a most important element in this economy, flourished out of the borderlands northward through the Plains. Indians traded horses to Anglo-Americans and to northern tribes; the Indians used them to hunt, for social status, and for trade while the Americans drove the animals to Westport, Missouri, to supply overlanders. From the East came guns, metal pots and pans, clothing, metal tools (for example, awls, which made sewing bison hides much easier than the older method of using bone needles), and whiskey. Besides bringing in trade goods for these middlemen tribes, horses also aided them in hunting bison, which added to their economic strength. The increases in dressed robes enhanced the bartering power of these hunters dealing with Anglo-American fur and horse traders.[13]

Horses became critically important to the tribes' economies. The Plains Indians exchanged mustangs with the tribes to their east and north for European and American goods. The Cheyennes also stole horses from New Mexico settlements and the Comanches in order to supply other tribes, while at other times they traded guns to the Comanches for their horses. The Comanches needed guns in order to stem the Anglo-American and relocated Indian tribes' advance into western Texas. The Spanish, and later the Mexican, government had effectively avoided any gun trade with the Comanches, thereby leaving northern and eastern tribes as the only source of firearms.[14]

The plains tribes used a great many horses in conducting their trade. George Bent, son of Owl Woman, who was the daughter of White Thunder, keeper of the Cheyenne Medicine Arrows, and of William Bent, who had established the most renowned trading post in the Arkansas Valley in 1833, knew well both the Anglo-American and Indian worlds. He was a careful observer and remembered as a boy seeing Comanches', Kiowas', and Plains Apaches' "pony herds grazing along the [Arkansas] [R]iver for fifty miles." Near present-day Fort Sill in Oklahoma, Lieutenant Wheelock recorded a Comanche village of two hundred lodges with three thousand horses in 1834. On tributaries of the Canadian River Lieutenant Marcy wrote of his encounter with three hundred Comanches who had two thousand horses and mules in 1849.[15] In 1855 agent W.D. Whitfield for the Upper Arkansas Agency enumerated the following numbers of horses owned by the tribes under his charge—the Upper Comanches: 20,000 horses, the Kiowas: 15,000, the Plains Apache: 2,000, the Southern Cheyennes: 17,500, and the Southern Arapahos: 15,000. Except for the Cheyennes who had 5.55 horses per capita, the other tribes had 6.25 horses per person (see Table 1).

Even though it was not the most difficult place on the High Plains to tend horses, the Arkansas River Valley proved a trying environment for its inhabitants to keep their herds. Precipitation, which averaged between fourteen and twenty inches annually depending upon locale, is the key to understanding the valley and its surroundings. Great variability marked the fall of rain and snow. During extremely hot days, heat radiating off the land evaporated rain showers before they ever touched earth. Sporadic and scattered thunderstorms could suddenly thrash the land and its inhabitants with gale, hail, and freshets. The unabated effect of dry, hot winds could leave little moisture in the upper layers of the soil, killing grass and drying water courses and springs.

The interaction of an erratic climate and High Plains Indian cultures, one aspect of the geodialectic, often caused suffering for horses and their tribal owners. During the fall the rising of the sun could melt away morning frost on dormant grasses revealing bluish-gray-colored blades stressed by drought, which provided poor forage, or nutritious golden-brown blades in years (like 1864) of plentiful rains. Balmy autumn days could suddenly turn windy and snowy and ensnare unsheltered people and horses in a deadly grip. In winter howling blizzards, or crystal clear blue skies and frigid temperatures, or perhaps mild sunny days could all follow in rapid succession and made trips across the exposed plains extremely hazardous for horses and people alike. In the spring, wet snows or rains, when and if they fell, nourished the grasses mantling the plains with a thick velvet-green carpet. During this time the Indians celebrated

TABLE 1
Per Capita Horse Ownership Per Tribe in the Upper Arkansas Agency, 1855

Tribe	population	# of horses	# of horses per capita
Comanches	3,200	20,000	6.25
Kiowas	2,400	15,000	6.25
Apaches	320	2,000	6.25
Cheyennes	3,150	17,500	5.55
Arapahos	2,400	15,500	6.25

SOURCE: W.D. Whitfield, Census of the Cheyenne, Comanche, Arapaho, Plains Apache, and the Kiowa the Upper Arkansas Agency, 15 August 1855, U.S. Department of Interior, Bureau of Indian Affairs, Letters Received by the Office of Indian Affairs, 1824–81, Record Group 75, M234, roll 878.

the return of the short grasses, which restored the strength of Indian ponies, returned the bison for the hunt, and renewed the tribes.

Plains Indians found the constraints to herding horses difficult enough in normal years but they found grazing conditions severely affected during droughts, for example those in the 1850s. In the western portion of present-day Kansas along the upper reaches of the Republican and Smoky Hill Rivers, where Southern Cheyennes and Arapahos often camped and hunted, the grass supply could prove luxuriant or scanty depending upon precipitation. In years when more than twenty inches of precipitation fell, according to one study, an acre produced around 3,000 pounds of short grass, whereas in years when ten inches or less fell then the same acre yielded less than 450 pounds. Moreover, this same acre required four years to return to predrought yields if the acre had not been overgrazed.[16]

In such a variable environment, High Plains Indians' horses required a great deal of care from their owners. A rough estimation of nutritional needs for an average mustang was a pound of salt a week, ten to twelve gallons of water daily, and an equivalent in grass of ten to twenty-five pounds of hay daily. Besides, mustangs were susceptible to both parasites and sunstroke.[17] Wherever Indians camped on the High Plains they had to remember the needs of their horses or else lose them. If a band had one thousand horses and camped in western Kansas during normal years, those horses would have needed the equivalent tonnage of around seven acres of grass per day during normal precipitation, and forty-two acres per day during drought (see Table 2).

Besides variable range conditions, the unfolding of the geodialectic on the

TABLE 2

Comparison of Horse Forage Needs on the High Plains During Normal and Drought Years

	Normal Year	Drought Year
precipitation	<20 inches	>10 inches
yield of grass per acre in lbs.	3,000	450
# of horses	1,000	1,000
required # of acres with assumed consumption of ca. 20 lbs. of grass per horse per day	7	42

High Plains in winter required sophisticated adaptation strategies on the part of the Indians. Maintaining large numbers of horses imposed trying constraints on Plains Indians, especially in winter. Alan J. Osborn, an anthropologist, has revealed a close connection to horse distribution throughout the High Plains and the increasing severity of winters from south to north. Given the Plains Indians horse-feeding practices, Osborn maintains, the farther north a tribe the fewer horses per capita because harsh winters precluded the maintenance of large herds.[18]

Not only would the relative severity of winters on the plains influence the distribution of horses, varying winter conditions also produced a wide range of health problems for Indian ponies. By spring and the return of the short grasses tribal herds had endured deprivation in the Arkansas region. One study reveals some of what High Plains Indian horses must have suffered during winters. During several successive winters researchers compared two groups of horses, one fed on hay and grain, and the other fed primarily on native High Plains grasses. Horses fed on range pasturage, so the results showed, lost weight every winter; and even though the animals recovered most of their weight in the summer with the return of fresh grass they matured slower and weighed and grew less than the hay- and grain-fed horses. The experimenters also fed grains and hay to the open-range horses if the native grasses had been exhausted.[19]

In contrast, High Plains Indians could seldom supplement a horse diet with grains and hay when encountering a shortage of plains grasses. The reliance upon plains grasses meant certain obstacles in maintaining healthy horses in winter. For example, short grasses lose over half of their summer protein in winter.[20] Consequently, without recourse to hay or grain, healthy Indian horses in the fall suffered from malnutrition by the end of winter.

In adapting to the High Plains, tribes tried to avoid the difficulty of grazing horses on short grasses in winter by restricting their movements to timbered places along the rivers and springs. Tall bunchgrasses, for example side-oats grama (*Bouteloua curtipendula*) and big and little bluestem (*Andropogon geradii* and *Schizachyrium scoparium*, respectively), grew in these areas.[21] These tall grasses had an advantage over short grasses; they yielded greater volume to the acre than did short grasses. According to the calculations of R.V. Boyle and J.S. McCorkle, "if [riparian grasses] are in good condition they can produce 2 to 20 times more forage than do the adjacent uplands." A serious disadvantage attended grazing horses on tall grasses: they retained less protein in winter than did short grasses.[22] Still, horses could live on the riparian grasses supplemented with cottonwood and willows, which for the herders meant fewer camp relocations in winter, and better shelter and security for people and beasts.

The tribes used numerous riparian ecosystems throughout the High Plains. An area along the Arkansas called "Big Timbers" served as one popular winter camping ground. When Zebulon Pike marched along the river in 1806 he reported a continuous stand of plains cottonwood (*Populus occidentalis*) along the north bank of the river from about twenty miles west of the present-day Colorado–Kansas state line to a termination point around the mouth of the Purgatoire River. Indians quite far removed from this area knew of it. In 1821 the Osage Indians, near present-day Winfield, Kansas, described Big Timbers for Jacob Fowler's band of explorer-traders. In places riparian flora growth followed the Purgatoire River as well as the Timpas, Huerfano, Apishapa, and Sand Creek Valleys. Besides these rivers and creeks, springs punctuated the High Plains and some of these were lined by trees. On the High Plains, Waldo R. Wedel once figured, there were "few districts where springs, either primary or as creek bed water hole, were more than 25 to 30 miles apart, with 10 to 15 miles probably a much likelier average." Ponds or playa lakes, he also noted, abounded throughout the High Plains. These depressions, varying as widely as a few feet to several thousand yards in diameter, filled with runoff from rains or snows. Generally, though, they remained dry throughout much of a year. The Plains Indians, realizing the importance of these areas to their herds, kept an accurate mental map of their locations. For example, George Bent clearly knew the locations of important watering sites between the Arkansas River and the headwaters of the Smoky Hill and Republican Rivers.[23]

Indians in winter camps could keep their horses relatively healthy so long as a supply of riparian grass remained. Once the Indian herds had grazed the dormant grasses, they needed some other source of forage. The upper Comanche avoided some of the horse wintering problems faced by the tribes to their north. Captain R.B. Marcy's insightful observation of these people's adaptation strategies revealed the reason why. By 1836 the Comanche's range encompassed a great extent of territory—all lands between the Arkansas River south to the Mexican settlements in Chihuahua, and from the Cross Timbers west to the Grand Cordillera. According to Marcy this enabled them to hunt bison as far north as the Arkansas River during the summer months. With the onset of autumn, the bands returned south as far as the Brazos and Colorado Rivers in Texas. The grasses, Marcy noted, "remaining green during the winter, afford pasturage for their animals until the following spring, when they return again to the north."[24] Marcy was mistaken about grasses remaining green in these southern areas throughout the year; but the winters in these locations were less severe and usually shorter than those nearer the Arkansas River. A killing frost might not arrive until December, and the greening of the short grasses in the southern climes could arrive several months before the plains greened around the Arkansas River.

The Cheyennes and Arapahos, who lacked the freedom to move their herds south, often fed their animals cottonwood limbs and bark, or small trees and seedlings, often along with willows after the grass supplies failed. Many Anglo-American travelers through the Arkansas Valley expressed surprise when first encountering this practice. In August 1845, along the Purgatoire River, Lieutenant J.W. Abert reported: "We were astonished at seeing great numbers of fallen trees, but afterward learned that the Indians are in the habit of foraging their horses in winter on the tender bark and young twigs of the cottonwood." Significantly, Abert noted "great numbers" of these trees cut down, which undoubtedly affected cottonwood growth in the valley. During the winter months Plains Indian women, claimed Colonel Dodge, cut down "acres upon acres of young cottonwood" for their horse herds. One homesteader along the South Platte near present-day Denver once recalled a stretch of cottonwoods all topped off by Arapahos to feed their horses. Topping cottonwoods would have had a beneficial side-effect: in the following year the cut branches would sprout numerous small twigs, which would provide forage.[25]

Besides securing forage and shelter, the Indians could keep their animals, and themselves, more secure in these riparian woods. As evident at Sand Creek in 1864, the short grasses farther removed from a winter camp might offer a more nutritious alternative to riparian grasses, but out on the plains both shelter and protection could prove troublesome. Plains Indians faced considerable security problems even during optimal grass conditions when encampments were large and well protected. In June 1848, Captain R.B. Marcy observed and commented on the Upper Comanches' (called the "buffalo eaters," *KutsuEka*) herding techniques. The camp Marcy encountered numbered around three hundred people and had a herd of nearly 2,500 horses and mules. When grazing these animals on the short grass the band leaders made daily assignments of 150 animals to young boys. The guards, Marcy noted, oversaw their charges "with the strictest vigilance and attention" since these Comanches never knew when another plains tribe might raid their horse herds.[26]

In winter the normal problems of protecting the herd became compounded once the horses had grazed the dormant grasses around the camp. Each passing day herders took their small group of horses farther from the main camp in search of grass, leaving their herds increasingly exposed to the dangers of theft or military attack. Besides, venturing out onto the High Plains proper left the animals and their caretakers exposed to the elements. The longer a small group of winter-camping Indians could keep their horses near to them in the riparian woods, then the better their herd protection.

Winter and drought brought difficulties in supplying horses with critical levels of vitamin A, without which the animals were subject to anorexia, night blindness, respiratory difficulties, reproductive failure, convulsive seizures, blindness, bone lesions, and impaired disease resistance. The Indians had nothing in their adaptation strategies to compensate for this deficiency, as did Anglo-Americans and Hispanics, who relied on alfalfa and Timothy hay supplemented with grains. These hays provided good quantities of vitamin A but these plants were not introduced into the Arkansas region until after 1870. At Bent's Fort, Anglo-Americans compensated by harvesting riparian grasses for hay, but Indians failed to lay aside these same grasses for winter forage. Yellow corn had small amounts of vitamin A, which of course was better than none at all, but until 1843 large amounts of corn were not grown in the valley and the closest source of it existed in the settlements in northern New Mexico. Consequently, the Indians had only a few alternatives to compensate for the lack of vitamin A in winter riparian grasses.[27]

Some of the Indians used novel adaptation strategies to overcome the constraints of vitamin A deficiencies in winter. By 1843 the inhabitants at the trading fort of Pueblo had begun planting corn on irrigated fields. In the spring of 1846 several thousand Arapaho, so Francis Parkman noted, arrived when the corn ripened. At this time of year the Arapahos' horses were recovering from the hardships of winter. Really having no recourse, the traders offered the harvest to the tribes who helped themselves "most liberally." The Indians though had enough foresight "to leave enough of the crops untouched to serve as an inducement for planting the fields again for their benefit in the next spring."[28]

The traders at Pueblo did not always lose in their grain exchanges with the tribes. Apparently the farmers grew more than enough corn to supply visiting Arapahos and themselves, because during the winter of 1846–47 they kept quite busy supplying other Indians. William Tharp, a trader at the fort, and others exchanged numerous wagon loads of corn for bison robes of the Cheyennes wintering along the Arkansas River at "Big Timbers." Some dealers took in more robes than they had in any year previous. This mutually beneficial arrangement flourished until the fall of 1854 when the Utes, irritated by misconduct toward them, raided and killed the inhabitants at Pueblo.[29]

Along with vitamin deficiencies, Plains Indians had to overcome horse herding difficulties in dealing with parasites, a problem compounded by winter camping practices. Strongyles, a nematode of the family *Strongylidae*, have been a historical nemesis of horses. These intestinal worms are passed in manure and hatch and develop into ineffective larvae in as few as five days. Under harsh conditions these larvae can survive as long as three months on a pasture without a host. Consequently, any horses remaining in a camp for more than five days or returning to the same place within three months would be exposed to infection. Since Indians remained in a winter camp for some time, their horses certainly encountered strongyle infection. Moreover, malnourished horses were susceptible to two species of lice, the blood sucking louse, *Hematopinus asini*, and the chewing louse, *Damalania equi*.[30]

Over time, Plains Indians unintentionally began degrading their winter camps, especially those like Big Timbers, which were used frequently. These places initially posed no infectious threat, but with each passing season worms became an increasing problem with no remedy. Also, these Indians were partly responsible for the reduction of these riparian woods like Big Timbers. Jacob Fowler's troop encountered Big Timbers in November 1821 near present-day Lamar, Colorado. He met about nine hundred lodges of Comanches, Kiowas, Arapahos, and Cheyennes along the banks of the river at the eastern terminus of Big Timbers. He estimated the tribes' combined horse and donkey herds in excess of twenty thousand animals.[31] Such a herd required great quantities of fodder. Based on the estimates in Table 2, this many animals would have needed over four hundred tons of grass per day. Depending upon the quality of the grass, these herds would have grazed enough plants to have covered between 135 and 915 acres. Either the herders would have taken their animals farther from the camps each passing day for grazing, or fed their horses cottonwood and willow. By feeding their herds on tree saplings, twigs, bark, and small branches, the tribes could keep their animals near their camps and provide better herd security even though the dormant grass had been grazed.

In time, such grazing practices coupled with the fuel needs of the tribes, and in conjunction with American travelers' need for the fuel and emergency fodder for their draft animals, greatly reduced the extent of timber in winter campgrounds like Big Timbers. It is impossible to quantify the exact toll each culture, American and Plains Indian, exacted on Big Timbers. Clearly, though, each culture made its greatest demands on the area at different times of the year. When Americans began plying the Santa Fe Trail, which passed through Big Timbers, in ever-increasing numbers after 1821, most of the traveling occurred from late spring into late summer. The grass was good at this time and these Americans needed timber only for fuel. The Indians used Big Timbers in winter and took the trees both for fuel and browse. The combined American and Plains Indian effect on Big Timbers was a thinning of the trees, a clearing of the saplings and most of the underbrush, and an areal shrinking of Big Timbers.

Historical accounts of Big Timbers confirm this conclusion. The shrinkage of the area was well under way by 1850. In 1847 Colonel Emory encountered Big Timbers nearly ten miles to the west of where Jacob Fowler entered it in 1821. By the mid 1850s the area had lost most of its undergrowth and young trees. In 1853 Lieutenant E.G. Beckwith described the grove as follows: "The trees are scattered over the bottom, in numbers, not unlike those of the new cotton-fields of Georgia and Alabama, with inviting shades." The trees were "not thick enough to obstruct the view," he further observed, "and the opposite bank of the river discovers the same dry hills as heretofore." During the 1850s and early 1860s, both Indian and American traffic, especially so with the Pikes Peak gold rush, quickened through what remained of the trees. In 1864 Army reports indicated the cutting of the last trees of Big Timbers, and a pre-1864 photo of Fort Lyon, located in the same area, reveals a landscape nearly devoid of trees.[32]

The reduction of major winter camping areas like Big Timbers placed increasing stress on the smaller riparian wooded areas like the campground at Sand Creek. By the late 1850s Indians and American overlanders both began frequenting the upper portions of the Republican and Smoky Hill watersheds. In less than one season George Bent recalled the disappearance of the riparian trees surrounding small lakes and creeks in this formerly little-traveled region of the High Plains.[33] For the Plains Indians, the disappearance of these nourishing shelters meant difficult, if not impossible, conditions for grooming their winter herds. The tribes lacked a recourse to the barns, hay, corn, and grains that sustained American horses in the same region. In the 1850s and early 1860s, a deteriorating environment weakened their horses in winter in the High Plains Arkansas River Valley. But the return of spring promised an end to suffering. With the first green blades of buffalo grass Plains Indians' winter camps broke into smaller units. In little groups these people ventured out onto the plains and grazed their horses on the protein-rich grasses.[34] This arrangement had the benefits of keeping each grazing unit small so horses could eat without competition and avoid parasite problems.

Tribal movements out onto the plains in early spring, an adaptation strategy designed to strengthen winter-weakened horses, also made each small camp extremely vulnerable to horse theft. For example, the Cheyennes, like the noted Yellow Wolf, would head out on foot to northern Comanche camps. He knew his foe were few in number and if skillfully approached the guards could be easily overcome and the horses taken. Yellow Wolf knew time was precious in reaching the herds before the horses recovered from their winter travails. On one particular Comanche horse raid in 1828, he, as the "leader" of the party, prohibited ritualistic, and time consuming, practices accorded, and expected from, a leader. Yellow Wolf instructed his fellow warriors to ignore tradition and to follow him quickly. On this foray, and others, Yellow Wolf returned successful, and ceremony, represented by the "leader's" rituals, had been devalued as many Cheyennes came to revere Yellow Wolf's abilities and to respect his leadership.[35] This one recorded incident may be indicative of the cultural transformation occurring among the Cheyennes as they adapted to the High Plains.

Traditionally, High Plains Indians regrouped in early summer for the preparation of their tribal bison hunts. Their horses had recovered strength by eating the vitamin- and protein-rich spring grasses and were at peak vigor. The Indians marked this time by celebrating their tribal renewal in elaborate ceremonies like the Sun Dance. Although they observed their restoration with the return of the short grasses, strong horses, and bison, the onset of summer could pose many difficulties in tending their horses.

Giving chase to bison fatigued Indian horses quickly. Even in the best of condition, these animals were highly susceptible to sunstroke or heat exhaustion. For example, in May 1849 while west of Cross-Timbers and on the divide between the Canadian and Red Rivers, Captain R.B. Marcy recorded the debilitating effects one chase had on his well-groomed horse. A ten-mile run, Marcy noted, "was very severe upon my horse, and I have no doubt it injured him more than three weeks' traveling. Poor fellow!" After the chase Marcy fed the wearied beast corn, rather than grass, a nutritious alternative possessed by few Indians.[36]

Marcy's observation is useful in comparing the probable condition of Army horses to Plains Indians' horses. Most of the tribes especially valued what they called a "five-mile" horse, an animal capable of running hard for a distance of five miles.[37] Marcy rode his well-groomed horse for ten miles. Horses like Marcy's were certainly in far better shape than Plains Indian horses if the best of the Indian horses were of the five-mile variety.

Considering the fine condition of army horses, the question arises as to how Indians escaped capture so often? An explanation lies in the different tactical uses of horses by Indians and the army. When the army attacked an encampment the Indians would flee in every direction with as many horses as possible, leaving the army to pursue small groups in unfamiliar territory. This problem plagued Chivington and other commanders such as General Robert Mitchell, who chased Sioux and Cheyennes along the Republican River in the winter of 1865. Unless the army could encircle a camp and take the whole tribe, then the Indians would simply escape to join other villages. Even the Indians at Sand Creek escaped *on foot* and were later rescued by warriors from other camps. Cavalrymen, like Chivington and his men, never ventured far from their supply wagons, and once forage reached a critically low level they would return to their post.[38]

Still, even the better-fed American horses doing ordinary chores had difficulties on the short-grass prairies. The journals of American explorers are replete with examples of fatigued horses in the short-grass region. In July 1834 in the Washita River Valley Lieutenant T.B. Wheelock recorded "[s]eventy-five horses and mules disabled: rapid marching in the heat of the day, and poor grazing at night, are supposed to have been the causes." In September 1846 along the upper Washita River Valley Lieutenant J.W. Abert watched as "loose animals, rendered wild by their suffering for water, were running in advance of the party, in spite of the efforts of their drivers." In 1848 Colonel W.H. Emory noted when past the 99th meridian in the Arkansas River Valley that "[h]orses occasionally fed on grain become very weak feeding on grass alone, and should never in that condition be subjected to quick work." During August of the same year Emory reported his horses "falling away in an alarming manner" while he marched his army along the Timpas River on his way to Santa Fe.[39] If army horses suffered from the heat then surely Plains Indians' horses endured similar distress.

At times in summer even finding water for horses could prove difficult for people on the southern High Plains. The rivers in the region flowed intermittently, not perennially. The Arkansas River, for example, simply served as a great drainage system for the melting snow pack on the eastern slopes of the Greenhorn, Sangre de Cristo, and Rocky Mountains. From May through June the river generally brimmed until the sun had finished melting the snowpack in mountain meadows, forests, and rocky slopes. Winters of unusually deep snowpack caused the river to flow longer into the summer. Frequent mountain rains and infrequent plains thunderstorms might also cause an uninterrupted flow into the subhumid plains. The region's precipitation, though, usually failed to produce a surface flow throughout the entire year.[40]

The river bed beyond present-day Pueblo, Colorado, often absorbed the flow during the hot summer months. In July 1853 at Fort Atkinson, Lieutenant E.G. Beckwith observed Comanches swimming in the Arkansas; however in 1851, the soldiers informed him, they had dug trenches in the riverbed in order to water their stock. Dry riverbeds also occurred in winter. In November 1863 the army freighter Edward L. Berthoud found the riverbed dry from Fort Lyon east to Fort Larned and encountered great difficulty in keeping his stock groomed. The Comanches, Cheyennes, and other tribes were just as susceptible to the watering problems faced by the army and freighters.[41]

The difficulties in keeping horses on the southern High Plains led many Indians, like those at Sand Creek, to use mules. Any experienced traveler, freighter, and hunter knew the value of mules over horses. Mules proved more resilient to heat, could go longer without water, and had greater stamina than horses. In 1834, when Lieutenant T.B. Wheelock summarized the Dodge summer expedition into the Comanches' rangeland he noted, "It is worthy of remark, that the mules of the command look better than when we started on the campaign, while it would be difficult to select *ten horses* in good order."[42] In 1821 Fowler recorded "donkeys" in the Indian herds in and around Big Timbers, which is a good indication of mule breeding. In adapting to the southern High Plains the Indians were finding mules more reliable than horses, and came to realize the great utility in using mules.

The geodialectic on the southern High Plains resulted in many obstacles impeding the Indians' adaptation to the region. Variability in precipitation made finding good pasturage difficult from one year to the next. While the wet year of 1844 provided superb grass conditions, a series of dry years following, especially 1846, 1851, and 1854, led to deteriorating range conditions.[43] These dry years coincided with a large increase in Santa Fe Trail traffic, which undoubtedly worsened an already bad situation. Indians found alarming the continuing degradation of Big Timbers and other riparian woodlands. Reoccurring drought, intermittent river flows, Indian grazing practices, the influx of American traders, soldiers, and travelers and their herds, weakened the environment for miles on either side of the Santa Fe Trail and the alternate routes through the Republican and Smoky Hill River basins.

From around 1820 through the 1860s the Plains Indians' adaptation strategies worked less well in a region undergoing rapid environmental change each passing year. Success, in terms of adaptation theory, carries with it a maintenance of material culture and population regardless of environmental flux, and in this light the tribes had failed to adjust to the region of the Arkansas River Valley. Comparing figures (even if they are rough estimates) from the 1855 and 1861 census of the Arkansas Agency reveals tribes undergoing calamity. In a short six-year time span, population and the numbers of horses fell

TABLE 3
Per Capita Horse Ownership Per Tribe in the Upper Arkansas Agency, 1861

Tribe	population	# of horses	# of horses and mules per capita
Comanches	[?]	[?]	[?]
Kiowas	1,520	3,500	2.30
Apaches	415	500	1.20
Cheyennes	1,380	2,000	1.45
Arapahos	1,400	2,600	1.86

SOURCE: A.G. Boone's census of the Upper Arkansas Agency, 1861, U.S. Department of Interior, Bureau of Indian Affairs, Letters Received by the Office of Indian Affairs, 1824–81, Record Group 75, M234, roll 878.

precipitously among all of the tribes (see Tables 1 and 3). In 1864 the Cheyennes and Arapahos at Sand Creek had about one horse per capita.

Plains Indian horse herding adaptation strategies had failed to work well. These people had little time to work out their adjustments to the southern High Plains, and besides, many aspects giving rise to the High Plains culture also worked to undermine it. High Plains Indian adaptation strategies depended on many nonindigenous factors. Indian use of the horse, an animal introduced to the High Plains and included as part of a large economic system encompassing not only other Plains Indians but also Americans and Hispanics, illustrates how this animal both strengthened and weakened the material culture of these tribes. With injurious consequences, these people clung to certain maladaptive horse maintenance strategies in an environment undergoing rapid alteration.

Nature, it is best to remember, provides no preferential regard for human life. With or without humans, life will continue on this planet for some time to come. Scholars can miss the mark when they argue about a culture's ecological ideology. A culture may be "ecologically" minded and still miss understanding the geodialectic around it and suffer severely as a result. Horses proved both an innovative addition, and a vexation, to High Plains Indians. Regardless of how Indians viewed their place in nature, environmental flux, caused both by people and other forces, rendered many of their adaptation strategies ineffective. By November 1864 the Sand Creek massacre was not only tragic in terms of human misery, but also in how it represented the frailties of High Plains Indians' environmental adaptation. Indeed, during the first half of the nineteenth century the Indians of the southern High Plains never fully adjusted their horse-tending practices to their variable environment as the geodialectic worked to their disadvantage.

ACKNOWLEDGMENTS

The author would like to thank Dan Flores, Richard White, Janet Lecompte, Gary Anderson, Willard Rollings, Dwight G. Bennett, Robert Righter, and Scott Slovic for reading and critiquing various portions or all of this essay. Their invaluable insights helped me considerably in formulating my own thinking on this subject. Of course, only I, not they, should be held accountable for the ideas presented in this piece.

NOTES

1. I have relied upon U.S. Senate, *Sand Creek Massacre*, 39th Cong., 2d sess., Ex. Doc. 26, 1867, 1–228; U.S. House, *Massacre of Cheyenne Indians*, 38th Congress, 2d sess., 1865, 1–108; U.S. War Department, "Engagement on Sand Creek, Colo. Ter." in *The War of the Rebellion: A Compilation of the Official Records of the Union and Confederate Armies*, vol. 41, part 1 (Washington, D.C.: Government Printing Office, 1880–1901), 949–72; and George E. Hyde, *Life of George Bent Written from His Letters*, ed. Savoie E. Lottinville (Norman: University of Oklahoma Press, 1968), 110–63, for my description of the Sand Creek massacre.

2. See Dan Flores, "Bison Ecology and Bison Diplomacy: The Southern Plains from 1800 to 1850," *Journal of American History* (September 1991): 465–85, for an excellent description of the decline of the bison herds on the southern High Plains.

3. James Lovelock, *The Ages of Gaia: A Biography of Our Living Earth* (New York: W.W. Norton,1988), 236. Gaian theory is not without its critics and perhaps the best critique is by Daniel B. Botkin, *Discordant Harmonics: A New Ecology for the Twenty-First Century* (New York: Oxford University Press, 1990), 133–51, 222–23. Still, both agree on two essential points to this essay: people are a part of not apart from, nature, and that life is the environment. The term *geodialectic* takes shape in the same manner as other terms formed with Gaia as a prefix: e.g., geography, *ge* or *gaia*, earth, plus *graphien*, to write. Murray Bookchin's, *The Philosophy of Social Ecology: Essays on Dialectical Naturalism* (New York: Black Rose Books, 1990), has shaped my recent thinking on the dialectic. I like to think of reality as Bookchin summarized it, "developmental—of Being as an ever unfolding Becoming."

4. Donald Worster, "Seeing Beyond Culture," *Journal of American History* (March 1990): 1142.

5. For a good discussion of "awareness" see Bookchin, *The Philosophy of Social Ecology*, 39–42.

6. John Bennett, *Northern Plainmens Adaptive Strategy and Agrarian Life* (Chicago: Aldine, 1969), 14. Also see John Bennett, "Human Adaptations to North American Great

Plains and Similar Environments" in *The Struggle for the Land: Indigenous Insight and Industrial Empire in the Semiarid World*, ed. Paul A. Olso (Lincoln: University of Nebraska Press, 1990), 41–80.

7. See J. Donald Hughes, *American Indian Ecology* (El Paso: Texas Western Press, University of Texas at El Paso, 1983); J. Baird Callicott, "American Indian Land Wisdom," in Paul A. Olson, ed., *The Struggle for the Land*, 255–72 (Lincoln: University of Nebraska Press, 1990); and especially John H. Moore, who argues the Cheyenne had perfected their ecological adaptation to the High Plains, in *The Cheyenne Nation: A Social and Demographic History* (Lincoln: University of Nebraska Press, 1987), 174. Useful for understanding the debate over the "Indian as ecologist," see Christopher Vecsey and Robert W. Venables, eds., *American Indian Environments. Ecological Issues in Native American History* (Syracuse, NY: Syracuse University Press, 1980); Richard White, "Native Americans and the Environment," in *The Scholar and the Indian*, ed. by William Swagerty (Bloomington: University of Indiana Press, 1984); and Willard Rollings, "In Search of Multisided Frontiers: Recent Writing on the History of the Southern Plains," in *New Directions in American Indian History*, ed. Colin G. Calloway, (Norman: University of Oklahoma Press, 1987).

8. While I fully realize, and sympathize with, the importance of culture in shaping landscape, the approach of prominent historians like William Cronon, Carolyn Merchant, Richard White, and many others, I have intentionally focused the subject of this essay on material culture. I take this approach in an attempt to avoid what Donald Worster calls the "nostalgia for that old narrow focus on the self-referential history of ideas, society, and culture, with its tendency to dismiss nature as a mere epiphenomenon." See Worster, "Seeing Beyond Culture," 1142.

9. Richard White, who has discussed the horse-herding practices of the Pawnees, portrays them as having come to terms with their horses on the prairies, but encountered grief as their culture became dependent upon whites, both in political and material conditions. See Richard White, *The Roots of Dependency: Subsistence, Environment, and Choctaws, Pawnees, and Navajos* (Lincoln: University of Nebraska Press, 1983), 204. It might be possible to cast the history of the tribes around the upper Arkansas River in the same light; however, a dependency model places the cause for the tribes' deteriorating well-being beyond their own responsibility, and makes whites their sole victimizers. The history of these southern Plains tribes' adaptation strategies reveals a more complex story, one in which both the actions of the tribes and whites together in a violable environment resulted in a disastrous outcome for the horse practices of the tribes. Also see Richard White, "The Cultural Landscape of the Pawnees," *Great Plains Quarterly* 2 (winter 1982): 31–40. For traditional studies of Plains Indians and their horses see John C. Ewers, *The Horse in Blackfoot Indian Culture, With Comparative Material from other Western Tribes* (Washington, D.C.: U.S. Government Printing Office, 1955); and Frank Gilbert Rose, *The Indian and the Horse* (Norman: University of Oklahoma Press, 1955).

10. Charles L. Kenner, A *History of New Mexico-Plains Indian Relations* (Norman: University of Oklahoma Press, 1969), 23–77.

11. Zebulon Montgomery Pike, *Exploratory Travels through the Western Territories of North America* (Denver: W.H. Lawrence & Co., 1889), 199–207; Jacob Fowler, *Journal of Jacob*

Fowler, ed. Elliott Coues (New York: Francis P. Harper, 1898), 59–65; and Donald J. Berthrong, *The Southern Cheyennes* (Norman: University of Oklahoma Press, 1963), 4–23.

12. Hyde, Life of George Bent, 33.

13. Hyde, *Life of George Bent*, 71–72; David Lavender, *Bent's Fort* (Lincoln: University of Nebraska Press, 1954), 154–59; and Jacob Jablow, *The Cheyenne in Plains Indian Trade Relations, 1795–1840* (New York: Monographs of the American Ethnological Society, 1951), 58–60.

14. Hyde, Life of George Bent, 37–40.

15. Hyde, *Life of George Bent*, 37; and U.S. House, *Journal of Colonel Dodge's Expedition from Fort Gibson to the Pawnee Pict Village*, by Lieutenant T.B. Wheelock, 23rd Cong., 2nd sess., 1835, H. Doc. 2, 86.

16. Robert Coupland, "The Effects of Fluctuations in Weather upon the Grasslands of the Great Plains," *Botanical Review* 24 (May 1958): 288.

17. B.F. Beebe and James Ralph Johnson, *American Wild Horses* (New York: David McKay Company, Inc., 1964), 86–87; J. Warren Evans, Anthony Borton, Harold F. Hintz, and L. Dale Van Vleck, *The Horse* (San Francisco: W. H Freeman and Co., 1977), 28–32, 290–311; and A.G. Boone to A.M. Robinson, 7 March 1861, U.S. Department of Interior, Bureau of Indian Affairs, Letters Received by the Office of Indian Affairs, 1824–81, National Archives, Record Group 75, roll 878.

18. Alan J. Osburn, "Ecological Aspects of Equestrian Adaptations in Aboriginal North America," *American Anthropologist* 85 (1983): 563–91.

19. W M. Dawson, R.W. Phillips, and S. R Speelman, "Growth of Horses under Western Range Conditions," *Journal of Animal Science* 4 (1945): 47–54.

20. D.A. Savage and V.G. Heller, *Nutritional Qualities of Range Forage Plants in Relation to Grazing with Beef Cattle on the Southern Plains Experimental Range*, U.S. Department of Agriculture Technical Bulletin No. 943 (Washington, D.C.: U.S. Government Printing Office, 1947), 25. The authors reported: "Thus, buffalo grass is found to have lost during the winter only 47.3 percent of its average summer protein; side-oats grama, 52.7 . . . and blue grama, 57.1."

21. Cornellia Fleischer Mutel and John C. Emerick, *From Grassland to Glacier: The Natural History of Colorado* (Boulder, CO: Johnson Books, 1984), 27–33.

22. U.S. Senate, *Report of the Secretary of the Interior*, 36th Cong., 2d sess., 1860, S. Ex. Doc. 1, 453; Hyde, *Life of George Bent*, 46–47 and 158–63; and R.V. Boyle and J.S. McCorkle, U.S. Department of Agriculture, "Guard First the Bottom Land," in *Grass: Yearbook of Agriculture, 1948* (Washington, D.C.: U.S. Government Printing Office, 1948), 233.

23. Pike, *Exploratory Travels*, 203; Fowler, *The Journal of Jacob Fowler*, 16; and Waldo R Wedel, "The High Plains and Their Utilization by the Indian," *American Antiquity* (29 July 1963): 5–7.

24. Ernest Wallace and E. Adamson Hoebel, *The Comanches: Lords of the South Plains* (Norman: University of Oklahoma Press, 1952), 12; and U.S. House, Report of Exploration and Survey from Fort Smith, Arkansas, to Santa Fe, New Mexico, Made in 1849, 31st Cong., 1st sess., 1850, H. Ex. Doc. 45, 45.

25. U.S. Senate, Report of an Expedition Led by Lieutenant Abat, on the Upper Arkansas and through the Country of the Comanche Indians, in the Full of the Year 1845, by Lieutenant J.W. Abert, 29th Cong., 1st sess., 1846, S. Doc. 438, 9; Richard Irving Dodge, Our Wild Indians: Thirty-Three Years' Personal Experience among the Red Men of the Great West (New York: Archer House, 1959), 588; and Cragin Notebooks, notebook IV, p. 12, Western Collections, Denver Public Library.

26. U.S., Report of Exploration and Survey from Fort Smith, Arkansas, to Santa Fe, New Mexico, Made in 1849, 31st Cong., 1st sess., 1850, H. Ex. Doc. 45, 37, 45.

27. Evans, *The Horse*, 275–76.

28. Frands Parkman, *The Oregon Trail Sketches of Prairies and Rocky-Mountain Life* (Boston: Little, Brown, 1900), 304.

29. Janet Lecompte, *Pueblo, Hardscrabble, Greenhorn: The Upper Arkansas, 1832–1856* (Norman: University of Oklahoma Press, 1978), 159–60.

30. Dwight G. Bennett, DVM, Ph.D., personal communiation to author, 2 August 1990; and Lydden Polley, "Strongylid Parasites of Horses: Experimental Ecology of the Free-Living Stages on the Canadian Prairie," *American Journal of Veterinary Research* 47 (August 1986): 1686–93.

31. Fowler, The Journal of Jacob Fowler, 59–65.

32. U.S. Senate, *Notes of a Military Reconnaissance,* 32nd Cong., 1st sess., 1848, S. Ex. Doc.41, 13; and U.S. House, Report of Exploration for a Route for the Pacific Railroad, by Capt. J.W. Gunnison, and Topographical Engineers, near the 38th and 39th Parallels of North Latitude, from the Mouth of the Arkansas River, MO., to the Sevier Lake, in the Great Basin, by Lieutenant E.G. Beckwith, 33rd Cong., 2nd sess., 1855, H. Doc. 91, 27–28; and Bent-Hyde Papers, file folder 22, Western Historical Collections, Norlin Library, University of Colorado, Boulder.

33. Hyde, *Life of George Bent,* 108.

34. George Bent to George Hyde, 15 October 1904, George Bent letters, Colorado Historical Society, Denver, Colorado.

35. George Bird Grinnell, *The Cheyenne Indians: Their History and Ways of Life* (Lincoln: University of Nebraska Press, 1972), 2:13–14.

36. U.S. House, Report of Exploration and Survey from Fort Smith, Arkansas, to Santa Fe, New Mexico, Made in 1849, 31st Cong., 1st sess., 1850, H. Ex. Doc. 45, 36.

37. Tom McHugh, *The Time of the Buffalo* (New York: Alfred A. Knopf, 1972), 76.

38. Eugene F. Ware, *The Indian War of 1864*, with introduction and notes by Clyde C. Walton (1911; reprint, New York: St. Martin's Press, 1960), 307–57.

39. U.S. House, *Journal of Colonel Dodge's Expedition from Fort Gibson to the Pawnee Pict Village, by Lieutenant T.B. Wheelock,* 23rd Cong., 2d sess., 1835, H. Doc. 2, 73; U.S. Senate, *Report of an Expedition led by Lieutenant Abert, on the Upper Arkansas and through the Country of the Comanche Indians, in the Fall of the Year 1845,* by Lieutenant J.W. Abert, 29th Cong., 1st sess., 1845, S. Doc. 483, 483; U.S. Senate, *Notes of a Military Reconnaissance, from Fort Leavenworth, in Missouri to San Diego, in California, including part of the Arkansas, Del Norte, and Gila Riders,* by Lieutenant Colonel W.H. Emory, 32nd Cong., 1st sess., 1848, S. Ex.

Doc. 41, pt. II, 11, 17; and U.S. House, *Report of Exploration and Survey from Fort Smith, Arkansas, to Santa Fe, New Mexico, Made in 1849*, by Lieutenant James H. Simpson, 31st Cong., 1st sess., 1850, H. Ex. Doc. 45, 36.

40. *The State of Kansas v The State of Colorado et al.*, U.S. Supreme Court, October Term, 1901, No. 3, Original Testimony and Proceedings in the Above Entitled Cause before Special Commissioner Granville A. Richardson, Defendant Colorado's Evidence, L.G. Carpenter's testimony, vol. 5, 2060–70, location #47936, State Archives, Denver, Colorado.

41. U.S. House, *Report of Exploration for a Route for the Pacific Railroad, by Capt. J.W. Gunnison, Topographical Engineers, near the 38th and 39th Parallels of North Latitude, from the Mouth of the Kansas River, MO., to the Sevier Lake, in the Great Basin*, by Lieutenant E.G. Beckwith, 33rd Cong., 2d sess., 1855, H. Doc. 91, 25–28.

42. U.S. House, *Journal of Colonel Dodge's Expedition from Fort Gibson to the Pawnee Pict Village*, by Lieutenant T.B. Wheelock, 23rd Cong., 2d sess., 1835, H. Doc. 2, 90.

43. For a good description of the High Plains climate see Merline Paul Lawson, *The Climate of the Great American Desert: Deconstruction of the Climate of Western Interior United States, 1800–1850* (Lincoln: University of Nebraska Press, 1974).

SECTION III

If man is destined to inhabit the earth much longer, and to advance in natural knowledge with the rapidity which has marked his progress . . . , he will learn to put a wiser estimate on the works of creation, and will derive not only great instruction from studying the ways of nature in her obscurest, humblest walks, but great material advantage from stimulating her productive energies in provinces of her empire hitherto regarded as forever inaccessible, utterly barren.

GEORGE PERKINS MARSH, Man and Nature *(1864)*

The late-nineteenth and early-twentieth centuries marked an important turning point in Americans' attitudes toward their environment. George Perkins Marsh, probably more than any other American, presaged these views. In his book, *Man* and *Nature*, he called into question the environmental harm increasingly done through industrialization. Americans, he feared, were destroying the stability of nature, and weakening the natural resource base of their culture. He believed America could solve its problem through the scientific study of nature, and by using the knowledge gained manage its natural resources in a sustainable manner. American society would need the skills of experts who understood forestry and hydrology, and of governmental planners who could manage the use of these resources in the public's interests. The American conservation movement had its origin with Marsh.

Environmental historians have approached the study of the American conservation movement in many different ways. Some view it as a healthy social movement, while some harshly criticize it for unleashing governmental centralization of power or for being too much the handmaiden of strong economic interests. Some historians see conservationists as understanding their environmental surroundings while other portray these reformers as sowing the seeds of environmental problems just as problematic as unfettered industrial capitalism.

Environmental historians seldom study urban environments. Professor Martin Melosi, the former president of the American Society for Environmental History, has stood out as one of the exceptions. He, along with a few others such as Joel Tarr of the University of Pittsburgh, have led the way in the study of urban environmental history, especially concentrating upon how urban reformers fit into the conservation movement. Still, only a few precious studies of western urban environments exist.

Two landscape architects, E. Gregory McPherson and Renee A. Haip, wrote the article in this anthology depicting the changes in the historical environment of Tucson, Arizona, from the 1870s to the 1970s. They portray how culture, both material and otherwise, shaped the city. Part of the conservation movement was the "City Beautiful," or the attempt to transform an industrial environment into a garden or park. The people of Tucson employed a similar notion in their attempt to transform the environment of their city. The question arises: how have people in western desert communities adapted to the environments in which they live? Have they devised "sustainable" strategies? In what way has xeriscaping, desert landscaping, been dependent upon technological advances, and what have been the environmental liabilities of this approach? Has xeriscaping proven any better than the earlier planting of trees and grass?

Cultural values always act as force shaping any given environment. Initially, Euro-American settlers arrived in the Kennewick Valley in the state of Washington hoping to "create a prosperous community in the 'bunchgrass waste.'" Dorothy Zeisler-Vralsted, a historian at the University of Wisconsin, La Crosse, describes how the officials of the Northern Pacific Railway were finally able to stabilize irrigated agriculture in the Valley. The first companies, with idyllic-sounding names like "Dell Haven," failed miserably in transforming the arid, volcanic soils and their ecosystems into fruit-tree orchards. The railroad officials, like Thomas Cooper, came to the valley with different notions, so Zeisler-Vralsted agrues, and their understanding paved the way for successful irrigation. How did the views of the early settlers and the railroad officials differ? How did the settlers see the land differently from the railroad officials? Who, possibly, had a more accurate notion of the environment of the valley? What were the environmental difficulties in building a successful irrigation company? and at what cost to the Northern Pacific, farmers and their families, and the land?

Thomas Dunlap, now at Texas A & M University, shows how animal populations have varied in national parks as a result of administrators' values. He explains carefully how shifts in ecological theory during this period of conservation reform changed the way park service personnel managed animal populations, especially those of large predators. Dunlap also addresses the sticky question of how managers defined a "natural system" for the parks. Dunlap's queries lead to some unanswered questions. For example, can "natural systems" be managed without, in effect, becoming "artificial"? Is holding park environments static in a state resembling their appearances in the 1870s a refusal to recognize the power, and perhaps right, of an environment to change itself?

Richard Sellars, a historian with the National Park Service, shows how cultural values shaped the environments of National Parks, especially Yellowstone. Stephen Mather, the first superintendent of the National Park Service, oversaw the creation of highways, trails, communications services, lodges, and campgrounds, all the while believing that he was keeping the parks "natural." How could Mather rationalize this? He also opposed other parks developments as being "unnatural." What kind of logic was Mather using to make these distinctions? How did these policies affect the park environments? In what way does Sellars's findings corroborate those of Dunlap?

CHAPTER SIX

EMERGING DESERT LANDSCAPE IN TUCSON

E. GREGORY MCPHERSON AND RENEE A. HAIP

In 1875 there were only three trees growing in Tucson, Arizona.[1] By 1910 thousands of exotic trees had been planted in an effort to transform the desert city into a garden spot of the Southwest. An equally dramatic change is now occurring throughout the city: the lush green vegetation of trees is being replaced by desert landscaping. The transition from a desert city to a garden city and the current return to the former reflect shifting attitudes of the populace toward the environment. What compelled this change, and what are the implications for urban dwellers in the future? Is the emergence of desert landscaping another example of history repeating itself, or does it express an evolutionary process that points to a more symbiotic relationship between man and nature? In this essay we examine urban vegetative changes in Tucson with the goal of answering these questions. We focus specifically on the geographical processes and natural-resource constraints that influenced attitudes toward tree planting and house landscaping during the past century.

Urban vegetation reflects both the cultural milieu and the physical environment.[2] In Tucson vegetative patterns are linked to climate, water resources, cultural heritage, urban morphology, and the values of the population. Tucson is located in a Sonora Desert basin surrounded by four mountain ranges. Average annual rainfall of eleven inches arrives during two seasons, summer and winter. The seasonal rains and hot, arid climate support abundant and diverse native desert flora that include many arboreal species such as saguaro cactus (*Carnegiea gigantea*), paloverde (*Cercidlum microphyllum*), and mesquite (*Prosopis velutina*). Many plants from temperate and humid subtropical climates thrive with consistent irrigation. Hispanic culture is an important aspect of the city's heritage, because non-Native Tucson originated as a Spanish settlement along the Santa Cruz River in 1775. As in many other western American cities, the horizontal, low-density developmental pattern reflects the post–World War II population expansion and the importance of the automobile as the dominant mode of transportation. The population of the city tripled between 1950 and 1970. Currently, more than 650,000 people live in metropolitan Tucson, and the population is projected to reach 1.6 million by 2025.[3]

The portion of Arizona in which Tucson is located became part of the United States through the Gadsden Purchase in 1854. By then, the vegetation of the small town was a mixture of native American species and Spanish-Mexican imports. After the Gadsden Purchase, Anglo settlers arrived and the adobe townscape expanded. The form of this Spanish-Mexican townscape consisted of buildings with facades directly on the narrow streets, so that little space was available for trees. The central portions of blocks were left vacant for domestic animals and gardens. Plants were located in interior courtyards similar to the mission style. Although shade trees were not abundant, plantings did include chinaberry (*Melia azedarach*), Mexican paloverde (*Parkinsonia aculeata*), Arizona ash (*Fraxinus velutina*), and peppertree (*Schinus molle*).[4] Most trees in early Tucson were probably volunteers that settlers nourished rather than intentionally planted.[5]

By 1875 Tucson was still a desert city, although civic leaders began planting trees to shade their houses and to beautify the city. The greening of Tucson began a year later when a handful of Bermuda grass (*Cynodon dacrylon*), brought from San Diego, turned a dusty corner lot into a cool, verdant oasis.[6]

THE DESERT TRANSFORMED

The late 1870s marked the beginning of great horticultural experimentation and effort to transform Tucson from a dusty desert city into a garden oasis. The amelioration of the inhospitable desert climate and the enhancement of the place as a winter health resort were driving forces behind a forty-year-long afforestation effort. Abundant water supplies made possible this change.

It is not widely known that the primary impetus for pumping groundwater in Tucson was to irrigate decorative landscapes. Around 1880 the first windmill was imported from Indiana to provide water for an experimental garden.[7] Its introduction alleviated the burden of drawing water by hand from the fifty-foot-deep well. Several years later new steam-powered water pumps were installed, and the storage capacity was enlarged to irrigate a double-row planting of cottonwoods (*Populus fremontii*) that extended a half-mile along both sides of the Southern Pacific Railroad track.[8] The purpose was to create an attractive entry for visitors. The new plantings were not always successful. An unusual shortage of irrigation water resulted in the loss of transplants throughout the city in 1892. That year stress-tolerant species like the chinaberry or the mulberry (*Morus alba*) were recommended as the best shade trees for the region, and planting of the water-thirsty cottonwood was discouraged.[9]

City leaders actively promoted tree planting. In 1888 the city council negotiated a contract with the local utility, Tucson Water, to provide free water for street trees.[10] Local plant enthusiasts and botanists associated with the University of Arizona and the Carnegie Desert Laboratory generated increased interest in horticulture through exemplary gardens and articles written for the local newspapers. The former provided an important medium for the dissemination of ideas and information about urban vegetation. For example, in 1907, "it has been declared by several local botanists and tree experts that it is impossible to grow oak and eucalyptus in this vicinity with any degree of success. A trip to the home of A.R. MacDonald in the north eastern part of this city, will, however, soon disprove of that thought."[11]

The editors of the *Arizona Daily Star* and other respected civic leaders mounted a citywide street-tree-planting campaign during the last part of the nineteenth century. Tree planting was considered a civic duty and was promoted on the basis of shading and beautifying streets. Furthermore, one editor encouraged public participation by extolling the healthful effects of trees on urban climate and air quality. He noted that tree planting "will result in greatly reducing the temperature of the summer months, as vegetation absorbs the heat, and more growing trees absorb many kinds of poisonous gases and thus they are not liable to be inhaled by the people."[12]

The passage of Arbor Day legislation by territorial lawmakers in 1901 fueled the tree-planting movement. Arbor Day was specifically geared toward education, so "that children may be trained to take an interest in the planting and caring for gardens and trees."[13] Editorials exhorted residents to celebrate Arbor Day, and Tucsonans responded by planting trees. According to the *Arizona Daily Star*, residents planted ten thousand trees in 1907 and 1908, or approximately one tree for every other resident. A politician was quoted as saying, "The people in the east have an idea that all the vegetation that can survive in Arizona soil is cactus and soap weed, but I expert to see the day when there will be no other city in the country that will be beautified with trees as will the Old Pueblo."[14]

New architectural styles and developmental patterns influenced the transformation from desert to garden city. In March 1880 the railroad reached

Tucson, and people with their eastern and midwestern landscape images and values as well as access to different building and plant materials began arriving. Residents abandoned traditional Sonoran architectural and landscape styles for the ones imported from the East. The structure of residential neighborhoods changed dramatically, as setbacks for houses resulted in large front and back yards as well as space for planting trees along streets.[15] Planting yards with Bermuda grass and winter rye (*Lolium* sp.) created an aesthetic setting for houses. Foundation plantings were used to soften the harsh environment and to embellish the architecture. Exotic species were collected from around the world. For example, in 1895 "several palms, magnolia trees, and varieties of shrubbery, arrived at the depot yesterday for the university, from Riverside, Cal."[16] The ornamental qualities of exotic species such as oleanders (*Nerium oleander*) and roses (*Rosa* spp.) were important in these new horticultural plantings.

Ironically, while Tucsonans were planting a host of exotic trees, the natural riparian woodland was being destroyed as the water table was lowered, as land was cleared for farming, and as timber was harvested for fuel. Flow in the Santa Cruz River became intermittent prior to 1890, and between 1886 and 1890 floods cut a deep channel and washed away vegetation.[17] The water table continued to drop as pumping increased and smaller amounts of runoff recharged the groundwater. Thus the creation of a mythical Arcadian garden at Tucson occurred at the expense of its natural riparian forests. Today, riparian woodlands and mesquite bosques that lined the Santa Cruz River and other watercourses in Tucson account for less than 10 percent of their original abundance.[18]

During the initial two decades of the twentieth century, public participation in tree planting remained high. Residents planted trees to improve the quality of their environment; however, interest in the activity gradually diminished. In 1916 Arbor Day was no longer a school holiday.[19] An editorial in the *Arizona Daily Star* early in 1920 recommended that the day be dropped as a state holiday because people planted trees anyway.[20] Although interest in tree planting waned, the urban forest created by one generation was a legacy. "Now there are thousands of street trees, several parks, hundreds of homes beautified with plant life of all kinds, so that Tucson today, seen from a nearby mountain looks like a young forest where 30 years ago there were but three trees."[21]

SUN-DRENCHED OASIS

By the early twentieth century, Tucsonans had created an oasis in the desert. Having established a new image for the city, the business community began promoting it as a winter resort. The Sunshine Climate Club was established in 1922 to advance tourism.[22] Furthermore, the mild winters attracted the health industry. The construction of a principal veterans' hospital in 1928 brought additional residents.

Landscape design became an important means of enhancing the image of the city as a winter-resort community. For example, a professor at the University of Arizona proposed that a new boulevard be lined with plantings of "palms, peppers, olives, and oleanders . . . illustrating the semitropical climate of Tucson to our visitors."[23] Landscape-design experts recommended overseeding winter lawns with Australian rye grass and using evergreens such as pyracantha (*Pyracantha* sp.), eucalyptus (*Eucalyptus* sp.), African sumac (*Rhus lancea*), and silk oak (*Grevillea robusta*) to accentuate the illusion of a subtropical climate.[24] After 1950 the few new residential areas with tree-lined streets had species like sour orange (*Citrus aurantium*) and palms symbolic of a subtropical climate rather than deciduous trees that connoted the bleak, cold winters that winter visitors were fleeing.[25]

At the same time that Tucsonans were maximizing winter sun and promoting tourism, the advent of evaporative cooling reduced the need for summer shade. Sweltering summertime heat was the impetus for planting trees and escaping to the nearby mountains. Tree shade was a necessity because summers were spent outside prior to the availability of mechanical cooling. However, effectively engineered evaporative cooling was developed in the early 1930s, and by 1940 "the average home and workshop on the desert are more comfortable places in June and July and August than the dwellings in milder zones where air-cooling has not been accepted as essential."[26] Abundant energy supplies fueled the postwar boom of the air-conditioned desert and the gradual demise of tree-lined sidewalks and streets.

A tremendous surge in housing demand after World War II resulted in many subdivisions laid out beyond the city corporate limits. Building codes and zoning ordinances were not implemented in Pima County until the mid-1950s, and the first general land-use plan was approved in 1960.[27] Tract houses were typically ranch-style buildings surrounded by Bermuda-grass lawns. Foundation plantings were infrequent, although developers sometimes planted fruitless mulberry or elm trees on each parcel.[28] Extensive landscaping was not characteristic of the suburban tract developments that arose east and north of the city center between 1940 and 1975. Hence the horticultural landscape was replicated as the city expanded, but fewer street and front-lawn trees were planted than previously. It was estimated that more than half of the new houses constructed after 1950 had no shade trees in the front yard.[29]

New streets were added and existent ones widened to accommodate the growing number of motorists. Main streets that were once lined with trees became starkly bare when they were widened or when old trees were removed and not replaced. Although medians of new boulevards were planted with pines, eucalyptus, palms, and grass during the 1960s, roadsides were seldom landscaped.

By about 1950 there was little public interest in tree planting, and a concern for shade and other functional uses of plank was virtually absent. A maturing urban forest covered older neighborhoods, and horticultural plantings in new subdivisions were ornamentally symbolic of the sundrenched oasis. However, between 1950 and 1970 an increased number of residents began converting their horticultural plantings to desert landscapes. This shift was attributed to a change in the value of leisure time and a heightened appreciation of the natural environment of the region.[30] Tucsonans were participating in a countrywide recreational mood. Golf, swimming, and tennis became more popular than gardening as leisure-time activities. Upper-class Anglos were the first to install desert plants, stone, and Mexican paving instead of grass lawns in the front yards. Desert-landscape precedents existed in the foothills and in a midtown subdivision called Colonia Solano, where lower-density development was accompanied by the preservation of existent vegetation.

RETURN TO A DESERT CIY

Although many residents began to appreciate the natural beauty of the desert and to incorporate its elements in garden design, the impetus for desert landscaping came when most Tucsonans perceived water scarcity as a problem. Tucson is wholly dependent on groundwater, and since approximately 1950 more of it has been pumped than has been naturally returned to the aquifer. However, a comparison of identical surveys administered in 1971 and 1977 revealed that not until 1977 did most residents consider the falling water table to be a serious problem.[31] This increased awareness coincided with a period of growing national concern about environmental issues as well as intense local debate and media coverage regarding water supply. Declining groundwater supplies, increased water prices, and water conservation programs have spurred an astonishingly rapid change in the city landscape. The traditional greensward and lush plantings have been replaced with rock mulches and low-water-use species at a steady rate since the late 1970s. Tucson began to shed its green mantle as the desert city reemerged.

Water conservation began to be a concern for many Tucsonans when the severe drought struck in 1974. For several days portions of the city at highest elevations were without water because pumping capacity did not meet demand. In 1976 a recently elected city council levied a 22 percent water-rate increase in July, one of the heaviest periods of water consumption. Four council members were eventually recalled, but the rate was never returned to its previous level.[32] Because 50 percent of the typical monthly water bill during summer can be for landscape irrigation, many homeowners replaced lawns and trees with rock and cactus. The local water utility successfully combined increased rates with conservation preachments to reduce demand. In 1976 the utility initiated a peak-demand-reduction program. Because landscape irrigation accounted for 30 to 50 percent of peak municipal demand, it hoped to promote water-efficient landscaping. A study was conducted to compare change in irrigated lawns and water use for two periods. Results of the study showed that a 1972-to-1976 trend of 1.7 percent increase in irrigated lawns was reversed to a decrease of 17.5 percent between 1976 and 1979. The largest change was on front lawns. Municipal water consumption decreased 20.6 percent in that period. Per-capita water use declined from 204 gallons in 1974 to 148 gallons in 1979.[33] Desert landscaping, which had been fashionable for some during the 1950s, became an economic necessity for others during the 1970s.

An unlikely coalition of business people and environmentalists joined the local utility in promoting water conservation and desert landscaping.[34] The Southern Arizona Water Resources Association (SAWARA), largely directed and funded by the business community, advocates desert landscaping, that is, xeriscape, through an annual design competition and educational conference. Many of its members view water scarcity as a threat to the continued growth of the region. Tucsonans are encouraged to conserve water so that there will be enough for its booming population, which is expected to reach one million before 2025. In contrast, local environmental organizations like the Arizona Native Plant Society advocate water conservation for ethical reasons. They believe that water conservation is an important mechanism for growth management and that both are needed to bring Tucson back to harmony with its natural desert environment. From their perspective, desert landscaping reflects a new ecological sensitivity toward the surrounding natural environment. Hence, conservationists regard desert landscaping as the manifestation of a new environmental ethic, while business leaders regard it as an economic necessity. Although their ultimate objectives vary, the combined voice has resulted in the wide acceptance of desert landscaping for water conservation during the past two decades.

City service departments began advocating desert landscaping through example and by providing informational materials. In the mid-1970s the parks-and-recreation department converted median strip plantings along main streets from lawn and large shade trees to desert landscaping. The state Groundwater Management Act now prohibits watering of plants that are not low in water use along public rights-of-way. In 1979 the city planning department released a publication entitled Landscaping in the Desert. In 1988 the city adopted a water-conservation landscape ordinance that restricts the use of turf and mandates planting of low-water-using species for required landscaping on private nonresidential properties.

The emerging desert landscape in the city bears little resemblance to earlier desert landscape. Now, many plants are drought-tolerant exotics from Australia and South Africa instead of native volunteers. Homeowners rake decomposed granite instead of sweeping packed earth, and they use sophisticated subterranean irrigation systems instead of buckets of water. Moreover, the emerging desert landscape contains relics of the Arcadian oasis that were not present in the original desert landscape. Giant Aleppo pines (*Pinus halapensis*) and Italian cypress (*Cupressus sempervirens 'Stricta'*) silhouette the skyline as testaments to a bygone era.

CLOSING COMMENTS

After more than a century, Tucsonans are changing to plantings that are compatible with the natural environment. Once the change began, the rapidity of its acceptance is striking. This rapid shift from horticultural to desert landscape illustrates how strong sociocultural traditions like a grassy front lawn can be modified if people are presented the right combination of incentives, mandates, and educational materials. The new landscape appearing in the city abandons the features of the Arcadian horticultural garden with its profligate waste of water, its illusion of subtropicality, and its monotonous anonymity. The new desert landscape has immense potential for both good and bad. It can strengthen a sense of place by creating a direct link between the urban environment and the natural desert contact. It can foster a greater appreciation of the city's history and cultural diversity by assimilating features symbolic of preview times. It can nurture growth, both personal and collective, by making the city a more livable and attractive place.

The ecological role of vegetation is more important today than during the previous desert landscape, when the environment was not so befouled with the by-products of a highly consumptive urban populace. Urban trees are one of the most cost-effective ways to reduce carbon-dioxide emission from power plants because their shade lowers the demand for cooling energy.[35] Urban plantings can also reduce storm-water runoff, noise along principal streets, and atmospheric pollutants.[36] Generally these ecological benefits increase as the amount of plant biomass increases. With continued growth of tourism in the city and with environmental quality an increasingly important issue, need for urban vegetation will expand.

For this study we sampled land-cover change by using aerial photograph from 1953, 1971, and 1983. The results indicate that vegetative cover in Tucson decreased from 37.3 percent to 28.6 percent between 1971 and 1983. Desert landscaping appears to be associated with an overall reduction in the amount of vegetation in the city. Additionally, because the mature size of most desert trees is less than that of the removed exotics, the potential amount of future tree-canopy cover is less for desert landscapes than for the replaced horticultural ones.

Are the goals of conserving water and increasing urban vegetation mutually exclusive? The answer is likely to be yes, if landscaping policies and ordinances continue to address only water conservation rather than the broad goal of managing the urban vegetative resources to make the city more livable. Homeowners in several urban places near Tucson receive landscape rebates if their designs meet water-conservation criteria. The ultimate water-conserving landscape is the zeroscape of decomposed granite devoid of plank, and it qualifies for the rebate. This landscape is not only unattractive and biologically sterile but also is a hot spot that increases demand for cooling energy by 20 to 30 percent.[37] If city government and other community leaders do not actively promote appropriate desert landscaping, zeroscapes may become the rule, not the exception.

The evolution of the new desert landscape in Tucson will be informative for other cities that may soon face similar water shortages. Inevitably, the rising cost of water will prompt additional abandonment of the luxury of verdure. Tucson is likely to become a hotter, drier, and dustier place than it has been previously. On the other hand, residents might be persuaded to reforest the city with desert-adapted species. A new spirit of environmental stewardship may arise as individuals take action to improve local environments and begin the long-term process of ongoing care for the vegetative resource. Collectively they can emphasize the nurturing milieu. The transformation from a garden city to a desert city may well be completed by the twenty-first century. Whether the new landscape will make the city more livable will be determined in the future.

ACKNOWLEDGMENTS

We are grateful for the comments on an early draft of this article from Donald Bufkin, Bernard Fontant, Thomas Sheridan, David Taylor, Mark McPherson, and Raymond Turner. This research is a contribution to the Hatch Project, entitled Impacts of Urban Forests in Arizona, sponsored by the University of Arizona Agricultural Experiment Station.

NOTES

1. *Arizona Daily Star*, 25 October 1908.
2. G.G. Whitney and S.D. Adams, "Man as Maker of New Plant Communities," *Journal of Applied Ecology* 17 (1980): 431–48.
3. *PAG Population Handbook: 1987*, Pima Association of Governments, Tucson, 1987.
4. W. Rogers, "Looking Backward to Cope with Water Shortages: A History of Native Plants in Southern Arizona," *Landscape Architecture* 60 (1979): 304–14.
5. *Tucson Preservation Primer: A Guide for the Property Owner*, ed. R.C. Geibner (Tucson: University of Arizona, College of Architecture, 1979).
6. *Arizona Daily Star*, 25 October 1908.
7. Ibid., 25 October 1908.
8. Ibid., 18 April 1893.
9. Ibid., 21 December 1892.
10. Ibid., 31 January 1888.
11. Ibid., 2 November 1907.
12. Ibid., 31 January 1888.
13. Ibid., 7 February 1902.
14. Ibid., 28 February 1908.
15. Tucson Preservation Primer.
16. *Arizona Daily Star*, 16 February 1895.
17. J.R. Hastings and R.M. Turner, *The Changing Mile* (Tucson: University of Arizona Press, 1972).
18. W.W. Shaw, J.M. Burns, and K. Stenburg, *Wildlife Habitats in Tucson: A Strategy for Conservation* (Tucson: Pima County Department of Transportation and Flood Control and University of Arizona School of Renewable Natural Resources, 1986.
19. *Arizona Daily Star*, 4 February 1916.
20. Ibid., 7 February 1920.
21. Ibid., 25 October 1908.
22. M.T. Parker, "Tucson, City of Sunshine," *Economic Geography* 24 (1948): 79–113.
23. *Arizona Daily Star*, 20 January 1920.
24. "Tucson's Leading Landscape Experts Give Their Own Views," *Magazine Tucson* (November 1949): 32–33, 52.

25. Melvin E. Hecht, "Climate and Culture, Landscape and Lifestyle in the Sun Belt of Southern Arizona," *Journal of Popular Culture* 11 (1978): 928–47.

26. J.H. Collins, "Cooling the Desert Air," *Desert Magazine* (May 1939): 29–31.

27. Donald Bufkin, "From Mud Village to Modern Metropolis—The Urbanization of Tucson," *Journal of Arizona History* 22 (1981): 63–98.

28. Personal communication from David Taylor, City of Tucson Planning Department, 14 April 1988.

29. Hecht, "Climate and Culture, Landscape and Lifestyle."

30. Melvin E. Hecht, "Decline of the Grass Lawn Tradition in Tucson," *Landscape* 19 (1975): 3–10.

31. T.F. Saarinen, "Public Perception of the Desert in Tucson, Arizona," paper presented at workshop entitled Social Implications of Environmental Problems, Rio Rico, Arizona, 9–11 March 1983.

32. W.E. Martin, H.M. Ingram, N.K. Laney, and A.H. Griffin, *Saving Water in a Desert City* (Washington, D.C.: Resources for the Future, 1984).

33. D.A. Mouat and M.C. Parton, "Assessing the Impact of the Tucson Peak Water Demand Reduction Effort on Residential Lawn Use: 1976–1979," Report to Tucson Water from the Office of Arid Lands Studies, University of Arizona, Tucson, 1980.

34. W.E. Martin and H.M. Ingram, *Planning for Growth in the Southwest* (Washington, D.C.: National Planning Association, 1985).

35. H. Akbari, J. Huang, P. Martien, L. Rainer, A. Rosenfeld, and H. Taha, "The Impact of Summer Heat Islands on Cooling Energy Consumption and CO_2 Emissions," Proceedings of the ACEEE Summer Study on Energy Efficiency in Buildings, Asilomar, California, 1988.

36. E.G. McPherson, "Functions of Buffer Plantings in Urban Environments," *Agriculture, Ecosystems, and Environment* (1988): 281–98.

37. E.G. McPherson, J.R. Simpson, and M. Livingston, "Effects of Three Landscape Treatments on Residential Energy and Water Use in Tucson," *Arizona, Energy and Buildings* 13 (1989): 127–38.

CHAPTER SEVEN

RECLAIMING THE ARID WEST
The Role of the Northern Pacific Railway in Irrigating Kennewick, Washington

DOROTHY ZEISLER-VRALSTED

As historians revise traditional interpretations of the West and reexamine the roles of women, Native Americans, and others, they are also looking anew at business institutions and their effect in the arid West. The railroad, an institution for many years portrayed as an oppressor, was in fact an agent for orderly development. Present-day Kennewick, on the lower Yakima River in southeastern Washington, offers evidence of the railroad's positive role. Early settlers intent on farming and orcharding in that dry country were dependent for water on entrepreneurs struggling to profit from irrigation schemes. Only the advent of the Northern Pacific Railway in 1901 broke the boom-and-bust cycles that had inhibited local growth: the railroad's irrigation subsidiaries made a long-range—albeit short-lived—commitment to the region, giving Kennewick its first period of steady, organized development.[1]

The Northern Pacific made its first appearance in the Yakima Valley in 1879 when it established the town of Ainsworth. Located on the Snake River, just north of where the Snake meets the Columbia River, Ainsworth was a shipping and construction center for the railroad's Pend Oreille Division. A line was to extend from the town northeastward, and by 1881 crews had reached Spokane Falls. Another part of the surveyed route would run from Ainsworth up through the Yakima Valley and then across the Cascades to Tacoma. News of this proposed venture provoked a flurry of activity in the Yakima Valley, as residents prepared for a land boom. But financial troubles at the Northern Pacific halted the project for several years. Reorganized in 1883, the NP resumed its plans to construct a trans-Cascade line. It made Pasco its local headquarters, and optimistic valley boosters and agriculturists awaited the prosperity that would surely come with the trains.²

The Kennewick area, like many other arid regions in the West, received only eight inches of rainfall a year. As a result, farming was impossible without a supplemental supply of water. By the 1880s Yakima Valley pioneers, having first tried themselves to irrigate their lands with makeshift ditches, turned to large companies, which were usually backed by eastern capital and had elaborate reclamation schemes. Following a familiar pattern, the corporations bought up land considered worthless and platted it for resale at much higher prices, promising to irrigate the tracts. Profits from land sales were to be used to build canals and water works, but all too often the irrigation system was never begun or was left unfinished. The Yakima Irrigating and Improvement Company (YI&I) illustrates this pattern.³

Organized in 1888, the YI&I was incorporated with a capital stock of $400,000. Its place of business was listed as Tacoma, but two of the trustees had addresses at Niagara Falls, New York. Once the company set up in the Kennewick area, in 1889, valley newspapers referred to the financial backers as the Oneida community because of their New York connections. Although its exact intentions were somewhat vague and the amount of construction it actually accomplished was negligible, the YI&I planned to create a prosperous community in the "bunchgrass waste." An irrigation system to water farms and orchards would bring settlers, and within a few years a town—Kennewick—would be established. Previously, various merchants had been drawn by the prospects of a town developing there with the arrival of the Northern Pacific; they set up businesses but soon left. After all, the area "was nothing but hot winds, and dust and sagebrush." Water would make the difference. Concerted efforts of YI&I entrepreneurs attracted a population from the Mississippi Valley and the Midwest. The company was "the first major project aimed at development of . . . a fruit-growing territory in the area now known as the Tri-Cities." From the outset, Yakima Irrigating and Improvement focused on small-scale orcharding on 10- to 20-acre tracts.⁴

One of the company's first acts was to buy land, 4,000 acres, according to one account; another source states that the company bought 20,000 acres, alternate sections, from the Northern Pacific, then set about acquiring the adjoining sections. Initially, the YI&I proposed to deliver water to an area of approximately 20,000 acres. Beginning in 1889, work started on a canal above the village of Kiona located on the north side of the Yakima River. Company officials ambitiously estimated that on completion the ditch could serve 4,000 acres of YI&I land as well as 40,000 acres downriver that did not belong to the company. Furthermore, the canal would double as a waterway for steamboats entering the Yakima. Promoters foresaw the possibility of participating in the freight trade at the Northern Pacific station in Pasco, through a system of river dams and terminals. Such prospects touched off a brief boom in the area. The population grew, and eager settlers bought land for orchards. In addition to the Kiona canal, the YI&I began to construct two other ditches: from Horn Rapids, the lower Yakima ditch was to extend almost as far as Riverside, or present-day Richland; the Kennewick canal was to start at Horn Rapids and run along the south bank of the Yakima through what would be the townsite of Kennewick to a point opposite Wallula. All three ditches claimed water rights from the Yakima River.[5]

Despite what appeared to be a large enough landholding to support development expenses, however, the YI&I was in financial trouble almost immediately. On 3 October 1889 it went into receivership, but with so many subscribers to the irrigation system, construction of the canals continued. By 1893 the Kennewick canal reached the Kennewick townsite, thirty-four miles from Horn Rapids, and pushed on. But work progressed slowly, and discontented area residents, still awaiting water, formed the Dell Haven Irrigation District in 1892 to purchase and operate the Kennewick canal. Still, the YI&I persisted in building its irrigation system, even while the Dell Haven group sought to sell its bonds. But misfortune befell the YI&I with the panic of 1893, and by 1895 local newspapers again reported that the company had gone into receivership. This time the firm did not recover.[6]

In the meantime, the situation in the Dell Haven Irrigation District worsened because the Kennewick canal could not deliver the subscribed amount of water. Like many other early irrigators, the YI&I lacked the engineering expertise needed to build the project; almost from the start the system was inadequate, and the irrigation works deteriorated rapidly. By 1896, after only three years, the district stopped operating, and in 1897, its members rejected a bid for reorganization because the outstanding indebtedness of $211,000 probably could not be repaid through operation of the present system.[7]

So ended the careers of the YI&I and the Dell Haven group. The lower Yakima, the Kiona, and the Kennewick gravity canals would remain idle until 1902, as one more phase of the irrigation movement in the lower Yakima Valley passed. Meanwhile, settlers who stayed on were undoubtedly forced to practice dryland farming or resort to digging crude ditches to water part of their holdings. By 1901, the enterprising James J. Hill was overseeing the affairs of the Northern Pacific. Under his supervision, the railroad renewed efforts to realize a profit from its land grants. Awarded twenty sections for each mile of track laid in the territory, the railroad had acquired a sizable interest in the Yakima Valley. The land would be valuable agriculturally—if it had water.

Thus, in March 1901, Thomas Cooper, one of the NP's western land agents, charged with investigating potential reclamation projects in the area, advised the company president, C.S. Mellen, of the "irrigation possibilities in [the] vicinity of Pasco." The railroad's crucial involvement in Kennewick irrigation had begun.[8]

In his report, Cooper outlined three potential irrigation projects that the railroad could take on with en eye to selling its excess land. The first, called the Snake River Pumping Project, would deliver water from the Snake to approximately 37,600 acres. The NP already owned over 15,000 acres in the vicinity to be irrigated. Costs of building the system were estimated at $523,100. The second enterprise was to take over the YI&I's Kennewick canal. Cooper reported that $100,000 would buy the canal, the water rights, 9,800 acres, the townsite of Kennewick, and the Kennewick hotel; he estimated that another $48,400 would be necessary to repair the canal. But railroad land under this project amounted to only 350 acres. The third irrigation system was the YI&I's lower Yakima canal. For $26,000 the NP could buy the ditch, along with almost 7,000 acres adjacent to the canal route. Finishing the canal would cost $36,900. Cooper concluded his report with a projection of operating expenses and potential profits for each irrigation system. In addition, he attached a map detailing the canals and the land involved in the projects.[9]

The most intriguing part of his investigation was the assessment of "traffic value." In this segment, Cooper listed the primary goals of the railroad and analyzed the failings of previous irrigation companies. He calculated the net profit in increased rail traffic that the company would realize if the land watered by the various canals were cultivated, figuring, for example, the freight rate per ton of alfalfa shipped to Puget Sound, the number of tons per acre, and the total acreage. Even conservative calculations pointed to sizable profit, yet Cooper worried that "irrigation is not the business of a transportation company." His advice, therefore, was to engage the railroad in the development activity without losing sight of its real goal:

> My idea is that the Company should not go into this business for the purpose of making a direct profit out of the irrigation enterprise, but should figure on selling the lands and water rights at the lowest price that will retire the capital invested. with interest, and look for . . . profit in the increased traffic earnings.[10]

With these objectives in mind, Cooper proposed that the NP purchase both the Kennewick and lower Yakima systems. The Kennewick canal should be repaired immediately, he said, and the land it would water should be offered for sale at the reasonable price of $25 an acre. Accompanying water rights should also be sold at a fair price. Cooper hoped to complete both rehabilitation work and sales in time for the 1902 irrigation season. Thereafter the company could finish the lower Yakima ditch and sell the adjacent land. If the two projects were successful, the NP might consider taking on the Snake River project. As for Kennewick, a townsite with a history of "long waiting and hope deferred," it would soon undergo a period of calculated. steady promotion.[11]

Back in St. Paul later that year, NP officials approved Cooper's proposal, and they adopted the policy of investing in irrigation with an eye toward eventual traffic profits rather than immediate gain. They gave him permission to begin. But he encountered difficulties immediately. He could not ascertain the amount of Dell Haven district warrants still outstanding. Furthermore, irrigation district water users had to dissolve the district before the railroad could proceed. Also, the Kiona canal was to be included in the purchase, which was not part of Cooper's original plan. Nonetheless, Cooper began negotiations with all those who had interests in the properties, including members of the defunct YI&I and Dell Haven Irrigation District representatives. On 30 December 1901, the NP's subsidiary, the Northwestern Improvement Company, bought the Kennewick holdings once owned by the YI&I

The purchase completed, NWI started repair work on the Kennewick canal without delay. Of the three former YI&I projects, it was the one Cooper thought offered the most promising returns with the least investment. The other two would wait. Unlike its predecessors in the irrigation business, NWI would not sell any land or water rights until Cooper was "satisfied that we will be able to run water through the canal next season."[13]

Banner headlines in valley newspapers reflected the community's enthusiasm for the project. One boasted, "Over 20,000 Acres to Be Reclaimed," with the subheading, "Northern Pacific Irrigation Undertakes Gigantic Irrigation and Settlement Scheme." Comparing the project to the Sunnyside system, a project begun in 1885 that irrigated 45,000 to over 100,000 acres of orchards and cropland farther up the valley, the article expounded on the virtues of NWI. Especially praiseworthy was the company's intention to sell acreage at a reasonable price. The report concluded with a prediction that 2,500 new settlers would be living on the land within twelve months after the repairs were made and the area was watered. The future of the small agricultural community of Kennewick had never looked so bright.[14]

In January 1902, Cooper announced that the repairs would take one to two months. It soon became apparent that he could not meet this schedule, but public relations remained harmonious despite citizens' impatience. Newspapers reported regularly on the status of the project. Railroad officials had contracted with one E.C. Burlingame to repair and enlarge the canal. Burlingame predicted a land boom and rising real estate prices in anticipation of NWI land being placed on the market. But his most significant remarks concerned the company's intention 'to settle as great a population as possible on its 5000 acres and to that end . . . not allow a tract of more than 40 acres to be sold to any one person, . . . [thereby restricting] sales to bona fide settlers on the land."[15]

In developing the Kennewick system, NWI sought to discourage the purchase of irrigated lands by speculators and nonagricultural interests. Although spawned by economic self-interest, the policy benefited developer and water user alike. Purchasers who acquired acreage for the purpose of reselling at a higher price once the irrigation project was built drove land prices up beyond the reach of settlers of modest means. Moreover, these nonagricultural landholders did not buy water (and thus help pay for the operation and maintenance of the irrigation system) and did not produce crops that would be shipped to market by train. So it was in the NP's and the settlers' best interests to restrict land purchases to farmers. orchardists, and horticulturists.

Yakima Valley residents quickly took an active part in supporting the Kennewick irrigation project. People throughout the arid West perceived irrigation as a panacea, and numerous small-town journalists prophesied that the land would "blossom like the rose" with the application of water. Technological advances contributed to the widespread optimism. Surely if humans could contemplate a massive undertaking like the Panama Canal, they could dig a ditch converting desert land into productive farms. Communities created booster clubs, such as the Yakima Commercial Club and the Kennewick Commercial Club, and horticultural societies. The new organizations promoted what they believed to be the exceptional soil, climate, and opportunities existing in the lower Yakima Valley. Kennewick, which had been nearly deserted after the YI&I failed, was "an old town and yet a new town," for it was growing: a flour mill was under construction in 1902, and the population was expected to reach four or five hundred. Prosperity seemed assured. The irrigation engineer Burlingame promised to build the best canal in the state of Washington."[16]

Although NWI officials were naive in believing that there was enough water to irrigate company acreage and that most of the land was suitable for irrigation, they nonetheless possessed a shrewd business sense about their Yakima holdings. In March, Cooper, who had his home base at the NP offices in Tacoma, arrived in Kennewick to inspect the rehabilitation work. Much to his disappointment, he found that there would be no water delivery for the 1902 season. Construction crews, seventy men and forty teams, were completing the repair and enlargement of the main ditch but encountering problems. Because the original builders had not allowed for seepage, the required amount of water could not get to the land at the farthest reaches of the system, which explained why the Dell Haven Irrigation District had found the canal too small. And additional restoration work, like the patching of rodent and other vermin holes that contributed to the seepage problem, was neccessary. Cooper realized that costs would exceed his original engineering estimate. Now he recalculated, but he assured Mellen that the extra expense would be recouped through land sales. Moreover, he said, it would be an "economy in the end to do a good job now." NP officials took the long view of the project, which early developers could not and did not emphasize.[17]

Other complications presented themselves. The Kennewick townsite had to be replatted because the YI&I had misplatted it after the first survey. Cooper intended to sell lots ranging in price from $50 to $200 in town. He even anticipated platting an addition. He also wanted to offer large "garden tracts," close to town and along the river. "This land is smooth, has a gentle slope to the river, excellent soil and [is] in every way desirable. When in cultivation, it will present a very attractive appearance from our trains." Cooper, at any rate, kept in mind the real business of the Northern Pacific and tried to shape the image the railroad presented, whether to potential buyers of NP tracts or to NP passengers.[18]

As for the irrigated land sales, Cooper suggested that NWI begin selling about 1 May 1902. For 5- and 10-acre parcels close to town the price would be

> $40 per acre upwards. All the other lands I propose to put at from $25 to $35 per acre, with a rebate of $10 per acre to actual settlers who comply with residence and cultivation requirements that we will prescribe, but with a minimum price of $25 per acre. This will . . . curb, but not prohibit speculation.[19]

Water rights would be sold at prices ranging from $15 to $20 an acre or exchanged "on a basis of four acres of water for six acres of land." Cooper set interest rates at 6 percent and recommended a 5 percent discount for cash sales. If the buyer decided to purchase on a six-year contract, he would have to pay one-sixth of the total price at the time of sale, and then for the first year pay only the interest, and for the remaining five years pay one-sixth of the total price annually. In conclusion, Cooper assured Mellen that the project was sound, especially the provisions to discourage speculators.[20]

Curbing speculation was crucial to all reclamation enterprises, at least those that aspired to longevity. After the initial investments, most irrigation systems supported themselves by means of annual operation and maintenance fees. If landowners insisted on keeping the land dry, the company would not be able to collect the fees that sustained the irrigation system. To avoid this situation, NWI deeds stated that a delinquent maintenance payment would result in a lien on the land that, if left unpaid, would permit company officials to foreclose on the property. Undoubtedly, the measure proved successful in discouraging speculators because a dollar-per-acre annual maintenance fee was a high price for merely retaining land.[21]

All of these features—moderate land prices, professional engineering investigations, thorough reconstruction of the system, and the deterrence of speculators—contributed to a nonexploitative image for the railroad's Northwestern Improvement Company. By March 1902, NWI had received four times the number of land applications it could fill, even though it did not advertise. One valley newspaper, praising the company's reclamation project, stated that it was not a "money making scheme."[22]

Meanwhile, Mellen and other NP executives worried. Prompted by concern over a perpetual obligation clause in the NWI contract, which would jeopardize other NWI holdings if the irrigation scheme failed, they advised Cooper to create a separate company to handle all irrigation matters. The new organization would have sole responsibility to deliver, in perpetuity, water to the land sold with water rights. Isolating the irrigation venture from the NP's other business in the area caused Mellen to contemplate the nature and extent of the railroad's own relationship to the project:

> It has never been my opinion that we should part with the canal or trust its operation either to the owners of the land or any other parties, for we are going into this enterprise as an incident to the development of the railway, and we do not wish, through lack of ability, financial or otherwise, the canal shall disappoint the parties to whom it has contracted to deliver water, and the tolls should be established sufficiently high to fully care for the canal, keep it at all times in first class condition, and furthermore establish such fund as will provide for its replacement.[23]

These sentiments revealed the extent to which the Northern Pacific linked its destiny with that of the area. Like the purveyors of other reclamation enterprises, the NP was motivated by economics, but unlike the others, it had a long-range interest in the Yakima Valley. The more goods that valley residents produced, the larger the railroad's shipping profits. Although there would be feuds over freight rates, the railroad needed a productive West to insure its own success. And the farmers and orchardists needed the water that the NP's subsidiary provided. Earlier reclamation companies, such as the YI&I, realized profits only from the sale of land and water and had no compelling reason to sustain irrigation projects.

Acting on instructions from Mellen, Cooper proceeded to organize a new firm, the Northern Pacific Irrigation Company (NPI). He transferred all NWI property that pertained to the Kennewick project to the new corporation. The system would operate as planned. The local press prophesied that the new company would inaugurate more irrigation projects on NP land.[24]

In May 1902, NPI began selling plats in the Kennewick townsite: irrigated land sales were postponed since, as Cooper had foreseen, the still uncompleted ditch would not be in operation until October—much too late for the 1902 season. But the delay did not dampen local enthusiasm. As town lots sold, newspapers gloated, predicting that Kennewick would thrive. An indication of future prosperity was the number of people applying for NPI land; according to one estimate, there were eight hundred bidders for one hundred irrigated tracts, and another source declared that each parcel had seven applicants. Company officials planned to select purchasers by choosing randomly from a ballot box, and boosters touted Kennewick as the next Sunnyside.[25]

Anticipating the day when the canal would be completed and irrigated land offered for sale, W.C. Sampson, one of the NP's land agents, mailed a form letter telling prospective buyers about the company, the area, land sales, and water delivery. In reviewing the advantages of the lower Yakima Valley, Sampson listed the long growing season, infrequent killing frosts because of low elevation, and the proximity of the railroad. He stated the terms for land sales and specified the amount of water guaranteed per acre. Finally, the letter prohibited the sale of more than forty acres to any one buyer, with a minimum purchase of ten acres. This stipulation, he explained, sought to encourage agriculture while avoiding both large farming enterprises and noncommercial residential gardeners.[26]

Despite precautions and meticulous groundwork, NPI began to experience difficulties similar to those of its predecessors. By September, Cooper was explaining to Mellen why costs continued to rise. The engineer, a Mr. Taylor, was a railroad man not familiar with irrigation works. His $48,000 rehabilitation estimate had, by 11 September 1902, been revised to a whopping $143,299. Furthermore, the canal was still unfinished, a fact Cooper attributed to a labor shortage and the scarcity of supplies. But still Cooper reassured Mellen that the high construction charges would be offset by land receipts.[27]

And in fact, by February 1903, the Northern Pacific Irrigation Company investment in Kennewick seemed to be showing a return. All of the platted lots in the Kennewick townsite had been sold. Furthermore, buyers clamored for more, and since any addition that the railroad would plat would be classified as residential rather than as unimproved agricultural, the railroad could realize a profit of $7,500 to $12,000. Even so, the NP's president still worried that the Kennewick system was costing more than anyone had anticipated. Problems continued to arise, and NP planners—like those in most reclamation enterprises—persistently misgauged the scope of the irrigation project. In a letter to M.P. Martin. who replaced Cooper as NPI general manager, Mellen mused:

> I shall be only too thankful if in the end we are able to pull out the money we have invested, with a reasonable rate of interest for the time we have had it tied up, looking only to the development of the country and the incidental railroad business that such development will produce, for our profit.[28]

The president's sentiments grew more firm, and in a few months, Mellen wrote Martin that the NP intended to "get our money out of the Irrigation Company as soon as we reasonably can; we do not stand to make any great profit out of the transaction and our only advantage is in the development of the territory." When Howard Elliott became president of the railroad later in 1903, he agreed with Mellen. Thus, after two years, the NP started looking for a way to unload what its officers now perceived as an unwise irrigation investment.[29]

In 1903, few understood all that building, operating, and maintaining an irrigation system entailed. Even irrigation engineers for the United States Reclamation Service often underestimated the cost of building a system and overestimated its reliability. Even though their interest lay in future traffic revenues, NP officials nonetheless must have been dismayed as the profits from land sales went back into the Kennewick irrigation system, which still showed no signs of self-sufficiency.

Contrasting with the opinions of NP executives was the outlook of the Yakima Valley press, which declared that the Kennewick ditch was "working like a charm" and that railroad officials were confident in the growth of the area. That the NP still had a certain measure of faith in Kennewick was demonstrated by its building a new depot there and completing the scheduled irrigation work. President Theodore Roosevelt, on a visit to Washington that year, predicted greatness for the state. Further reinforcing this rosy view was the Northern Pacific Irrigation Company's finally offering perpetual water rights for sale, although the first crop could not be planted until 1904. Another indication of prosperity was the planned sugar beet factory to be located nearby in Prosser. Although not an NP endeavor, the project had the assistance of NPI and the blessing of the railroad. Because of all these hopeful things, many Kennewick residents must have looked sanguinely to the future in late 1903.[30]

The outward evidences of success did not, however, alter the fact that the NP's irrigation project absorbed money as relentlessly as Yakima Valley ground absorbed water. Now Martin, like his predecessor, Cooper, was reassuring the railroad's president about increased tract value from hitherto unproductive acres even while admitting that profits from land and water sales could only recoup the company's initial capital outlay. And, echoing Cooper, Martin asked for approval to spend still more money—this time some $4,000 to $5,000 to set up a telephone line and build a small dam. The Kennewick system always seemed to need one more expenditure to insure success. Turning next to the other two canals, the Kiona and the lower Yakima, Martin reminded Elliott that repairs were mandatory. (Initially, Cooper had recommended postponing work on them.) More money would be required. President Mellen had once approved the sale of both canals if the company could get its asking price of $75,000; Martin thought the owners of the Sunnyside project might be interested.[31]

But NPI did not sell any canal. In fact, in 1904 Elliott directed the NP land department to stop all sales until a thorough examination had been made of the railroad's holdings. Further, he and Cooper, who had been promoted to the position of NP land commissioner, scheduled a trip to meet with Frederick H. Newell, director of the U.S. Reclamation Service, and other officials in Washington, D.C., reportedly to ascertain the government's reclamation plans. Thus began a working relationship between railroad officials and federal administrators regarding the sale of lands and water appropriations in potential federal reclamation projects. That association would extend throughout the first half of the century, and it would inform development of much more than Kennewick.

> One of the most important legislative measures for the development of the west was the enactment by Congress of the Reclamation Law. . . . Shortly following the passage of the Act the Northern Pacific Railway Company was asked by the Reclamation Service to cooperate with the Service in making the law a success. This co-operation took the form of selling its land under the various projects at prices and upon terms and conditions approved by the Secretary of the Interior. This co-operative arrangement has been carried out on the Lower Yellowstone Project in Montana and North Dakota, the Hermiston and West Umatilla Projects in Oregon, and the Tieton and Sunnyside Projects in Washington, in all of which the Company owns more or less lands.[32]

Again, the railroad's interest in the arid West, though neither philanthropic nor altogether realistic, contributed to an era of steady development.

Meanwhile, NPI was evaluating the status of the Kiona, lower Yakima, and Kennewick canals. If the company decided to enlarge the Kiona, it could place 10,000 additional acres of its land under irrigation. In the case of a rebuilt lower Yakima canal, 2,200 additional acres could receive water, although NPI no longer owned any of this land (it had sold to a private company in January 1904). Of primary interest, however, was the Kennewick ditch, where NPI had made its major investment. There, the problems resulting from the topography and climate of the region were most pronounced. Engineers observed that the soil was made up of "sand, gravel, volcanic ash, and volcanic rock," and soil composition made the ditch prone to seepage, which in turn caused canal breaks. As a result, NPI had incurred expenses for extensive puddling (compacting) of the ditch in order to diminish the amount of seepage. Despite efforts to control the fine gravelly soil that allowed so much water to escape the canal walls, washouts would continue to hamper operation of irrigation works throughout the area.[33]

Still, landowners prepared and cultivated the ground to be irrigated by the Kennewick canal. By the early summer of 1904, approximately 12,850 acres had been planted. An additional 1,400 acres had been cleared preparatory to seeding. Water users concentrated on fruit raising and had planted fifteen thousand fruit trees. Growers had shipped eight hundred crates of strawberries in the 1904 season, with a net value of $3.50 per crate, and NPI officials anticipated increases in yields. As to the future, M.P. Martin's optimism was tempered with realism when he told Howard Elliott that "the Canal will improve as the banks, and its weak spots are developed [in] time: but it is impossible to make an Eden out of a desert, or a perfect canal to water the Eden in one year, or in two."[34]

NPI continued to encounter setbacks. By early 1905, one newspaper reported that the Kennewick canal had "already cost more than the company can ever get out of it directly." The railroad did make an additional $10,000 each month in shipping receipts. Yet Martin now urged Elliott to quit the irrigation business. The canal continued to be a source of anxiety. The Northern Pacific, he argued, would enjoy increased traffic value whether or not it had irrigation investments in the lower Yakima Valley.[35]

No longer hopeful about the project's ultimate success, Martin wanted to sell NPI at a price that would retain the original investment, allow for a profit, and free the railroad of all liability. In justifying this recommendation, he reviewed the defects inherent in the Kennewick irrigation system. First, he considered the high maintenance costs, attributing these expenditures to the topography of the area and the lack of a clay material to maintain the canal.

Clay ditch walls would have been more stable and less permeable than volcanic ash and gravel. But even clay walls could not prevent the wind from filling the canal and flumes with rock, ash, "and the drifting sand from the sage brush." And breaks in the canal were inevitable:

> A break is liable to occur any time: we have had two since April 1st that cost $1557 to repair. At one of them a man was right on the spot and saw a tiny stream, which he tried to stop break out at the bottom of the slope, but in fifteen minutes, in spite of all his efforts, about 60 feet of embankment and eight hundred yards of material was carried away, bringing destruction to a considerable tract of the farm land lying below the ditch.[36]

The manager of the irrigation company did not see any way to recover maintenance costs. Even doubling the annual fee charged water users would not provide adequate funds for operation of the system. Because of these unresolvable difficulties, Martin concluded, Elliott should get the Northern Pacific out of the reclamation business.[37]

During the same period, the federal Reclamation Service had begun to investigate Yakima Valley lands. It considered, and rejected, a project that would have comprised NPI lands; 40 years later that project would—as the Kennewick Irrigation District—secure federal funding. Meanwhile, President Elliott, faced with more ditch repairs, commented to a subordinate, "These canals seem to be expensive playthings." Slightly alleviating the situation was the sale of the Kiona ditch in May 1907.[38]

Martin's 1905 advice and their own grasp of the irrigation situation had spurred railroad officials to seek a way to dispose of the Kennewick canal. By October 1907 there were several potential buyers, and by late 1908 the irrigation works were sold to two valley residents for $175,000, which according to railroad calculations netted a profit of $100,000. Cooper, who had negotiated the deal, had accomplished quite a feat, for the transaction was more lucrative than many had expected.[39]

Yet the Kennewick *Courier*, speculating on that sale, suspected that the railroad got $200,000 for the land and the canal. The reporter acknowledged that the Kennewick canal had not been meeting operation and maintenance expenses, but he also declared that the NP had slowed growth in the area since 1904 by postponing land sales in the hope that a combined land-and-canal package would attract buyers. Even before that, declared the reporter, the policy of selling only parcels large enough to farm and restricting land sales as much as possible to agriculturists who would irrigate had been a deterrent to development. Now at last NPI acreage would be placed on the market, spurring growth. The community could survive without the Northern Pacific.[40]

Despite the bravura, the situation was not a happy one. The new owners of the Kennewick system ran short of cash almost immediately. Determined water users struggled to eke a living out of the ground irrigated by the canal for the next forty years. Not until 1948 did the Kennewick Irrigation District secure a federal appropriation and establish a project that operates today, watering part of the NP tracts.

The prosperity that many predicted for Kennewick was not realized in the early 1900s, but the Northern Pacific did inaugurate a new era in the history of the area. Almost overnight the 1902 village of 50 became the town of Kennewick, population 500 in mid-1903. New businesses came in. Most important, for the first time people with an interest in the region were guiding Kennewick's destiny. The NP continued to help the community by trying to secure federal financing for the irrigation project. Until the 1930s the Northwestern Improvement Company was still the largest landholder under the Kennewick system. Thus, economics undoubtedly motivated railroad officials, but their strategy was different from that of earlier irrigators, because the railroad anticipated profits not from water and land sales but from increased traffic. Fortunately for the Kennewick inhabitants who produced the commodities that the railroad shipped, those monetary gains were renewable annually. Thus the railroad lent stability to the area and insured its existence in the first years of the twentieth century.[41]

NOTES

1. For two different perceptions of the railroad's development role, see Alfred Runte, *National Parks: The American Experience*, 2d. ed. rev. (Lincoln: University of Nebraska Press, 1987), and Donald J. Pisani, *From the Family Farm to Agribusiness: The Irrigation Crusade in California and the West, 1850–1931* (Berkeley: University of California Press, 1984). Yakima Valley lands affected by the Northern Pacific Railway are also part of the Columbia Irrigation District and the Kennewick Irrigation District, which is a U.S. Bureau of Reclamation project.

2. D.W. Meinig, *The Great Columbia Plain: A Historical Geography, 1805–1910* (Seattle: University of Washington Press, 1968), 268; Walter A. Oberst, *Railroads, Reclamation and the River: A History of Pasco* (Pasco, Wa, 1978), 16–18. General histories of the Yakima Valley include W.D. Lyman, *History of the Yakima Valley, Washington: Comprising Yakima, Kittitas and Benton Counties*, 2 vols. (Chicago, 1919), and *An Illustrated History of Klickitat, Yakima and Kittitas Counties* (Chicago, 1904).

3. Lawrence B. Lee has published extensively in the field of water resources, including

"Water Resource History: A New Field of Historiography?" *PGR* 57 (1988): 457–67. Other leaders in reclamation studies include Norris Hundley, Jr., *Water and the West: The Colorado River Compact and the Politics of Water in the American West* (Berkeley: University of California Press, 1975); Donald Worster, *Rivers of Empire: Water, Aridity, and the Growth of the American West* (New York: Pantheon Books, 1985); Marc R. Reisner, *Cadillac Desert: The American West and Its Disappearing Water* (New York: Penguin Books, 1986); and Pisani, *From the Family Farm to Agribusiness*. For a discussion of Yakima irrigation, see C. Brewster Coulter, "The New Settlers on the Yakima Project, 1880–1910," *Pacific Northwest Quarterly* (PNQ) 61 (1970): 10–21.

4. Yakima Irrigating and Improvement Company (YI&I) articles of incorporation, Washington Department of Conservation and Development Records, Washington State Archives, Olympia; Work Projects Administration, *Washington: A Guide to the Evergreen State* (Portland, Or: Binsfords & Mort, 1941), 370 (first quotation); interview with W.F. Sonderman by Augusta Eastland, 1936, Washington Pioneers Project, Kennewick Irrigation District (KID) Archives, Kennewick (second quotation); *An Illustrated History*, 227–28; Ted Van Arsdol, *Desert Boom and Bust* (Vancouver, Wa, 1972), 5 (third quotation); *Yakima Herald* (North Yakima), 2 February 1889.

5. Lyman, *History of the Yakima Valley*, 1:354–55; Emmett Kaiser Vandevere, "History of Irrigation in Washington," Ph.D. diss. (University of Washington, 1948), 46–47; Van Arsdol, *Desert Boom and Bust*, 3; *Yakima Herald*, 6 February 1890. Intended at one time to extend thirty miles, the Kiona canal was nine miles long in 1904; M.P. Martin to Howard Elliott, June 30, 1904–annual report, President Subject file, Northern Pacific Records, Minnesota Historical Society, St. Paul. According to Vandevere (p. 46), YI&I officials purchased 22,966 acres from the Northern Pacific.

6. *Yakima Herald*, 3, 24, and 31 October 1989, 23 July 1891, 4 July 1895, 19 November 1896; *Colfax Commoner*, 25 October 1889; *Franklin Recorder* (Pasco), 12 March 1897; and File 587, Yakima County Superior Court, Organization of Dell Haven Irrigation District, in Van Arsdol, *Desert Boom and Bust*, 8.

7. Van Arsdol, *Desert Boom and Bust*, 36–37; Lyman, *History of the Yakima Valley*, 1:355; *An Illustrated History*, 228; Thomas Cooper to C.S. Mellen, 11 March 1901, President Subject file, NP Records; *Yakima Herald*, 11 March 1894.

8. Cooper to Mellen, 11 March 1901, President Subject file, NP Records.

9. Ibid.

10. Ibid.

11. Ibid.; *An Illustrated History*, 227 (quotation).

12. Cooper to Mellen, 11 March, 23 September, 18 November, and 30 December 1901, and to Martin, 2 January 1902, President Subject file, NP Records; Frank H. Rudkin to Cooper, 28 October 1901, Northern Pacific Irrigation Company (NPI) Branch Lines file 13, President Subject files, NP Records; *Yakima Herald*, 17 December 1901.

13. Cooper to M.P. Martin, 2 January 1902, President Subject file, NP Records.

14. *Yakima Republic* (North Yakima), 3 January 1902; Rose M. Boening, "History of Irrigation in the State of Washington," *PNQ* 10 (1919), 22–23.

15. *Yakima Republic*, 24 January 1902, 31 January 1902 (quotation).

16. *Yakima Republic*, 16 May 1902 (quotation); Burlingame quoted in *Yakima Herald*, 4 February 1902. For an insight into promotional activities in eastern Washington, see G. Thomas Edwards, "Irrigation in Eastern Washington, 1906–1911: The Promotional Photographs of Asahel Curtis," *PNQ* 72 (1981), 112–20.

17. Cooper to Mellen, 13 March 1902, President Subject file, NP Records. Although Cooper did not criticize the quality of work performed by the YI&I, the condition of the Kennewick canal would indicate sloppy engineering.

18. Ibid.

19. Ibid.

20. Cooper to Mellen, 13 March 1902, President Subject file, NP Records. In considering the amount of discount for cash sales, Cooper mentioned 7.5 percent offered by the Sunnyside Company, which he believed was too high.

21. Ibid.

22. *Yakima Republic*, 14 March 1902.

23. Cooper to Mellen, 13 March 1902, and Mellen to Cooper, 20 March 1902, President Subject file, NP Records.

24. Cooper to B.S. Grosscup, 1 April 1902, President Subject file, NP Records. The new company incorporated with a capital stock of $250,000 and prominent NP executives listed as trustees and officers. In the same letter, Cooper, who would be promoted to company land commissioner, stated that he would be moving to St. Paul, but he did stay closely associated with events in Kennewick. *Yakima Herald*, 22, 29 April 1902.

25. *Yakima Herald*, 29 April, 24 June 1902; *Yakima Republic*, 16 May, 27 June 1902 (this article noted that the majority of applicants came from Tacoma, with the rest from North Yakima, Spokane, Walla Walla, and the Kennewick area).

26. W.C. Sampson to prospective buyers, 28 July 1902, NJPI Branch Lines file 13, NP Records.

27. Cooper to Mellen, 11 September 1902, President Subject file, NP Records.

28. Cooper to Mellen, 28 February 1903, and Mellen to Martin, 23 April 1903, President Subject file, NP Records.

29. Mellen to Martin, 24 June 1903, President Subject file, NP Records. And see also Mellen to unknown, 27 July 1903, ibid.

30. Kennewick Columbia Courier, 24 April, 19 June, 17 July 1903; *Yakima Republic*, 24 July 1903. Although crops were planted in 1904, full production would not occur for three to four years. See general sales agent to Otto L. Hanson, 23 Septtember 1903, NPT Branch Lines file 13, NP Records.

31. Martin to Elliott, 6 November 1903, and Cooper to Mellen, 30 December 1901, President Subject file, NP Records.

32. Martin to Elliott, 23 January 1904, Elliott to Charles Leavey, 16 April 1904, Cooper to Elliott, 28 December and 13 October 1905, President Subject file, NP Records. Memorandum for Judge Reid (ca. 1914), Yakima Irrigation Companies file, NP Records (quotation). Earlier correspondence between the NP sales agent and T.A. Noble, engineer of the U.S.

Geological Survey, regarding various irrigation data also pointed to cooperative efforts between the railroad and the government bureaucracy. See general sales agent to Noble, 12 August 1903, NPI Branch Lines file 13, NP Records.

33. Martin to Elliott, June 30, 1904—annual report, President Subject file, NP Records. Despite Martin's sober tone in the annual report, which was made public, the townspeople remained convinced of the superior quality of the canal, saying it was one of the finest in the country. See *Columbia Courier*, 4 November 1904.

34. Martin, annual report. In this report, Martin furnished an exact accounting of expenditures.

35. *Columbia Courier*, 24 March 1905; Martin to Elliott, 11 May 1905, President Subject file, NP Records.

36. Martin to Elliott, 11 May 1905, President Subject file, NP Records.

37. Ibid.

38. *Kennewick Courier*, 4 January 1907. Despite the financial drain, Elliott never questioned the necessity of repairing the canal, for he acknowledged that the ditches must be kept "in order"; Elliott to Leavey, 2 April 1907, President Subject files, NP Records. *Yakima Herald*, 31 May 1907.

39. Cooper to Martin and G.H. Plummer, 25 October 1907, and to Elliott, 2 May 1908, President Subject file, NP Records.

40. *Kennewick Courier*, 5 June, 23 October, 13 November 1908.

41. Cooper to Martin, 16 July 1908, President Subject file, NP Records. Today, the Highlands is the only land once irrigated by the Kennewick Irrigation District. The other land served by the Kennewick canal falls within the boundaries of the Columbia Irrigation District; *An Illustrated History*, 223, 228; Cooper to R.K. Tiffany, 21 January 1918, Yakima Irrigation Companies file 19 J-2, NP Records; "A Statement Submitted by Kennewick Irrigation to the Bureau of Reclamation Concerning the Kennewick Division of the Yakima Project in Washington" (ca. 1926), J.M. Hughes to Donnelly, 8 October 1930, President Subject file, NP Records.

CHAPTER EIGHT

WILDLIFE, SCIENCE, AND THE NATIONAL PARKS, 1920–1940

THOMAS R. DUNLAP

The values Americans have attached to national parks have been changing as long as the parks have been in existence, reflecting and sometimes anticipating society's shifting ideas about nature and its place in our culture. Nowhere has this been more pronounced than with regard to animals. The earliest park acts provided for protection for wildlife and some of the first game protection laws were applied to the parks, but until well into the twentieth century, park officials cared for only a few species, ignoring and even killing the others. Now wild animals are among the parks' major attractions, and the preservation of all species is an important part of their mission. The change came between the two world wars, when the new National Park Service broke with practices that went back to the earliest years of the parks. Declaring that it had a mission to save all native animals and the web of relationships among plants, animals, and land, the Park Service began to protect the formerly hunted predators, saving endangered species, and eliminating foreign animals from their parks.[1]

A small group, mainly mammalogists, was behind this change. Using the concepts of the new field of animal ecology and the prestige of science, they gave a different meaning to the idea of an authentic, primitive natural landscape. Their ideas had implications for wildlife everywhere; they first affected the parks because it was there they could most easily bring scientific information and preservation sentiment to bear on actions. The Park Service was small, nonpolitical, committed to guidance by experts, and just beginning to take on institutional form. Employed in the Service's Wild Life Division (established 1933), the scientists could work for the preservation of all species in an atmosphere free of long-established bureaucratic restraints and more conducive to seeing new ideas transformed into policy.

In the late-nineteenth century Americans began establishing national parks to preserve what they valued in nature—the spectacular scenery of mountains, chasms, and geological freaks. Though park acts provided for the protection of wild animals as well as scenery, for park officials wildlife meant only those creatures that added to the parks' appeal. The ideal animal was large and stood around in groups in the open—posing nobly in the middle distance against a background of mountain peaks—or entertained tourists with its "cute" antics. "Management" meant encouraging these species, killing their "enemies," and ignoring the others. Yellowstone superintendents, for example, kept the buffalo near the roads and stampeded them to show prominent visitors the "thundering herd." They fed elk through the winter and set up feeding stations for the bears at the park's garbage dumps—with bleachers for the tourists.[2] They shot, trapped, and poisoned wolves, coyotes, and mountain lions. "Up to about 1930," notes an in-house Park Service history, "a great majority of people, including National Park Service officials, took it for granted that complete protection of wildlife involved elimination or drastic control of all predators. The term 'extermination of predators' appears often in . . . official business . . ."[3] No one thought of the parks as preserves for all species in a balance dictated by natural forces.

People thought instead in terms of "good" and "bad" animals and of encouraging the good ones and getting rid of the bad ones. Even the Audubon Society recommended that bird-killing hawks be shot on sight. Its only reservation about public campaigns against the hawk was that most people could not distinguish the "good" from the "bad" raptors.[4] The Forest Service killed predators in reserves (like the Kaibab National Forest) to build up the deer herd, and in 1918 Aldo Leopold, editing the *Pine Cone* for the New Mexico Game Protective Association, called for getting the last predator scalp in the state.[5] In 1915 Congress had authorized the Bureau of Biological Survey to assist ranchers in controlling predatory animals, and by the mid-1920s the agency had killed the last breeding wolves in the West and was hunting down the stragglers. Even people who liked predators were often resigned to seeing them killed off. Their deaths were the price of progress.[6]

By the 1920s a few people began to see value in the despised varmints and to urge their preservation. The first organized protests came from academic zoologists, mainly westerners, who took aim at the Biological Survey's predator control program. At the 1924 meeting of the American Society of Mammalogists, a small group—Joseph Dixon, Lee Dice, Harold Anthony, and C.C. Adams—claimed that the Survey's liberal use of poison was wiping out local wildlife populations, threatening whole species, and taking an unacceptable toll on nontarget animals. They did not object to control, even to poisoning (this was not a protest on "environmental" or humane grounds), but to unnecessary slaughter.[7] The debate, which went on until 1931, centered around reconciling preservation with progress; "stock-killers," everyone admitted, could not be allowed where sheep and cattle grazed. The parks, which had, after all, been set up to preserve wild America, seemed to be the solution.[8]

Joseph Grinnell, director of the Museum of Vertebrate Zoology in Berkeley, California, led the opposition to predator control and was a strong voice for wildlife preservation. Born in 1877 in the Indian Territory (his father was a doctor at an Indian agency post), he had grown up on the reservations and in Pasadena, California, studying the land as an amateur naturalist at a time when it was rapidly changing. After 1908, when he became head of the new Museum of Vertebrate Zoology, he participated in numerous collecting expeditions that added a valuable perspective to his impressions about what stock raising, mining, and irrigation were doing to the land.[9] About the same time, he started faunal studies in Yosemite to establish the distribution and abundance of animals in the region. He realized that the West he had known was vanishing, and he wanted to save it.

Grinnell's sentiments were becoming more common; even the agents of progress were uncomfortable with their role. Vernon Bailey, the Survey's chief naturalist, had been involved in predator control since the agency's surveys of national forests in 1907, but he said of the wolves he was killing: "[f]ew animals are more devoted to their home life, braver, or more intelligent. Yes, they were cruel killers, but not half as cruel as we have been."[10] Stanley Paul Young, who made his Survey career in predator and rodent control, thought wolves wonderful creatures. Both he and Bailey wanted to keep them, but in areas without livestock.[11] A few went further. In 1929, Olaus Murie, then a predator and rodent control supervisor for the Bureau of Biological Survey, confessed to one of his colleagues that he was "very fond of native mammals, amounting almost to a passion," and that he "would make considerable sacrifice for the joy of animal companionship and to insure that other generations might have the same enjoyment and the same opportunity to study life through the medium of the lower animals." The Survey's predator control operations, he said, had no scientific backing, "and to tell the truth I could not see that the queries of the [mammalogists] had been met."[12]

The debate over poisoning crystallized sentiment in the Park Service—or provided an excuse for action. Certainly, park officials were ready. At their annual conference in 1925 the park superintendents reduced the money that rangers received from the sale of pelts and cut the list of predators to three species: wolves, coyotes, and cougars. Three years later—following a speech by Joseph Grinnell—they passed a resolution opposing trapping in the parks. In May 1931 Park Service Director Horace Albright pledged "total protection to all animal life," including predatory animals, which "have a real place in nature." They would be killed only where they were causing "actual damage" or threatening a species that needed special protection. Trapping would end and poison would be used only for rodent control in settled areas or emergencies. He proclaimed a wildlife mission that went well beyond the "scenic" ideal. "[A]ll animal life should be kept inviolate" and the parks should preserve "examples of the various interesting North American mammals under natural conditions for the pleasure and education of the visitors and for the purpose of scientific study."[13]

Albright echoed the sentiments of the wildlife experts who were finding a place, and influence, in the Park Service. No one had had direct responsibility for park animals in 1928, when George Wright, an independently wealthy part-time ranger, proposed and offered to pay for a survey of park animals and their habitat. The Service agreed and the results of the two-year study (by Wright, Ben H. Thompson, and Joseph Dixon) were so impressive that in 1931 the agency made Wright a field naturalist and, two years later, chief of the new Wild Life Division. Animal ecology and concern for the preservation of all animals were a part of the new bureau's ethos. As a forestry student at the University of California, Wright had taken a minor in vertebrate zoology under Grinnell and spent a summer making life history studies of the fauna of Mt. McKinley (now Denali) National Park with Grinnell's assistant, Joseph Dixon.[14] Three other members of the staff had worked or been trained at the Museum of Vertebrate Zoology, and another Grinnell student, Harold Bryant, had been on the committee recommending establishment of a permanent wildlife branch (and as head of the park education program would spread the gospel to visitors).[15] Carl P. Russell, who became chief after Wright died in a car accident in December 1935, studied under Lee Dice (also a Grinnell student) who had established a pioneer program in animal ecology at the University of Michigan. Russell's assistant, Victor Cahalane, had worked with Dice at the Cranbrook Institute of Science, and the man whose work was crucial to the defense of natural management of all species, Adolph Murie, had taken his Ph.D. under Dice at the University of Michigan in 1929.

Wright immediately published his survey as the first of a new monograph series—"Fauna of the National Parks."[16] All species, it said, "should be preserved, for each is the embodied story of natural forces which have been operative for millions of years and is therefore a priceless creation." Predators were included, for they find in the parks "their only sure haven . . . [and] are given opportunity to forget that man is the implacable enemy of their kind, so that they lose their fear and submit to close scrutiny." Boundaries should be adjusted to make each park an ecological unit where native wildlife should be saved or restored and exotics curbed or eliminated. Park officials needed to educate the public to appreciate wildlife in its natural setting, to thrill to the lone bear in the forest rather than the group at the garbage dump.[17]

Events were even then reinforcing Wright's case. The Yellowstone elk herd had been a problem since the early part of the century. Ranchers protested that the swollen herd was devouring range grasses outside the park; the annual slaughter as the migrating animals left the park appalled many; and no one liked the occasional, spectacular die offs during bad winters.[18] In the winter of 1924–25 the protected deer in the Kaibab National Forest, on the north rim of the Grand Canyon, began dying in the thousands because of disease and starvation. Herds across the country, many of them in the populated Northeast, went through the same cycle, and the spectacle of dead deer and stripped forests was a stark challenge to the idea of complete protection.[19] In *Game Management* (1933), Aldo Leopold had come far enough from his earlier views to suggest that predator control had to be proved useful in each instance; it could not be assumed to be needed or helpful.[20] As the decade went on his views would shift even further toward an ideal of restoring nature's harmony in the parks. Others would slowly follow his lead.

The Wild Life Division's scientists, though, did not argue, on the evidence of disasters, that all wildlife should be preserved, still less did they use aesthetic arguments about the beauty of nature. They grounded their appeals on scientific studies, justifying their love of wild animals with ecological theory and field research that saw each species as part of a system. Niches, ecosystems, food chains, and the pyramid of numbers replaced vague ideas about the "balance of nature"; quantitative studies undermined myths of "bad" animals. This kind of defense would not have been possible earlier, for it depended upon ideas just being developed. Victor Shelford's *Animal Communities of North America*, the first attempt to organize animal communities on ecological principles, had been published in 1917, and Charles Elton's Animal Ecology, which popularized the ideas of niche and food chain and provided a research program for a generation of scientists, only a decade later.[21]

The new defense of nature can be seen by comparing the studies that Adolph Murie did for the Park Service on the coyotes of Yellowstone (1937–39) with those of his brother Olaus conducted for the Biological Survey on the coyotes of Jackson Hole, Wyoming (1927–32).[22] The coyote populations overlapped, the areas were similar, both studies began in response to complaints that predators were killing off the more desirable species, and both aimed to see what effect coyotes had on elk under wilderness conditions. Even their conclusions were the same: individual elk were vulnerable, the population was not.[23] Their methods, though, were very different. Olaus analyzed stomach contents and scats to determine the "Food Habits of the Coyotes of Jackson Hole, Wyoming." Elk, he pointed out, were a small part of the coyotes' diet (and probably came from carrion, not kills). This suggested that elk and coyotes could live together. Adolph looked at many factors—weather, food, disease, cover, and historical patterns of wildlife abundance. He counted elk, studied calf survival, and reported on coyote behavior and the animal's relationships with species ranging from bison and moose to porcupines and squirrels. His work was as comprehensive as the title: *Ecology of the Coyote in the Yellowstone*, and his conclusions about the elks' ability to survive grounded in a different theory. The change in methods was even more apparent in Adolph's study of the wolves and Dall sheep of Mt. McKinley National Park (1939–41), the first ecological study of a large predator and its prey population. There he tested the argument that predators selected old, weak, ill, or otherwise defective individuals by collecting over eight hundred sheep skulls. He compared the living and dead populations in terms of age and sex, and looked at the numbers showing signs of disease or injury. He watched wolves at their dens over two summers, providing the first extensive observations on their behavior in the wild.[24]

It was not a matter of an unsophisticated approach versus a sophisticated one; Olaus worked under the direction of the Survey's W.L. McAtee, for years head of the Survey's Division of Food Habits Research, and his studies reflected faithfully the best methods of scat and food analysis available. Nor did the brothers have different ideas about wildlife preservation. Their research was different because they were of different professional generations, with different training and conceptual tools. Olaus, born in 1889, had graduated from college in 1912. He worked on field and museum jobs for a few years before joining the Biological Survey. He got an M.S. (from the University of Michigan in 1927), but he did it at his brother's urging, and the work was clearly an interlude in his career in the Biological Survey. His appreciation of ecology came later, when he was studying elk with Adolph.[25] Adolph, ten years younger, finished college in 1923, and though he did some natural history work—including a faunal survey in Alaska with Olaus in 1922–23—he received a professional education in wildlife ecology under Dice at Michigan, where he learned concepts and research methods developed since his brother had left school.

By the mid-1930s the Park Service had ended most predator killing and was moving toward management to achieve a natural state; it needed studies like Adolph's to fend off the considerable opposition to saving predators in the parks.[26] Ranchers complained that the parks were now sanctuaries for "stock-killers"; hunters were concerned about "their" trophies; and even some staunch park defenders were baffled by the idea of saving "vermin." In the late 1930s Horace Albright, by then retired from the Park Service, clearly thought that things had gone too far. Coyotes, he declared, should be reduced or eliminated. Visitors never saw them and they were worthless. "I find," he wrote in 1937, "that the impression is quite widespread that the National Association of Audubon Societies and perhaps other organizations are more interested today in saving the predatory species of birds and mammals than giving reasonable consideration to the species that are regarded as very important by the general public." He hoped that the Park Service would not take this attitude.[27] But it did, justifying the new policy by scientific studies describing all species as functional parts of a natural system.

The Division's point of view involved more than just saving the despised "varmints," more even than preserving natural systems. The scientists wanted a particular kind of natural system—North America as it had been before the whites came—and to get this they pushed for the elimination or curbing of introduced secies. This was less controversial but almost as novel as preserving predators. People had been importing and establishing new plants and animals since the first days of settlement. In the late nineteenth century there had been a flood of exotics. Homesick European immigrants imported songbirds from their native lands; the rich stocked estates with game animals and birds from around the world; and hunters set out new species for sport. Texas and California went through a "Mediterranean" phase, and California had its Australian days as well.[28] Nature was to be reshaped for human convenience and delight.

A reaction occurred in the 1890s, in part the result of disastrous introductions like the starling but also fueled by romanticism and nationalism that sought to preserve the pioneer experience. Grinnell's paean to the game birds of California is typical. "Conserve our native species! There are none whose qualities are superior: they are part of the natural heritage of our land, and have been serviceable in the past; we are responsible for their preservation."[29] By the 1920s debate was far enough advanced that John Phillips, reviewing the record of game bird introductions into the United States, could refer to two schools of thought "widely at varience. One of these, the conservative, represented by such eminent naturalists as Joseph Grinnell of California and many others, believes in preserving at all costs the present or rather the original status of native birds and harmless mammals, and points out the great dangers incurred in the importation of species in other parts of the world, and especially the danger of spreading new diseases." The other would try anything without considering the dangers of the suitability of the species to the country. (Phillips recommended a middle course that avoided both extremes.)[30]

As with predator control, advocates of native wildlife and native ecosystems (a word just coming into use) saw the parks as a sanctuary. In 1930 the Service took the first steps toward that goal; Albright sent out a notice calling for the preservation of native flora and fauna. Superintendents, he stated, should eliminate or curb nonnative forms and take steps to see that there were no introductions.[31] A decade later the Service told a U.S. Senate special committee on wildlife that:

> another phase of the Service program of wilderness protection is concerned with elimination of nonnative animals that at times have become introduced into various parks, often to the detriment of native fauna and flora. Any exotic species which has already become established in a park shall be either eliminated or held to a minimum provided complete eradication is not feasible, and the possible invasion of the parks by other exotics shall be anticipated and steps taken to guard against the same.[32]

The parks were to be refuges not just for nature, but for nature as it had been when the Europeans arrived.[33] A corollary was the preservation of native species. In 1930 Wright began a campaign for the trumpeter swan, and managed not only to protect the birds in Yellowstone but also to enlist the Biological Survey's help in creating a refuge (Red Rock Lakes Migratory Waterfowl Refuge) at the birds' breeding ground outside the park. The Service took more interest in the Yellowstone grizzlies, initiating a program that is now one of its most controversial. It also began shooting burros in the Grand Canyon to help the native wildlife. In still another move, it sought to increase public education on endangered species, in 1943 sponsoring a book, *Fading Trails*, to alert the public.[34]

The logical end of all this was a park in which wildlife would be the chief attraction and which would be managed to preserve the habitat in its "primitive" state. That was indeed one of the scientists' goals. Everglades, authorized as a national park in 1934, showed how public ideas were changing, but a better indicator was the Division's decade-long campaign for Isle Royale, the largest island in Lake Superior. The area had none of the spectacular displays of massed wildlife that made the Everglades a mecca for bird-watchers. It was, at best, an unspectacular northern forest. Nor was it pristine. Logged and mined in the nineteenth century, it was a maze of cut-over areas, mine dumps, and burned-out patches, with summer camps and fishing piers. Still, it could be restored and (because it was isolated) managed as an ecological unit. The scientists may have been enthusiastic because they knew it so well—C. C.

Adams had surveyed it in 1904 and Adolph Murie had studied the island's moose population in the early 1930s. Michigan had suggested making it a park as early as 1930 and the Wild Life Division pressed the case. In 1940 Congress finally authorized the park, and the Service had its (ecological) island.[35]

It is a long way from Yellowstone's geysers to the spruce forests of Isle Royale, from strychnine for predators to calls for reintroducing wolves, but the scientists traveled the road—and quickly. In twenty years, using ecological research that portrayed each species as part of a whole, and the whole as unique, irreplaceable, part of the American past, they changed the Park Service's policy toward animals and made preservation of all species a part of the agency's mission. A new public appreciation for wild animals and nature and the existence of the Wild Life Division were important (and the scientists helped form both). Still, the crucial element in the shift was the scientists' ability to use the cultural authority of science to define wilderness, to describe the romantic dream of the world as it had been in ways that could be converted to action on the ground.[36]

The strength of that dream became clear after World War II, as popular writing, movies, television, and public education carried ecological ideas to the public. More people came to see the world as a system and each species as an integral part of the whole, essential to its functioning. In 1935 when Harlow Mills, a Park Service naturalist, suggested that wolves and cougars be introduced into Yellowstone because they were "a vital part of the picture . . . which can never be the same in their absence," even the Wild Life Division gave that a quick burial.[37] The public, and in particular the stockman, would not understand. In 1980 the Northern Rocky Mountain Wolf Recovery Team, appointed under the Endangered Species Act of 1973 to produce a plan for the recovery of the population, found considerable opposition to reintroduction, but also much enthusiasm. When Canadian wolves crossed the border into Glacier National Park in 1986 the Park Service anxiously watched over them, and there have been suggestions that the Service has secretly reintroduced wolves into Yellowstone.[38] The public has also come to demand the application of ecological ideas outside the parks. The Endangered Species Acts of 1966, 1969, and 1973, the Marine Mammal Protection Act of 1972, and the 1973 Convention on Trade in Endangered Species are testimony to public interest in preserving as much of the ecosystem of the country as possible, even at the expense of some economic development. Laws are not always observed, and they may be changed, but current discussions on the preservation of wildlife in America are guided by ideas proposed and first implemented in the parks between the two world wars.

ACKNOWLEDGMENTS

Much of the research for this article was done with the financial support of the National Sciences Foundation (Grant Number SES 8319362) and the Virginia Tech history department. I owe a more personal debt to my colleagues, who criticized a seminar paper and earlier versions of this article.

NOTES

1. The role of scientists and scientific ideas in the nature conservation and presentation movements of twentieth-century America has not been investigated in any depth, nor have historians traced in any detail the changing place of wildlife in popular ideas about nature or the influence of field biology and ecology on these ideas. Alfred Runte, *National Parks: The American Experience*, 2nd ed. (Lincoln: University of Nebraska Press, 1987), chapters 6 and 7, deals with the administrative and policy aspects of changing views in the Park Service. On wilderness, see Roderick Nash, *Wilderness and the American Mind*, 3rd ed. (New Haven: Yale University Press, 1982), chapter 13. Donald Worster's *Nature's Economy* (San Francisco: Sierra Club Books, 1977) deals with scientific and popular ideas, but in such a wide context that individual scientists and their ideas vanish. He does not, in any event, concentrate on wildlife and its meaning as part of nature. Peter Mathiessen's *Wildlife in America* (New York: Viking, 1987) provides a popular treatment, but it is not well based in historical sources. The same might be said of Barry Holstun Lopez, *Of Wolves and Men* (New York: Scribner, 1978). There is a large literature on hunting, but this is hardly adequate for wildlife. The best recent studies on evolving attitudes toward animals are Keith Thomas, *Man and the Natural World* (New York: Pantheon Books, 1983) and James Turner, *Reckoning with the Beast* (Baltimore: Johns Hopkins University Press, 1980), but these deal with the general relation of animals to humans, not of wild animals to humans and human conceptions of nature.

2. Until the late-1920s, when the level of injuries and death forced a more cautious policy, park officials treated bears as tourist attractions. In 1922 Horace Albright, then superintendent of Yellowstone and later head of the Park Service, posed at a picnic table with several of his furry friends. See Horace M. Albright, *The Founding of the National Park Service* (Salt Lake City: Howe Bros., 1985), 182. Runte, *National Parks*, 168–169. Even today bear policy is not, with the exception of the grizzlies of Yellowstone, "wildlife policy," but "nuisance animal control," a matter of protecting people and property.

3. C.C. Presnall, "Discussion of Service Predator Policy" (1939), in "Condensed Chronology of Service Predator Policy," file 719, entry 7, Records of the National Park Service, Record Group 79, National Archives, Washington, D.C. (cited hereafter as RG 79, National Archives). Chronology and information on park service predator policy is found, unless otherwise attributed, in this file.

4. The Audubon magazine—*Bird-Lore* to 1940, *Audubon* thereafter—presents a fascinating picture of changing attitudes prior to 1940. On changes, see also Thomas R. Dunlap, *Saving America's Wildlife* (Princeton: Princeton University Press, 1988), chap. 6.

5. The Kaibab has been extensively treated. Thomas R. Dunlap, "That Kaibab Myth," *Journal of Forest History* 32 (1988): 60–68, is the latest review. On Leopold, see Susan Flader, *Thinking Like a Mountain* (Columbia, Mo: University of Missouri Press, 1974).

6. See, for instance, Ernest Thompson Seton's "The King of Currumpaw: A Wolf Story," *Scribner's*, vol. 16 (November 1894), 618–26, reprinted as "Lobo, King of the Currumpaw," in Ernest Thompson Seton, *Wild Animals I Have Known* (New York: C. Scribner's Sons, 1898), and comments in Seton, *Lives of Game Animals* (1925–28; reprint, New York: Literary Guild, 1953). Theodore Roosevelt felt the same way; this refrain runs through *The Wilderness Hunter: The Works of Theodore Roosevelt, National Edition* (1926), *Hunting Trips of a Ranchman* (1885), and *Hunting the Grisly and Other Sketches* (1893) (titles reprinted in *Hunting Trips of a Ranchman & The Wilderness Hunter: Sketches of Sport on the Northern Cattle Plains: An Account of the Big Game of the United States,* New York: Modern Library, 1996).

7. Thomas R. Dunlap, "Values for Varmints: Predator Control and Environmental Ideas, 1920–1939," *Pacific Historical Review* 53 (1984): 141–61. For a contrary view of the significance of this episode, see Worster, *Nature's Economy*, chap. 13.

8. Dunlap, "Values for Varmints," 147–58.

9. See, for example, Grinnell to C. Hart Merriam, 27 November 1907, in Rescued Correspondence, 1907, box 78, W.L. McAtee Papers; Joseph Grinnell, "The Methods and Uses of a Research Museum," *Popular Science Monthly*, vol. 77 (August 1910), 163–69, reprinted in Joseph Grinnell's *Philosophy of Nature* (1943; reprint, Freeport, NY, 1968). On feelings about the transformation of the West, see Lee Clark Mitchell, *Witnesses to a Vanishing America: The Nineteenth Century Response* (Princeton: Princeton University Press, 1981). The use of science in this enterprise was not unique. As Grinnell was mapping the distribution of cactus wrens and ground squirrels in California, Alfred Kroeber and his students were collecting the scattered fragments of Native American languages, rituals, and beliefs, intellectually reconstructing the world the whites were destroying.

10. Vernon Bailey, "The Home Life of the Big Wolves," *Natural History*, vol. 46 (September 1940), 120–22.

11. Vernon Bailey, "Wolves in Relation to Stock, Game, and the National Forest Reserves," *Forest Service Bulletin 72* (Washington, D.C., 1907), box 5, folder 12, Vernon Bailey Papers, Smithsonian Archives, Smithsonian Institution, Washington, D.C.; Stanley Paul Young, "The War on the Wolf," *American Forests* 48 (December 1942): 574. See also Stanley Paul Young and Edward A. Goldman, *The Wolves of North America* (1944; reprint, New York: Dover Publications, 1964); Stanley Paul Young and H.H.T. Jackson, *The Clever Coyote* (Washington, D.C.: Wildlife Management Institution, 1951); Stanley Paul Young, *The Last of the Loners* (New York: Macmillan 1970). The same holds true even for a later generation. See Charles L. Cadieux, *Coyotes: Predators and Survivors* (Washington, D.C.: Stonewall Press, Inc., 1983).

12. Murie to A. Brazier Howell, 7 May 1931, Department of Mammalogy, American Museum of Natural History, New York; Murie to W.C. Henderson, 9 January 1931, box 265, Olaus Murie Papers, Denver Conservation Library, Denver, CO. See also K.C. McMurray to J.N. Darling (head of the Survey), 18 December 1934, in file: Predatory—States, Michi-

gan, General Files Division of Wildlife Services. Records of the U.S. Fish and Wildlife Service Record Group 22, National Archives, Washington, D.C.

13. Presnall, "Condensed Chronology," file 719, entry 7, RG 79, National Archives; Horace M. Albright, "The National Park Service's Policy on Predatory Animals," *Journal of Mammalogy* 12 (May 1931): 185–86.

14. Ben H. Thompson, "George M. Wright, 1904–1936," *George Wright Forum* 1 (Summer 1981): 1–4.

15. The people associated with the Museum of Vertebrate Zoology included Joseph Dixon, E. Lowell Sumner, Adrey E. Borell, Theodore H. Eaton, Jr. All were listed on the first page of the memo that the Division in 1935 sent to the Bureau of Biological Survey asking that they be put on the mailing list for the Survey's new bibliographical guide to wildlife management research. Grinnell's commitment to ecology and the interests of his students may best be judged from his correspondence in the Museum of Vertebrate Zoology (hereinafter cited as MVZ), Berkeley. These collections cited by permission of Dr. David Wake, director, MVZ. Alfred Runte, personal communication, has suggested that while Grinnell's interest in wildlife was constant, his enthusiasm for ecological concepts developed by his students may have varied directly with the credit they gave him for his inspiration.

16. George M. Wright, Joseph S. Dixon, Ben H. Thompson, "Fauna of the National Parks of the United States," *Fauna Series Number One*, U.S. Dept. of the Interior (Washington, D.C., 1933). On policy, see Senate, *Report of the U.S. National Park Service*, 76th Cong., 2d sess., 1940, S. Rept. 1203, 350–52; Victor H. Cahalane, "The Evolution of Predator Control Policy in the National Parks," *Journal of Wildlife Management* 3 (July 1939): 229–37; and file 720, entry 7, RG 79, National Archives.

17. Wright et al., "Fauna of the National Parks," 2, 54–56.

18. Douglas B. Houston, *The Northern Yellowstone Elk: Ecology and Management* (New York: Macmillan, 1989).

19. On these irruptions see Aldo Leopold, Lyle K. Sowls, and David L. Spencer, "A Survey of Over-Populated Deer Ranges in the United States," *Journal of Wildlife Management* 11 (April 1947): 162–77.

20. Aldo Leopold, *Game Management* (New York: C. Scribner's Sons, 1933), 252.

21. Victor Shelford, *Animal Communities of North America* (Chicago, 1917); Charles Elton, *Animal Ecology* (1927; reprint, London, 1966)

22. Olaus J. Murie, "Food Habits of the Coyote in Jackson Hole, Wyo.," U.S. Dept. of Agriculture, Circular 362 (Washington, D.C, 1935); Adolph Murie, "Ecology of the Coyote in the Yellowstone," *Fauna Series Number Four*, U.S. Dept. of the Interior (Washington, D.C., 1940).

23. Both reports, to continue the similarities, raised opposition in the agencies they were intended for. Olaus's colleagues in predator and rodent control thought he had not sufficiently emphasized coyotes' appetite for mutton. Park Service officers who thought that "innocent" animals needed protection wanted to fire Adolph. See box 264, miscellaneous P file, Olaus Murie Papers, Conservation Center, Denver Public Library, Denver CO; and box 83, W.L. McAtee Papers, Library of Congress, Manuscript Division, for comments on Olaus's paper. On Adolph's troubles, see Olaus Murie to Harold E. Anthony, 5 December

1945, Murie file, Dept. of Mammalogy, American Museum of Natural History; Olaus Murie to Carl L. Hubbs, 30 July 1946, box 361, folder 7, Adolph Murie Papers, Conservation Center, Denver Public Library, Denver, CO; file 720, entry 7, RG 79, National Archives.

24. Adolph Murie, "The Wolves of Mt. McKinley," *Fauna Series Number Five*, U.S. Dept. of the Interior (Washington, D.C., 1944). On the importance of this work, see the dedication of L. David Mech's *The Wolf: The Ecology and Behavior of an Endangered Species* (Garden City, NY: Natural History Press, 1970). Durward Allen, an eminent wildlife biologist and Mech's teacher, confirmed Mech's views in a personal communication to the author.

25. On the relationship of the Muries, I am indebted to Professor Jim Glover. On ecology, see Olaus Murie to Marcus Ward Lyon, 30 March 1935, Murie file, Dept. of Mammalogy, American Museum of Natural History, New York.

26. Cahalane, "Evolution of Predator Control Policy." On the partial exception, in Denali National Park (formerly Mt. McKinley), see Samuel J. Harbo, Jr., and Frederick C. Dean, "Historical and Current Perspectives on Wolf Management in Alaska," in Ludwig N. Carbyn, ed., "Wolves in Canada and Alaska," *Canadian Wildlife Service Report Series Number 45* (Ottawa, 1983), 51–64.

27. Albright to A.E. Demaray, Nov. 24, 1937, file 720, entry 7, RG 79, National Archives.

28. See, for example, Robin Doughty, *Wildlife and Man in Texas*, College Station, Tex., 1983); T.S. Palmer, "The Danger of Introducing Noxious Animals and Birds," in U.S. Dept. of Agriculture, *Yearbook of Agriculture, 1898* (Washington, D.C., 1899), 87–110; and John C. Phillips, "Wild Birds Introduced or Transplanted in North America," U.S. Dept. of Agriculture, *Bulletin 61* (Washington, D.C., 1928); Woodbridge Metcalf, *Introduced Trees of Central California* (Berkeley, 1968).

29. Joseph Grinnell, Harold C. Bryant, and Tracy I. Storer, *The Game Birds of California* (Berkeley, 1918), 44.

30. Philips, "Wild Birds," 5.

31. Albright to Superintendent and Concessionaires, Nov. 11, 1930, file 720, entry 7, RG 79, National Archives.

32. Senate, *Report of the U.S. National Park Service*, 76th Cong., 2d sess., 1940, S. Rept. 1203, 352, 362.

33. The complexities and ambiguities in this definition have been part of park policy since.

34. Senate, *Report of the U.S. National Park Service*, 76th Cong., 2d sess., 1940, S. Rept. 1203, 352–365; U.S. Dept. of the Interior, *Fading Trails: The Story of Endangered American Wildlife* (New York: The MacMillan Company, 1943). On the grizzly controversy, see Alston Chase, *Playing God in Yellowstone* (Boston: Atlantic Monthly Press, 1986), an excellent example of the depth of emotion that the issue still rouses.

35. The island became an important wildlife research area when wolves colonized it in the early 1950s. In 1957 the Park Service contracted with Purdue University biologist Durward Allen to study the wolves and moose. The project, which still continues, was the training ground for L. David Mech, the current chair of the Wolf Specialist Group of the Species Survival Commission of the International Union for the Conservation of Nature. Information gathered on Isle Royale has been used in the preservation plan for the eastern timber

wolves in Minnesota. Durward Allen (personal communication, October 1988) thinks that the pack may not be viable; it is not breeding and may die out. L. David Mech, "The Wolves of Isle Royale," *Fauna Series Number Seven*, U.S. Dept. of the Interior (Washington, D.C., 1966); Durward L. Allen, *Wolves of Minong: Their Vital Role in a Wild Community* (1979; reprint, Ann Arbor: University of Michigan Press, 1993). On mainland wolf management, see wolf files, Office of Endangered Species, U.S. Dept. of the Interior, Washington, D.C.; and Eastern Timber Wolf Recovery Team, *Recovery Plan for the Eastern Timber Wolf* (Washington, D.C., 1978).

36. The development of this alliance is seen in the 1963 report of the Secretary of the Interior's Advisory Board on Wildlife Management on park wildlife, which placed ecology, ecosystem preservation, and native wildlife at the center of the parks' mission. Advisory Board on Wildlife Management, "Wildlife Management in the National Parks," in Wildlife Management Institute, *Transactions of the Twenty-Eighth North American Wildlife and Natural Resources Conference* (Washington, D.C., 1963), 28–44.

37. Harlow Mills to Ben H. Thompson, 21 June 1935, box 6, entry 34, RG 79, National Archives.

38. U.S. Fish and Wildlife Service, Northern Rocky Mountain Wolf Recovery Team, *Northern Rocky Mountain Wolf Recovery Plan* (Denver, 1980). See also *Agency Review Draft, Revised Northern Rocky Mountain Wolf Recovery Plan* (Denver, 1984); Chase, *Playing God in Yellowstone*, 129–41; Jim Robbins, "Wolves Across the Border," *Natural History* 95 (May 1986): 6–15.

CHAPTER NINE

MANIPULATING NATURE'S PARADISE
National Park Management under Stephen T. Mather, 1916–1929

RICHARD WEST SELLARS

Controversies over national parks continually make headlines as the National Park Service encounters public debate on fire control, restoration of wolf populations in Yellowstone, intensive development of the Yosemite valley, and other issues. Such debates raise fundamental questions concerning the purpose of the parks: Why were they set aside by Congress? Were they primarily meant to be vacation areas for public enjoyment? What role should science play in caring for the parks, and how much effort should be made to preserve their natural conditions? When the National Park Service came into being in the early twentieth century the answers to these questions seemed quite clear. They were heavily weighted in favor of public use and enjoyment rather than preservation, and they fostered management practices that altered the national parks forever.

After establishing Yellowstone National Park in 1872 and subsequently other parks such as Sequoia, Yosemite, and Glacier, Congress created the National Park Service in August 1916 to oversee this growing system of scenic reserves. The 1916 legislation, known as the park service's Organic Act, mandated that the national parks be treated differently from other public lands where extraction of natural resources was encouraged to enhance America's economic growth. The park service was given a truly distinctive mandate:

> to conserve the scenery and the natural and historic objects and the wild life therein and to provide for the enjoyment of the same in such manner and by such means as will leave them unimpaired for the enjoyment of future generations.[1]

The charge to leave the parks "unimpaired" became the principal standard for national park management. Today, some observers assume that this mandate has from the beginning meant that the National Park Service's primary purpose has been to preserve park resources such as wildlife, fish, and forests in their natural conditions.² But a close look at early park service management discloses substantial flaws in that assumption.

In fact, the new mandate to leave the parks unimpaired changed the direction of national park management very little from the pre-1916 focus on preserving the parks' scenery and popular game animals and providing good fishing opportunities for public enjoyment. Stephen T. Mather, the National Park Service's principal founder and first director, generally adhered to the policies and practices set by the United States Army and civilian superintendents who had overseen the parks in the late-nineteenth and early-twentieth centuries before Congress created the park service. Instead of setting new directions for park management, the immediate effect of the Organic Act was to strengthen national park leadership. It brought the parks under efficient, coordinated supervision and enabled Mather to accelerate park development for public use and enjoyment.

National park management under Mather's direction from 1916 to 1929 should be considered the park service's initial administrative interpretation of its legislative mandate, reflecting the service's perception of Congress's true intent regarding the parks. As the Organic Act's chief proponent, Mather surely had a clear understanding of congressional intent. And, indeed, widespread support of his administration of the parks suggests that the service met its mandate much as Congress and the public wished.³

Nevertheless, extensive alterations of the parks' natural conditions occurred during Mather's tenure. The National Park Service did not follow a policy of noninterference with nature, nor did it limit its operations to park protection activities such as patrolling boundaries or attempting to stop vandalism to natural features. Instead, under Mather the service manipulated nature in the parks in two fundamental ways—through development and construction to accommodate tourism and by direct interference with flora and fauna.

Stephen Mather, a wealthy Chicago businessman educated at the University of California had joined the campaign for a national parks bureau early in 1915. Polished and at ease with the rich, powerful and famous, he was also frenetically enthusiastic. His biographer referred to him as the "Eternal Freshman" who was "almost pathologically fraternal."⁴ Mather developed an intense loyalty to the national parks, and soon after passage of the Organic Act his friend, Secretary of the Interior Franklin K. Lane, appointed him as the National Park Service's first director. Fervently idealistic in his drive to preserve America's great, scenic areas. Mather devoted his energy, and much of his own money, to effective promotion of the national park cause, using finely honed business skills that had enabled him to amass personal wealth as head of a borax company with mines in the West.

As one of his principal goals, the aggressive new director sought public acceptance and support for the parks by opening them to greater tourism to increase their popularity. He asserted that with national parks, "the greatest good to the greatest number is always the most important factor determining the policy of the service."[5] In his crusade to popularize the national parks he courted their chief constituencies, the tourism industry and the traveling public. Railroad companies had for several decades campaigned to attract vacationers to fill the hotels they built in or near Yellowstone, Glacier, Mount Rainier, and other parks.[6] And the emerging fascination with automobile travel fit perfectly Mather's plans to develop the parks and thereby increase their accessibility and popularity.

Working closely with his top assistant, Horace M. Albright, who had a pervasive influence on national park policy during this period and who became the park service's second director, Mather succeeded in his transformation of the parks, which he had proclaimed in 1916 to be "greatly neglected."[7] By the time his health forced him to resign in January 1929, the parks had undergone extensive development involving virtually every type of construction needed to support tourism and park administration. Shortly after Mather's resignation, Albright summed up park development that occurred before and during Mather's tenure by reporting that the service was responsible for "1,298 miles of roads, 3,903 miles of trails. 1,623 miles of telephone and telegraph lines." Furthermore, national parks had campgrounds, museums, park office buildings, and many other facilities.[8]

In effect, during the Mather era the service came to regard national parks as being "unimpaired" as long as their development was restricted to that which supported tourism and was fitting to the natural scenery. That is, parks must be appropriately developed for tourism. With highways winding through majestic park landscapes and with rustic log and stone buildings and bridges, much of the development was designed to harmonize with the scenic beauty and thus seemed not to impair the parks. For instance, the service in the 1920s constructed rustic-style headquarters buildings in parks such as Sequoia, Grand Canyon, Glacier, and Mount Rainier. Ranger stations were built in many parks, some in a "trapper cabin" style suggestive of the fur trade era and particularly favored by Mather. And park concessionaires often erected splendid rustic structures including Sequoia's Giant Forest Lodge, the Phantom Ranch in the depths of the Grand Canyon and the spectacular Ahwahnee Hotel in the Yosemite Valley.[9]

These attractive facilities seemed to complement the parks' scenic majesty and helped draw increasing numbers of visitors each year. Yet they rejected numerous tourism-related proposals, sometimes after considerable internal debate over what constituted appropriate park development—in essence, what would impair the parks and what would not. A plan to suspend a cableway to carry visitors across the Grand Canyon, for example, had Horace Albright's support but not Mather's and was ultimately defeated. Similarly, a proposal for an elevator alongside the thirty-foot lower falls of the Yellowstone River enjoyed even less support and was never built.[10]

Although much of the construction allowed in the parks was designed to blend with the natural settings, some development necessarily consisted of less aesthetically pleasing facilities including parking areas, water storage and supply systems, electrical power plants, sewage systems, and garbage dumps. Moreover, tourism development often went beyond basic accommodations to embrace such amenities as a golf course and even a race track in Yosemite, which was used for several years in the park's "Indian Field Days" celebrations. Mather himself encouraged public golf courses in Yosemite and Yellowstone, believing that tourists would stay in the parks longer if they had more to entertain them (Yellowstone never got its golf course).[11] He also advocated that where feasible the parks should be developed for winter sports; and beginning during the last weeks of Mather's directorship Horace Albright led an aggressive campaign to host the 1932 Winter Olympics in Yosemite. Although the service lost out to Lake Placid in this effort, Yosemite initiated a winter sports carnival in 1931 using facilities developed during Mather's time, such as a toboggan run and ice rink, to attract hundreds of off-season tourists.[12]

While the service developed the parks for tourism as it deemed appropriate, it gained acclaim on the other hand by opposing commercial and industrial development proposals that did not relate to tourism and that were considered inappropriate in national parks. Mather especially feared such potential intrusions, believing they would spoil the majestic beauty and dignity of park landscapes and diminish the public's enjoyment. Perhaps most abhorrent of all the schemes for large developments such as dams, power plants, mines, irrigation projects, were plans to divert water from Yellowstone's lakes and streams to irrigate farms in Idaho and Montana, which Mather successfully fought in the 1920s.[13] In effect, intrusive proposals like those in Yellowstone were seen as potentially serious impairments that would frustrate the National Park Service's efforts to meet its basic mandate from Congress. By excluding such commercial and industrial intrusions, the service saved the parks' natural resources from great harm.

Beyond overseeing development and use, the National Park Service was required to care for the parks' flora and fauna. In managing these resources the service did not, however, attempt to maintain truly natural conditions. Rather, it sought to present to the public an idealized setting of tranquil, pastoral scenes with wild animals grazing in beautiful forests and meadows bounded by towering mountain peaks and deep canyons. This vision did not allow for violent disruptions like raging forest fires that blackened the landscape or flesh-eating predators that attacked popular game animals. Such an approach amounted to a kind of "facade management" in the parks, intended to preserve scenery, the resplendent facade of nature. And, to many, maintaining a serene, verdant paradise seemed to mean that natural conditions were being kept "unimpaired."

Also, in seeking to present a romanticized version of nature, the National Park Service from the very first saw no need for in-house scientific expertise to help manage the parks. As indicated in the 1918 "Lane Letter," the major national park policy statement of Mather's time, the service was to use scientists of other bureaus. It should "utilize their hearty cooperation to the utmost."[14] The service borrowed scientists from bureaus such as the Biological Survey, Bureau of Fisheries, Bureau of Entomology, Bureau of Plant Industry, and United States Forest Service, all of which practiced traditional utilitarian management, emphasizing resource manipulation and consumption. Mather apparently believed that the mandate to leave the parks unimpaired could be fulfilled through strategies similar to those used in tradtional land and resource management.

Indeed, under Mather the National Park Service steadily built its own landscape architecture and engineering capability to develop the parks for tourism. Thus, its willingness to rely on biologists from other bureaus to manage national park flora and fauna suggests how much greater was the service's interest in recreational tourism than in fostering innovative strategies in nature preservation.[15] With the park service committed to using the scientific expertise of other federal bureaus, natural resource management under Mather was, to a large extent, imitative rather than innovative. Nearly always intended to assure public enjoyment of the parks, the service's manipulation of nature was mainly an adjunct to its tourism management.

For example, the National Park Service accepted almost without question the policy, long established by the United States Army and civilian park superintendents, controlling predators to ensure the safety of popular game species such as elk and deer. Efforts to kill predators in the parks were also in accord with the ongoing, nationwide campaign to control carnivorous enemies of domestic livestock, as demanded by farmers and ranchers and promoted by the Biological Survey. Indeed, with extensive support from Biological Survey experts, the National Park Service practiced predator control with thoroughness, killing the species seen as detrimental to public enjoyment, from wolves, cougars, and coyotes, to bobcats, otters, martens, and foxes. At one point even pelicans were hunted in Yellowstone to protect both native and introduced trout populations. Predator control was particularly effective on some species. By the mid-1920s wolves and cougars had been virtually eradicated from several parks.[16]

Determined to keep the national parks unimpaired, the service acted as though the predators themselves were impairments, threats to be dealt with before they destroyed the peaceful natural scene it wished to maintain. Bloodthirsty predators had no place in the beautiful, pastoral parks, at least not in large numbers.

The National Park Service changed its predator control policy only gradually during the Mather era. Under pressure from the American Society of Mammalogists, the Boone and Crockett Club, the New York Zoological Society, and other organizations, the service reconsidered its policy of killing carnivores. By the mid-1920s aggressive predator control had diminished in the national parks. And by the end of Mather's administration in early 1929, the service had narrowed its predator policy to one of limited control. Only major predators were to be targeted. They were to be killed only under certain conditions and were not to be fully eradicated.[17] Although observed in varying degrees by different park superintendents, this modification of predator control was the most substantive change in natural resource management policy to occur during Mather's directorship, yet the change came only after populations of major predators had been eliminated or seriously reduced in some parks.

While the park service sought to eliminate predators, it used ranching methods to sustain other species, especially spectacular game animals. Perhaps nowhere was ranching more apparent than with management of Yellowstone's bison. Aware that bison had been near extinction in the United States, the service continued the intensive bison management program begun during the army's administration of the park. With activities centered at the Buffalo Ranch in the Lamar Valley of northeastern Yellowstone, the park's "Chief Buffalo Keeper" allowed the animals to roam the valley under the watchful eye of a herdsman during the months when forage was readily available, then rounded up and corralled the herd at Buffalo Ranch for a period of time in the winter. The confined buffalo were fed hay harvested from several hundred acres of plowed, seeded, and irrigated park land. In the corrals, the keepers separated the calves, castrating many of the young bulls to control the breeding population. Surplus bison were slaughtered for market or given to nearby Indian tribes. Moreover, the park held roundups and even occasional stampedes for the public's enjoyment.[18]

The park service, following precedents set by army and civilian superintendents, gave constant attention to other large mammals the public enjoyed such as elk, deer, and antelope. Somewhat like domestic livestock ranchers, superintendents sought to maintain proper big-game populations, with population size based upon how many animals the range would sustain without becoming overgrazed. Indeed, the service's concern for grasses in national parks focused not so much on grassland as an aspect of nature to be preserved but more on its value as rangeland, that is, as a food source to support game animals. To augment the food supply, the park service continued the army's practice of feeding hay to the animals during winter when snow-cover limited pasturage available for grazing and browsing.[19]

Fearing an excessive increase of some big game populations in the parks, however, the service obtained authority from Congress to ship "surplus" animals to state and federal preserves and to city parks and zoos across the country. By the early 1920s Yellowstone had become, in Mather's words, a "source of supply" and a "distributing center" for certain species, with the park having shipped game animals to twenty-five states, plus various locations in Canada.[20] Again adopting traditions established by its army and civilian predecessors, the service set up park zoos to ensure that tourists saw the more popular animals. Yosemite had zoos, and, in addition to its other wildlife displays, Yellowstone established a zoo at Mammoth Hot Springs, exhibiting bison, bears, coyotes, and even a badger.[21]

Perhaps of all large mammals, bears were most subjected to management practices specifically aimed at public enjoyment. Especially popular were the "bear shows" at park garbage dumps, where visitors, seated in bleachers and protected by armed rangers, could enjoy watching numerous bears feeding at close range. The shows had originated in the earliest decades of the national parks when dumps were established near newly constructed hotels. Aware that bears were probably the parks' most popular animals, the service continued the shows. But as tourism increased, so did conflicts between bears and people in campgrounds, near hotels, and along roadsides. In some instances problem bears were sent to zoos. Yet park rangers sometimes shot the most recalcitrant animals, as in Sequoia, where by the 1920s this practice had become common.[22]

The park service altered fish populations more than that of any other wildlife in the national parks. Although hunting game in the parks was illegal, the service aggressively promoted sport fishing, making it a premier national park attraction. To ensure successful fishing, hatcheries were built in Yellowstone, Glacier, Mount Rainier, Yosemite, and other parks.[23] Lakes and streams were stocked with both native and non-native fish. Some national park waters, Crater Lake for instance, had been barren of fish before army and early civilian superintendents, followed by the park service, undertook fish-planting programs. To operate the hatcheries and manage the stocking, the service relied upon the Bureau of Fisheries, an agency devoted to assisting the nation's commercial-fishing industry and sport-fishing enthusiasts. Also, game and fish commissions in such states as California, Colorado, Washington, and Oregon worked closely with the park service in fish management.[24]

In addition, the park service extensively altered the conditions of the forests under its jurisdiction. It sought to protect timbered areas from two major threats, fire and insects, of which, fire, the "Forest Fiend" as Mather called it, seemed the greater threat.[25] Mather continued forest management practices begun in the nineteenth century, and he relied heavily on United States Forest Service expertise. Even though the forest service had a fundamentally different mandate for land management, providing for harvest of a variety of resources, the park service unhesitatingly accepted the forest service policy of total fire suppression. In accord with the thinking of the time, and seeking to keep its forests green and beautiful, the park service viewed suppression of all park forest fires, including those begun by natural causes, as fully compatible with its mandate to preserve the national parks unimpaired.[26]

Moreover, the park service called upon the Bureau of Entomology to help combat forest insect infestations. In many instances insects killed trees throughout extensive areas of the national parks, affecting the scenic beauty. Spraying and other methods for reducing insect populations in the national parks were often focused on areas most important to the visiting public, including special scenic areas and road and trail corridors. Extensive spraying to control forest diseases, such as blister rust, did not get fully under way until after Mather resigned. Insect infestations also threatened vast areas of public lands adjacent to the parks. Therefore, similar to fire control, insect control frequently spread across boundaries between national parks, national forests. and other public lands, even though the mandate to leave national parks unimpaired differed substantially from the policies governing adjacent areas.[27]

As it implemented its mandate to leave the parks unimpaired and to provide for public enjoyment, the National Park Service faced only limited and infrequent protests over its management, an indication that the service was treating the parks in a manner generally satisfactory to Congress and the public. In fact, much of the park service's natural resource management seemed clearly sanctioned by the Organic Act. For instance, the act permitted the service to "control the attacks of insects or diseases" that might threaten a park's scenic features and to "provide . . . for the destruction" of animals and plants "detrimental" to park use. Moreover, the act made it clear that the parks were to be managed for the perpetual enjoyment of the public. Among other things, the National Park Service could grant "privileges, leases, and permits for the use of land for the accommodation of visitors."[28] And far from objecting to Mather's administration of the parks, Congress created new national parks and provided increased funding for development, especially near the end of Mather's tenure as director when Congress launched a ten-year, $51 million program to improve park roads and build new ones.[29]

Mather's aggressive promotion and development of the national parks reflected his belief in their great social value as sources of public recreation, health, and welfare. Upon his death in 1930, tributes poured in from Congress, conservation groups, businessmen, and friends across the country.[30] Such national respect further suggests that Mather's emphasis on developing parks and opening them to public use had met with widespread approval.

During the Mather years, objections to the service's management of nature, even as expressed by professional scientific organizations, were infrequent. Both the Ecological Society of America and the American Association for the Advancement of Science passed resolutions in the 1920s against introduction of non-native species in national parks.[31] Yet these two statements stood virtually alone and hardly represented a sustained effort to criticize the service's treatment of natural resources. The protests that built up in the mid- and late–1920s against the killing of predators were likely the most severe criticism from conservationists that national park management faced during Mather's time.

Perhaps the most penetrating critique of the park service's treatment of natural resources came from Charles Adams, a biologist at Syracuse University who had conducted research in Yellowstone. Following an examination of natural conditions in several national parks, Adams published an article in *The Scientific Monthly* in 1925 urging that park management come in line with the emerging science of ecology. He examined such programs as fire, wildlife, and fish management, concluding that the service must develop an ecological understanding of its natural resources. As Adams put it, if the service was to preserve the parks "in any adequate manner . . . there must be applied to them a knowledge of ecology."[32]

Adams referred to the "theoretical" policy of maintaining the parks as wilderness, a policy to which the service had "not adhered." The service was not meeting what he believed to be its true mandate, the preservation of natural conditions in the parks.[33] Adams's critique was important as an early effort to promote ecologically based management of the national parks and thus to interpret the Organic Act in that regard. Yet perhaps even more important was that it had little, if any, effect on national park policies. The National Park Service under Mather was strongly set on a different course.

Mather and other founders of the National Park Service have sometimes been identified as "aesthetic conservationists, concerned about preserving lands for their great scenic beauty as opposed to the "utilitarian conservationists," exemplified by Gifford Pinchot and the United States Forest Service, which sought sustained consumptive use of natural resources.[34] Certainly through its determined efforts to preserve the scenic facade of nature, the park service under Mather focused on aesthetic conservation. But as practiced during the early decades of the National Park Service, the nurturing of forests and certain game species that contributed most to public enjoyment had a strongly utilitarian cast and was, to a degree, even commodity oriented, such as with fish management. During the Mather years natural resource management was indeed practiced as a kind of ranching and farming operation in that it was intended in part to maintain the productivity and presence of favored species. Thus, just as it was virtually impossible to separate the basic idea of national parks from tourism development and economics (a connection going back to the Northern Pacific Railroad's support for the Yellowstone National Park legislation in the early 1870s), so was it also difficult to separate the treatment of specific park resources (bears, fish, and forests, for example) from the promotion of public enjoyment of the parks that fostered tourism and brought economic benefits.

The basic concept of setting lands aside as national parks, the development of the parks for tourism, and the detailed management of nature in the parks—none of these ran truly contrary to the American economic system. The establishment of national parks prevented a genuinely free-enterprise system from developing in these areas and required a sustained government role in their management. But this was done in part as a means of protecting recognized scenic values, which through tourism also had clear economic value. With the national parks, aesthetic and utilitarian conservation overlapped to a considerable degree. Frequently the differences between the two were not distinct. The national parks, in fact, represented another cooperative effort between government and private business, notably railroad, automobile and other tourism interests, to use the resources of publicly owned lands, particularly in the West.

Through the park service, the federal government collaborated with business to preserve places of great natural beauty and scientific interest, while also developing them to accommodate public enjoyment, thereby perpetuating an economic base through tourism.

Furthermore, with no precedents and no understanding of how to keep natural areas unimpaired, the newly created National Park Service could believe that it was truly preserving the parks. Especially because use and enjoyment of the parks were unmistakably intended by the Organic Act, harmonious development of public accommodations became a means of keeping the parks "unimpaired" within the essential context of public use; and during Mather's time, public use was far less intensive than it would later become.

It is also important to note that, in the context of the times, in comparison with other public and private land management policies encouraging traditional consumptive use of resources, the national parks were much more oriented toward the preservation of nature. Generally perceiving biological health in terms of attractive outward appearances, the service seemed to believe that it could fulfill what Mather called the "double mandate" for both preservation and public use. The service could preserve what it considered to be the important aspects of nature while promoting public enjoyment of the parks. For instance, the 1918 Lane Letter, the principal national park policy statement of the Mather era, embraced these two goals without any suggestion of contradiction. It asserted that the parks were to remain "absolutely unimpaired," but also stated that they were the "national playground system."³⁵

Similarly, the park service's faith in the importance of development and its compatibility with maintaining natural conditions in the parks found expression on no less than the bronze plaque honoring Stephen Mather, cast shortly after his death and placed in many national parks and monuments. The plaque's inscription noted that in laying the "foundation of the National Park Service" Mather had established the policies by which the parks were to be "*developed* and conserved unimpaired" for the benefit of future generations (emphasis added).³⁶ This assertion—in effect restoring the Organic Act's principal mandate and affirming the belief that developed parks could remain unimpaired—would epitomize park service rationale and rhetoric from Mather's time until at least the end of the first half century of the service's history. By that time (the mid-1960s) increased postwar tourism and a rising concern for ecology would have revealed much more clearly the inherent conflicts between park development and preservation.

Biologist Charles Adams recognized in 1925 that the United States Forest Service had gotten under way with the advantage of a forestry profession already developed in Europe in the late-nineteenth century. On the other hand, he believed that the national parks were a "distinctly American idea," with European precedents for parks being limited to "formal park design rather than large wild parks," such as those in the United States. Adams noted also that there had been "no adequate recognition" that "these wild parks call for a new profession, far removed indeed from that of the training needed for the formal city park or that of the conventional training of the forester."[37] In effect, America's national parks required more than "facade management"—more than the conventional landscape architecture and game and forestry practices applied by the National Park Service.

Not until the 1930s would a small group of wildlife biologists within the service begin to shift park management toward ecologically sound practices. Yet the shift would be erratic and very slow. At all times it would have to contend with the emphasis on recreational tourism that Mather firmly established and that ever since has remained the most influential factor in national park management.[38]

Indeed, the National Park Service during the Mather era excelled in park development, building on precedents it found in landscape design and in tourism and recreation management to make the parks enormously inviting. And, although operating under a unique and farsighted mandate to keep the parks unimpaired, the newly established bureau relied on precedents of traditional forest, game, and fish management. The park service practiced a selective kind of preservation, promoting some elements of nature, opposing others—altering natural conditions largely in an attempt to serve the other part of its mandate, the public's enjoyment of the parks.

NOTES

1. Hillary A. Tolson, *Laws Relating to the National Park Service, the National Parks and Monuments* (Washington, D.C.: National Park Service, 1933), 10.

2. Historians of the national parks take differing views of the Organic Act's primary purpose. John Lemons and Dean Stout argue in "A Reinterpretation of National Park Legislation," *Environmental Law* (Northwestern School of Law of Lewis and Clark College) 15 (Fall 1984): 53, 65, that the Organic Act's primary mandate is to preserve park resources: "the purpose of natural parks is to preserve pristine ecological processes" (p. 53); and that "the most basic fiduciary duties of the [National Park Service] are to reduce development and promote preservation of resources" (p. 65). Similarly, Robert B. Keiter, "National Park Protection: Putting the Organic Act to Work," in *Our Common Lands: Defending the National Parks,* ed. David J. Simon (Washington, D.C.: Island Press, 1988), 81, argues that the Organic Act "sets forth an impressive, unambiguous resource preservation mandate." And Alston Chase, in *Playing God in Yellowstone: The Destruction of America's First National Park* (Boston: Atlantic Monthly Press, 1986), 6, comments that the National Park Service's "sole mission is preservation." By contrast Alfred Runte, in *National Parks: The American Experience,* 2nd ed. rev. (Lincoln: University of Nebraska Press, 1987), 82–105, emphasizes the parks' tourism and economic potential as key motivating factors for the legislation establishing the National Park Service.

3. Although Mather's appointment as director became official in April 1917 when the park service was formally organized, as assistant secretary of the interior he was in effect already in charge of the parks when the Organic Act was passed in August 1916. On the meaning, interpretation, and implementation of the Organic Act to date see Richard West Sellars, "The Roots of National Park Management: Evolving Perceptions of the Park Service's Mandate," *Journal of Forestry* 90 (January 1992): 16–19; and Richard West Sellars "Science or Scenery? A Conflict of Values in the National Parks," *Wilderness* 52 (Summer 1989): 29–38.

4. Robert Shankland, *Steve Mather of the National Parks* (New York: Alfred A. Knopf, 1951), 20, 36.

5. Stephen T. Mather, "The Ideals and Policy of the National Park Service Particularly in Relation to Yosemite National Park," in *Handbook of Yosemite National Park: A Compendium of Articles on the Yosemite Region by the Leading Scientific Authorities*, ed. Ansel Hall (New York: B.P. Putnam's Sons, 1921), 80.

6. Alfred Runte, in *Trains of Discovery: Western Railroads and the National Parks*, rev. ed. (Niwot, CO: Roberts Rinehart, 1990), 9–61, gives an extended discussion of railroad interest in national parks.

7. See quote in *Annual Report of the Superintendent of National Parks for the Fiscal Year Ended June 30, 1916* (Washington, D.C.: Government Printing Office, 1916), 1.

8. Horace M. Albright to Ray Lyman Wilbur, 5 March 1929, entry 6, Record Group 79 (hereafter RG), National Archives, Washington, D.C. (hereafter NA).

9. William C. Tweed and Laura Soulliere Harrison, "Rustic Architecture and the National Parks: The History of a Design Ethic," typeset manuscript, 1987, chap. 3, pp. 2–12, chap. 4. pp. 6–9, copy courtesy of authors; Ned J. Burns, *Field Manual for Museums* (Washington, D.C.: Government Printing Office, [1941]), 6–10.

10. Shankland, *Steve Mather*, 207–8.

11. Alfred Runte, *Yosemite, the Embattled Wilderness* (Lincoln: University of Nebraska Press, 1990), 144, 156. On Mather's interest in golf courses see Stephen T. Mather to E.O. McCormick, 16 December 1920, entry 6, RG 79, NA; and John Ise, *Our National Park Policy: A Critical History* (Baltimore: Johns Hopkins Press, 1961), 198. Neither golf course was built. However, when the service acquired lands in the Wawona area of Yosemite, a golf course was already there and remains in use today. For a time, Yellowstone employees golfed on an abandoned army rifle range.

12. On Horace M. Albright's support for the Winter Olympics in Yosemite, see Horace M. Albright to James V. Lloyd, 13 February 1929, entry 17, RG 79, NA. See also Runte, *Yosemite*, 152–53.

13. On the Yellowstone dam proposals see *Report of the Director of the National Park Service to the Secretary of the Interior for the Fiscal Year Ended June 30, 1920* (Washington, D.C.: Government Printing Office, 1920), 21–30. See also Ise, *Our National Park Policy*, 307–16; and Horace M. Albright, as told to Robert Cahn, *Birth of the National Park Service: The Founding Years, 1913–1933* (Salt Lake City: Howe Brothers, 1985), 100–102, 105–7, and 113–14. Schemes to tap the waters of Yellowstone Lake would recur in the 1930s, but were unsuccessful. See Aubrey L. Haines, *The Yellowstone Story: A History of Our First National*

Park, 2 vols. (Yellowstone National Park, WY: Yellowstone Library and Museum Association in cooperation with Colorado Associate University Press, 1977), 2:344–46.

14. Albright actually drafted this policy statement, but it became known as the "Lane Letter" after Secretary of the Interior Franklin K. Lane issued it under his signature. Franklin K. Lane to Stephen T. Mather, 13 May 1918, entry 17, RG 79, NA. The statement is reproduced almost in entirety in Albright and Cahn, *Birth of the National Park Service,* 69–73.

15. Professional engineers and landscape engineers (later landscape architects) made up two of four divisions in the service's first formal organizational chart approved in July 1919. Russ Olsen, *Administrative History: Organizational Structures of the National Park Service, 1917–1985* ([Washington, D.C.: National Park Service], 1985), 36–37. By contrast, wildlife biologists had no voice in national park management until after Mather's career ended. Even then, the wildlife biology program was privately funded at first. Lowell Sumner, "Biological Research and Management in the National Park Service: A History," *The George Wright Forum* 3 (Autumn 1983): 5–7.

16. Victor H. Cahalane, "The Evolution of Predator Control Policy in the National Parks," *Journal of Wildlife Management* 3 (July 1939): 230–31; National Park Service, "Policy on Predators and Notes on Predators," 1939, typescript, various pagination, Central Classified File 715 (hereafter CCF), RH 79, NA.

17. National Park Service, "Policy of Predators and Notes on Predators," CCF 715, RG 79; Albright to Wilbur, 5 March 1929; Cahalne, "Evolution of Predator Control Policy," 235.

18. Curtis K. Skinner et al., "History of the Bison in Yellowstone Park," [with supplements], 1952, typescript, 6–10, Yellowstone National Park Archives, Mammoth, Wyoming; National Park Service, "The Bison in Yellowstone National Park," 13 November 1943, entry 19, RG 79, NA; Margaret Mary Meagher, *The Bison of Yellowstone National Park,* National Park Service Scientific Monograph Series, no. 1 (Washington, D.C.: U.S. Department of the Interior, National Park Service, 1973), 10–27.

19. On the park service's management of large grazing and browsing animals in Yellowstone see Don Despain, Douglas Houston, Mary Meagher, and Paul Schullery, *Wildlife in Transition: Man and Nature on Yellowstone's Northern Range* (Boulder, CO: Roberts Rinehart, 1986), 14–57, 72–110. On elk management in Yellowstone see Douglas B. Houston, *The Northern Yellowstone Elk: Ecology and Management* (New York: Macmillan, 1982).

20. "The Yellowstone as a Source of Supply," in *Report of the Director of the National Park Service to the Secretary of the Interior for the Fiscal Year Ended June 30, 1921 and the Travel Season 1921* (Washington, D.C.: Government Printing Office, 1921), 38; Report of the Director of the National Park Service to the Secretary of the Interior for the Fiscal Year Ended June 30, 1923 and the Travel Season, 1923 (Washington, D.C.: Government Printing Office, 1923), 24.

21. Shankland, *Steve Mather,* 269; *Report of the Director of the National Park Service, 1921,* 39; Runte, *Yosemite,* 130–34; Skinner, "History of the Bison in Yellowstone Park"; Horace M. Albright, "Our National Parks as Wildlife Sanctuaries," *American Forests and Forest Life* 35 (August 1929): 507.

22. Paul Schullery, *The Bears of Yellowstone,* rev. ed. Boulder, CO: Roberts Rinehart/National Park Foundation, 1986), 81–99; Runte, *Yosemite,* 136–40; Larry M. Dilsaver and

William C. Tweed, *Challenge of the Big Trees: A Resource History of Sequoia and Kings Canyon National Parks* (Three Rivers, CA: Sequoia Natural History Association, 1990), 145–46.

23. *Report of the Director of the National Park Service to the Secretary of the Interior for the Fiscal Year Ended June 30, 1922 and the Travel Season 1922* (Washington, D.C.: Government Printing Office, 1922), 39; Report of the Director of the National Park Service to the Secretary of the Interior for the Fiscal Year Ended June 30, 1925 and the Travel Season 1925 (Washington, D.C.: Government Printing Office, 1925), 6–7.

24. David H. Madsen, "Report of Fish Cultural Activities," 5 April 1935, typescript, CCF 714, RG 79, NA; *Report of the Director of the National Park Service to the Secretary of the Interior for the Fiscal Year Ended June 30, 1929 and the Travel Season, 1929* (Washington, D.C.: Government Printing Office, 1929), 26–27.

25. Mather, "Ideals and Policy of the National Park Service," 79.

26. Ansel F. Hall to the director, 29 October 1928, containing "A Forestry Policy for the National Parks," entry 17, RG 79, NA; John D. Coffman, "John D. Coffman and His Contribution to Forestry in the National Park Service," typescript, n.d., 34–35, National Park Service Archives, Harpers Ferry, West Virginia; Stephen J. Pyne, *Fire in America: A Cultural History of Wildland and Rural Fire* (Princeton: Princeton University Press, 1982), 296–97.

27. *Annual Report of the Director of the National Park Service, 1925,* 7; Ansel F. Hall, "Minutes of Meeting of the Regional Forest Protection Board, San Francisco, California," 16 February 1928, entry 17, RG 79, NA. The policy of concentrating on areas important to the public was re-emphasized in the service's 1931 forest policy statement that "remote areas of no special scenic value and not of high fire hazard, little used or seen by the public . . . may be omitted from insect control plans. . . . " National Park Service, "A Forestry Policy for the National Parks," typescript, 1931, p.7, entry 18, RG 79, NA.

28. Tolson, Laws Relating to the National Park Service, 10.

29. Annual Report of the Secretary of the Interior for the Fiscal Year Ended June 30, 1928 (Washington, D.C.: Government Printing Office, 1928), 169.

30. Shankland, *Steve Mather,* 287–91.

31. For the Ecological Society's resolution and the park service's response see Charles C. Adams, "Ecological Conditions in National Forests and in National Parks," *Scientific Monthly* 20 (June 1925): 569–70. For the association's statement see "A Resolution on the National Parks Policy of the United States," *Science* 58 (January 29, 1926): 115. See also R. Gerald Wright, *Wildlife Research and Management in the National Parks* (Urbana: University of Illinois Press, 1992), 37.

32. Adams, "Ecological Conditions," 563.

33. Ibid., 584.

34. These differing concepts are discussed in Samuel P. Hays, *Conservation Movement, 1890–1920* (Cambridge: Harvard University Press, 1959), 189–98; and Roderick Nash, *Wilderness and the American Mind,* 3rd ed. (New Haven: Yale University Press, 1982), 129, 135–39, 180–81.

35. Lane to Mather, 13 May 1918—the "Lane Letter."

36. The inscription is quoted in Shankland, *Steve Mather,* 291. Additional castings of this

plaque were placed in many other units of the national park system in 1991 to commemorate the park service's seventy-fifth anniversary.

37. Adams, "Ecological Conditions," 567–68; 589–90.

38. On the 1930s park service science program, see Richard West Sellars, "The Rise and Decline of Ecological Attitudes in National Park Management, 1929–1940," forthcoming in *The George Wright Forum* 16 (1993).

SECTION IV

The urbanized nature of Santa Fe, the virtual disappearance of commercial agriculture in the area and the rise of public utilities as the most common source of water tends to trivialize important [community-based] water rights. I believe that the preservation of these water rights is important to the vitality of a culture over three centuries old. The people, the land and the water are inextricably bound together and will be until Santa Fe is entirely paved over. It is this culture which is our greatest pride and not without considerable value, though not measurable directly in dollars.

U.S. DISTRICT JUDGE ENCINIAS, *from his decision of* Anaya v. Public Service Company of New Mexico *(1990)*

One of the most difficult problems facing westerners is who controls the region's resources and how does this control affect its society and environment. Have minorities, especially Hispanics, Indian peoples, and African Americans found themselves in worsening environments? How has the influential political and economic pressure of middle-class environmentalism shaped resource and pollution issues? Do scientists and environmental activists have the answers for solving mounting environmental problems related to resource exploitation and degradation? The answers to these historical questions are not simple, yet more clearly than ever before, the solutions involve solving more than soil-erosion, timber-cutting, wilderness-preservation, air-pollution, and water problems. Issues of race, gender and culture are inextricably a part of environmental problems in the West, and to address any of them well means addressing all of them.

The consumer middle class has played a defining role in shaping modern environmentalism, so Samuel Hays, Professor Emeritus at the University of Pittsburgh, stresses in his *Beauty, Health, and Permanence*. This class played an exceptionally important role in the preservation of Echo Park in Dinosaur National Monument during the 1950s. Mark W. T. Harvey, a historian at North Dakota State University, surveys the clash between a post–World War II middle-class America that believed in environmental or wilderness preservation and an older middle-class whose values defined the earlier conservation movement. Harvey argues that the differences between these two generations "lies at the birth of the modern American West," and that they "established a major theme of environmental conflict in the West for the next two generations." How did the post–World War II preservationists differ from the earlier conservationists? Were the preservationists able to stop the construction of Echo Park Dam singlehandedly; and if not, then how did they accomplish their goal? What role did scientists, especially hydrologists, play in preserving the monument? In what manner did social change work as a force shaping the environment of Dinosaur Monument?

Daniel Pope, a historian at the University of Oregon, echoes the findings of Hays in his piece on the nuclear power protest in Eugene, Oregon, in the early 1970s. Other factors besides a middle class bent on reform, Pope argues, entered the picture. Global, regional, and local economic trends affected voter behavior alongside their desires for a clean environment. What groups do the activists like Jane Novick, Chris Attneave, and Joseph Holaday represent? How would one characterize these nuclear opponents by race, education, and income? What political action groups came to support Novick, Attneave, and Holaday? Why were minorities not active in opposing nuclear energy in Eugene?

Racism has always figured into the development of natural resources in the West, and by implication, the manner in which environments are formed. An old adage in the West goes: "Water flows uphill to money." F. Lee Brown, an economist at the University of New Mexico, and Helen M. Ingram, a professor at the University of California, analyze in their article how "water runs from the poor and powerless." They make clear how Hispanics and Indian peoples in the Southwest place different cultural values on water than do Anglo-Americans in the same region. In a sense, Hispanics and Indian peoples share a topophilia with water, whereas most Anglo-Americans do not. How do Hispanics and Indian peoples address water differently than do Anglo-Americans? What solutions do Ingram and Brown propose to address Hispanic and Indian peoples' difficulties in determining their own fates and the control of water?

Environmental racism is a new issue on the agendas of many environmental organizations. At the heart of environmental racism is the fact that minorities often lack the resources and the knowledge to address the environmental problems confronting them. Moreover, affluent environmentalists, frequently accused of having only their own interests at heart, have often ignored the plight of minorities and their squalid landscapes. In response, several groups, including the Sierra Club, have recently worked to confront environmental racism, and have provided opportunities for leadership to minorities.

Mary Pardo, a graduate student in sociology at UCLA when she wrote her article, describes how Hispanic mothers in East Los Angeles opposed public policies threatening their environment. The women effectively challenged detrimental policies without any help from mainstream environmental groups. Clearly, the issues motivating middle-class Americans to fight for clean air and water are just as important to minorities and the poor. What were the effective tactics used by the women of East L.A. to achieve their goals? How did they derive support from within their communities and from their families? Are there any clues in Pardo's essay that might address why the mothers of East L.A. might not have the same interests as the environmentalists who belong to the Sierra Club?

Many contemporary ecologists emphasize the holistic working of environments. Everything is connected to everything else: culture, plants, animals, air, water, physical forces—all connected. John Opie, an environmental historian at the New Jersey Institute of Technology and one of the founders of the American Society of Environmental History, accepts holism in his discussion of greenhouse warming, drought, center-pivot irrigation, and agricultural production on the High Plains. He also explores how some elements of chaos theory define the limits of people's adaptations to any historical environment. He criticizes how modern economics and resource laws have encouraged an unsustainable form of agriculture. In what way does Opie believe environmental degradation goes beyond simple economic accounting? In what manner have Anglo-American adaptation practices been ill suited to surviving on the High Plains? How does Opie believe Americans should address the historical problems facing farmers?

CHAPTER TEN

BATTLE FOR WILDERNESS
Echo Park Dam and the Birth of the
Modern Wilderness Movement

MARK W.T. HARVEY

The best-known conflict over public lands and scenic preserves in the United States in the mid-1950s centered on a proposal by the Bureau of Reclamation to build a large dam near Echo Park inside Dinosaur National Monument. First suggested by the bureau in 1946, the dam threatened the remote and relatively unknown section of the national park system that spans the border between Utah and Colorado. The bureau wanted to build this dam on the Green River in a high-walled canyon near the center of the monument. Echo Park, a picturesque valley just upstream from the dam site, along with miles of nearby Lodore Canyon and Yampa Canyon, would have been buried beneath several hundred feet of water.

Beginning in 1949 wilderness enthusiasts and conservation organizations joined forces to defend the scenic preserve by mounting a national campaign against the dam. Determined to prevent the magnificent scenery from being flooded, they sought to reaffirm one of the founding principles of the national park system: scenery in the national parks should be left unimpaired for future generations. Led by Howard Zahniser of the Wilderness Society, Ira Gabrielson of the Emergency Committee on Natural Resources, Fred Packard and Sigurd Olson of the National Parks Association, and David Brower of the Sierra Club, conservationists badgered Congress to find an alternate site for the dam. Finally in 1956 Congress agreed. Echo Park Dam would not be built. By forcing the Bureau of Reclamation to back down, conservationists secured a major victory on behalf of the national park system and wilderness preservation. As Roderick Nash and other historians have recognized, the defeat of

Echo Park Dam proved to be a milestone for the fledgling wilderness movement and a critically important episode prior to the Wilderness Act of 1964.[1]

While historians have been aware of the significance of the Echo Park battle for some time, they have not examined it thoroughly in the context of the wilderness movement, national park history, or the history of the American West after the Second World War. In the broadest sense, the controversy over Echo Park revealed a generational clash in American conservation and environmental history, a clash that became a prominent theme in the West in the postwar era.

Earlier in the twentieth century, federal conservation agencies like the Bureau of Reclamation, Forest Service, and Bureau of Land Management had become deeply enmeshed in the economy of westerly states, setting the stage for a series of sharp conflicts with preservationists who became increasingly more powerful and effective after 1950. The Echo Park controversy was the first of these conflicts in the postwar era, an early sign that the West would be a constant battleground for two distinct generations, those who came before World War II and those who came after the war, in the nation's environmental history.

The Bureau of Reclamation was deeply rooted in the Department of the Interior and was part of the bedrock of the nation's conservation establishment. Created in 1902, the agency had been a keystone in the Progressive Era conservation movement. By building dams, canals, and irrigation and power facilities, the Bureau of Reclamation sought to make efficient use of the West's water resources. By the end of World War II, the agency had a substantial presence in the region with offices in Billings, Denver, Salt Lake City, and other cities. Its dams and power plants were part of the fabric of many western state economies. Backed by steadily increased appropriations from Congress, the bureau expanded across the West throughout the 1940s with major projects along the Columbia River, the tributaries of the Missouri River, and the Sacramento and San Joaquin Rivers. The agency also had a major presence along the lower stretches of the Colorado River, where it had erected giant Hoover Dam to control flooding and generate hydroelectric power for the Southwest; Parker Dam, connecting the Colorado River Aqueduct to greater Los Angeles; and the All American Canal, which transported Colorado River water west to the Imperial Valley.[2]

With the lower Colorado basin taking the lion's share of the great river, states upstream became increasingly restless and eager to use their own share of the water, to which they were entitled under the Colorado River Compact of 1922.[3] World War II brought increased population growth, federal spending, and industrial development to states in the upper Colorado basin, including Wyoming, New Mexico, Colorado, and Utah. These states now demanded that the bureau provide dams and power plants to encourage their growth.

The bureau responded with the proposed Colorado River Storage Project, a series of dams and power plants to be constructed along the upper Colorado River and its tributaries. As part of this project (known as CRSP), large storage and power dams were slated for Glen Canyon, south of the Utah–Arizona border, and at Echo Park, just below the junction of the Green and Yampa Rivers within Dinosaur National Monument. Although Glen Canyon Dam was to provide substantially more water storage and power, Echo Park Dam took on special importance because of its function in capturing Yampa River water for Utah and because of its role in the Central Utah Project, a part of CRSP.[4]

The Bureau of Reclamation announced plans for CRSP with a two-year flourish of publicity beginning in 1947. Joined by the Utah Water and Power Board and state water engineers from Utah's upper basin neighbors, bureau officials circulated throughout the region, touting the merits of CRSP for ranchers, farmers, rural electrical cooperatives, and urban dwellers. Many residents close to Dinosaur National Monument anticipated a business boom with the construction of Echo Park Dam and a further boost to their economy from tourists visiting the reservoir.[5] Newspapers such as the *Denver Post*, *Salt Lake Tribune*, and *Deseret News* heralded the dam and CRSP regularly. Soon Echo Park Dam became a much proclaimed instrument of economic growth and diversification, a symbol of progress and prosperity to one of the West's emerging subregions.[6]

While most Americans in the early Cold War era believed in dams and the numerous benefits they provided, some people were dubious about the bureau's concept of managed rivers. World War II and the postwar period proved to be a watershed in the evolution of popular attitudes toward nature. Increased income, rising living standards, and higher levels of education all worked to usher in a postindustrial society and an "advanced consumer economy." These trends contributed to a shift in public attitudes toward preservation of the environment. With the middle class less worried about fulfilling basic living needs and having greater amounts of leisure time, this growing segment of the populace expressed a desire to enjoy what one historian has called "amenities" of life, including vacations to parks, forests, and wilderness areas.[8]

These changes underpinned the growth and activism of the postwar generation of preservationists, represented by such groups as the National Parks Association (NPA), Wilderness Society, and Sierra Club. Keen to promote their agendas in the western states, preservationists sought to take advantage of the great travel boom that followed the war, when millions, ready to celebrate the nation's victory with a long-delayed vacation, took to the highways. Americans now could afford to travel to places distant from their homes. National forests and especially national parks were primary destinations, and Americans descended on them in record numbers.[9]

National parks and monuments in the West took on special meaning to preservationist groups. The NPA had special regard for the large scenic parks in the West, partly out of fear that the National Park Service was lavishing funds on historic sites, presidential birthplaces, cemeteries, and battlefields in the eastern United States, many of which had been absorbed into the park system in the 1930s. With the NPA anxious that the agency now favored "cannonball parks," *National Parks* magazine began to highlight "primeval parks" in the West to distinguish them from the gamut of eastern sites. The NPA lauded the scenic qualities of Mount Rainier National Primeval Park, Glacier National Primeval Park, and other scenic parks, primarily in the West.[10] "Primeval parks" demonstrated that Americans could protect the most significant portions of the country as samples of the "original" North American landscape.[11] The NPA's focus on "primeval parks" coincided with the growing activism of the Wilderness Society in the American West. In 1947 the Society first announced plans to begin a campaign for creating a national wilderness system, which fueled its desire to maintain primitive areas in national forests as well as safeguard the western scenic parks from private interests that sought access to them.[12]

Such rising hopes of the NPA, Wilderness Society, and other groups quickly faded in the postwar years, as logging, mining, and grazing escalated on public lands. From the mid- to late–1940s, preservationists witnessed numerous threats to Forest Service primitive areas and to several national parks and monuments in the West. Timber farms, ranchers, and other private interests sought access to the resources within Olympic and Kings Canyon National Parks as well as to Jackson Hole National Monument.[13]

From the preservationists point of view, an even more dangerous threat came from the Army Corps of Engineers and Bureau of Reclamation, federal dam-building agencies that introduced plans to dam rivers in several national parks in the West. The Corps of Engineers, responding to heavy flooding along the Columbia River, considered a dam along the north fork of Montana's Flathead River, an upstream tributary that formed Glacier National Park's western boundary. Water backed up by Glacier View Dam, according to a special commission studying the national parks, threatened to inundate "20,000 acres of the most primitive portion" of the park, including winter range for elk and white-tailed deer. After much wrangling, the corps agreed to a dam at Hungry

Horse, outside Glacier Park, but similar proposals by the Bureau of Reclamation followed. In 1948 the bureau indicated an interest in building dams on the Colorado River within the Grand Canyon, threatening both Grand Canyon National Park and Grand Canyon National Monument.[14]

On the heels of these proposals the bureau announced the Colorado River Storage Project, with dams slated along the Green River below Echo Park and at Split Mountain Canyon in Dinosaur National Monument. Conservationists regarded the proposal as a grave threat to the national park system. Echo Park Dam would submerge the lower portions of the Green and Yampa Rivers for some forty miles upstream from their confluence and, they contended, the heart of Dinosaur Monument would be transformed from a "primeval" wilderness into an artificial lake. Fred Packard of the NPA wrote that "many of the outstanding geological and scenic features of the monument, including Pat's Hole, Echo Park, Castle Park, Harding's Hole, and the famous Canyon of Lodore would be destroyed."[15]

The NPA asserted that the dam would establish a precedent for using natural resources in national parks for commercial purposes and thus provide an opening wedge into other national parks and monuments. The association warned that the Echo Park and Split Mountain Dams must be deleted from CRSP or the beauty of Dinosaur Monument and the foundation of the national park system would be lost.

When taken alongside other plans for dams in the Grand Canyon, Glacier, Kings Canyon, and Mammoth Cave National Parks, preservationists viewed the plan for Dinosaur as obstreperous, overbearing, and insensitive to a tradition of preservation as embodied in the national park system. At the core of their concerns was the belief that national parks were supposedly the best-protected type of public land preserve. They sprang from the National Park Service Act of 1916, which stipulated that the parks be conserved "unimpaired" for future generations.[16] National parks had been dedicated for preservation. If they were unsafe from intrusion, there could be no further advances in the preservation of public lands. Howard Zahniser, a leader in the Wilderness Society, wrote that the Echo Park controversy centered on "the sanctity of dedicated areas."[17]

Mounting pressures to use resources in the national parks and monuments greatly influenced National Park Service Director Newton Drury, who stood at the center of the Echo Park controversy from 1949 until early 1951.[18] Drury fiercely disputed the proposed dam with Bureau of Reclamation Commissioner Michael Straus. At the center of their argument was the role of the National Park Service in managing "national recreation areas," such as reservoirs behind the bureau's dams. The park service had taken administrative control over these areas, including Lake Mead behind Hoover Dam, in the 1930s, and Drury had further expanded that role in the war years, in part because the agency wished to aid local economies.[19]

After the war, the Wilderness Society, NPA, and other preservationists made clear to Drury their distaste for the agency's management of reservoirs. They considered such a task incompatible with the agency's responsibility to preserve and protect the "primeval parks."[20] Drury now found himself pressured to combat the Echo Park proposal, and he did so in several letters to Commissioner Skaus in the Department of the Interior in Washington, D.C. Their hostile exchange soon came to the attention of Secretary of the Interior Oscar Chapman.[21]

Chapman sympathized with the National Park Service, but his primary loyalties lay with the Democratic Party and with the Truman administration's interest in expanding federal management of western rivers. Furthermore, Chapman was under pressure from the Atomic Energy Commission to help supply electrical power reserves in Utah and the Mountain West, reserves needed to aid in the testing of the atomic bomb. These factors weighed heavily in his decision, announced in June 1950, to support the bureau's dam below Echo Park.[22]

Infuriated conservationists told Chapman he had taken a step to encourage "the raiding of the national park system."[23] Drury, for his part, was not surprised with the decision. He later recalled that "the great Bureau of Reclamation was . . . like the state of Prussia in the German empire, where everything was weighted in its favor." As for Chapman, Drury told an interviewer long afterward that the secretary was "very much in the position of the mahout who rides the elephant and thinks he's guiding it but is really being carried along. That wasn't true of men like [Harold] Ickes, but it surely was true of Chapman as Secretary of the Interior."[24] The director soon became completely alienated from Secretary Chapman, and under pressure from the latter, resigned his position on 1 April 1951. Drury's ouster provoked another outcry from conservationists, who were now galvanized against the Echo Park Dam and determined to protect Dinosaur National Monument.[25]

Bernard DeVoto launched the conservationists' national campaign against the dam in July 1950, with a hard-hitting article in the popular *Saturday Evening Post*. A native of Utah and a historian and writer based in Cambridge, Massachusetts, DeVoto had recently been converted to the conservation cause. His article—"Shall We Let Them Ruin Our National Parks?"—cast the showdown over Echo Park in dramatic terms and brought the controversy into living rooms across the country. Should the dam be constructed, he warned,

"Echo Park and its magnificent rock formations would be submerged. Dinosaur National Monument as a scenic spectacle would cease to exist."[26]

During the next fire years, DeVoto, along with leaders of the NPA, Wilderness Society, Sierra Club, Izaak Walton League, and other conservationists, denounced the bureau's plan as a threat to the integrity of the national park system. If the dam were permitted, it would not only severely intrude into Dinosaur Monument, it would greatly weaken barriers safeguarding all national parks and monuments. Any weakening of the park system, so they argued, would hamper efforts to establish a national wilderness system.[27]

Led by U.S. Grant III, a retired engineer from the Army Corps of Engineers, preservationists suggested that dams be built at alternate sites outside of Dinosaur. They argued that such sites were equally efficient if not more efficient for capturing water and producing power.[28] Grant, along with David Brower and Richard Bradley of the Sierra Club, also questioned the bureau's argument that inclusion of the Echo Park Dam in CRSP would minimize evaporation from the big reservoirs. Through careful scrutiny of the methods of calculating evaporation, they forced the bureau to concede that the amount of water it hoped to save by erecting the dam was not as great as it had originally deduced and that alternate sites could be used. As one possible alternative, conservationists suggested increasing the height of Glen Canyon Dam, a suggestion that later proved embarrassing to their side.[29]

Besides questioning the bureau's arrangement of dams and its evaporative statistics, preservationists found that publicizing the beauties of canyons along the Green and Yampa Rivers in Dinosaur National Monument was their most effective strategy. With DeVoto's article as inspiration, others produced essays and illustrations about Echo Park for the *Sierra Club Bulletin, Living Wilderness, National Parks* magazine, and other publications. David Perlman of the *San Francisco Chronicle* joined a Sierra Club float trip in 1954 before writing several articles that testified to the historic and scenic resources that would be flooded. Editorials against the dam appeared in the *Washington Post, Milwaukee Journal, San Francisco Chronicle,* and *New York Times.*[30]

In the summer of 1952 Harold Bradley and two of his sons floated the Yampa River to Echo Park. Long-time wilderness activists and members of the Sierra Club, the Bradleys produced a short film highlighting the unique beauties of the rivers and high-walled canyons. "The experience," Harold Bradley later wrote, of "threading our was through this superb gallery of matchless pictures displayed in ever-changing vistas, left us aghast at the thought that Bureau of Reclamation engineers are calmly planning the destruction of the Monument."[31] Bradley's film offered scenes of his family running their kayaks through rapids, eating lunch on quiet beaches, and gazing up at massive cliffs of Weber sandstone. The movie proved vital to the Echo Park campaign. It revealed the excitement of a new sport to a coming generation eager to experience the wilderness.[32]

The film sparked great interest within the Sierra Club. Club members, enchanted with scenes of swift rivers and sheer rock walls, discovered an entirely new kind of wilderness—strikingly different from the Sierra Nevada—but to them equally compelling and deserving of greater exploration. The Yampa and Green Rivers and the canyons of Utah and Colorado offered a unique landscape of white, pink, and red cliffs, rushing white water, and desert vegetation. Bradley's film awakened Sierra Club members to a region unknown to most of them and to the majority of wilderness lovers across the country.[33]

His film also helped to inspire Sierra Club river trips through Dinosaur during the following summers. Bradley knew that floating the rivers would acquaint club members with breathtaking scenery and provide them with a taste of the great wilderness experience that he and his two sons had had the summer before. He also believed that successful float trips would demonstrate that river running was not a dangerous sport—clearly not as risky as the proponents of Echo Park Dam wanted everyone to think. Some two hundred club members, guided by local river runner Bus Hatch, took a float trip through the monument in the summer of 1953 without incident, helping to refute the argument that "the river run is exceedingly dangerous, while the lakes will presumably be safe."[34] Participants of the river trips considered them a great success, with Harold Bradley concluding that "practically all came out filled with wonder and enthusiasm."[35]

The river trips also gave birth to another film of Dinosaur, *Wilderness River Trail*. Shot by professional photographer Charles Eggert, it was the first film produced by the Sierra Club that featured both sound and color.[36] By the end of 1953 the Sierra Club had made copies of Eggert's film available across the county, courtesy of the American Alpine Club in New York, the Appalachian Mountain Club in Boston, the Izaak Walton League in Chicago and Denver, and the Mazamas in Portland. *Wilderness River Trail* caught the attention of thousands of conservation and wilderness enthusiasts around the country.[37]

No one enjoyed the Sierra Club float trips or felt as moved by the scenic beauty as David Brower, recently appointed the club's executive director. Brower first learned of the Dinosaur matter from Martin Litton, who wrote articles on the controversy for the *Los Angeles Times* and told Brower that he thought the Sierra Club was doing too little to fight the dams.[38] Brower had assured Litton that the club's role would change. Brower exhibited an intensity and uncompromising manner reminiscent of John Muir's during his long and ultimately unsuccessful effort to save Hetch Hetchy, the valley in Yosemite National Park inundated by a dam early in the twentieth century. Having just attained leadership in the Sierra Club, Brower's sense of Muir and the club's past loomed large in his mind. Muir's passionate appeals to save Hetch Hetchy had been drowned out by more powerful voices, and as Brower first became aware of the threat to Echo Park he seemed to hear echoes of Muir resounding in the canyons of the Green and Yampa Rivers, calling him and the club to action. Hetch Hetchy was an old and deep wound in the Sierra Club, and in Echo Park Brower saw an opportunity to make it heal.

Brower's ability to captivate the public with images and descriptions of wilderness areas turned out to be his greatest gift. Influenced by photographer Ansel Adams, Brower understood the enormous power of still pictures and films in sparking public awareness of places worthy of being preserved. His publicity skills also emerged brilliantly in a book *This Is Dinosaur*, published by Alfred A. Knopf in 1955.[39] Brower believed that while films and articles in various publications had been of great value in publicizing the monument, an illustrated book would bring the spectacular beauty of Dinosaur to a larger audience. The "primary object of the book," he wrote, "is to let the people know what's there, [and] whether it is to be a milestone or a headstone—this question is for the people and Congress to decide."[40] In Alfred A. Knopf, Brower found a publisher who had a deep love for the national parks and for conservation. Knopf had long supported the parks and decided that a volume of essays about Dinosaur would be a fitting way to demonstrate his long love affair with them.

Brower approached Wallace Stegner to edit the work. Having just published *Beyond the Hundredth Meridian*, a biography of John Wesley Powell, Stegner seemed a natural choice. Stegner had immersed himself in western history for nearly a decade while writing the book on Powell. He had written about the canyons through which Powell had gone, and, having living part of his youth in Utah, he had a feel for the Colorado River and Plateau.

Stegner agreed with Brower that the book should seek to reveal the vast size and enormous beauty of the monument and demonstrate its national significance in the park system. Stegner insisted in his preface that the authors deliberately chose not to "make this book into a fighting document" because too much "bad feeling and bad prose" had already been generated by Echo Park Dam. *This Is Dinosaur* had a different aim, to provide a survey of "the national resource [and] its possibilities for human rest and recreation and inspiration."[41]

This Is Dinosaur became the first book-length publication in conservation history that sought to publicize a park or wilderness preserve and contribute to a preservation campaign. Filled with colorful pictures of Echo Park and "its magic rivers," the book was equally impressive for its excellent essays. Knopf himself furnished an essay, insisting that "the very special purposes of recreation, education, refreshment, and inspiration for which Parks and Monuments have been set aside prohibit many economic uses which are thoroughly legitimate elsewhere."[42] To help ensure the book's effectiveness, Knopf agreed to donate a copy to each member of Congress. Brower, wanting to leave no doubt about the book's message, included a foldout brochure inside the back cover displaying a picture of mud flats at Lake Mead created by Hoover Dam. A quote from Interior Secretary Douglas McKay served as the photograph's caption: "What We Have Done At Lake Mead is What We Have In Mind For Dinosaur."[43]

The Echo Park campaign had been launched to reaffirm the nation's commitment to the national park system—to strengthen the National Park Service Act of 1916 with its mandate to conserve the parks unimpaired for future generations. But over the course of the battle the fate of the monument became inextricably tied to wilderness preservation, thanks to all the intensive publicity of Echo Park and the canyons of Dinosaur, where the spectacles of rock and white water captured the essence of "wild" land to a new generation of preservationists. *This Is Dinosaur* offered the clearest exposition of the relationship between parks and wilderness, owing largely to Stegner's eloquence. A memorable passage foreshadowed the language of Stegner's "Wilderness Letter," written in 1960, and that of the Wilderness Act of 1964. In *This Is Dinosaur* Stegner wrote:

> It is legitimate to hope that there may [be] left in Dinosaur the special kind of human mark, the special record of human passage, that distinguished man from all other species. It is rare enough among men, impossible to any other form of life. It is simply the deliberate and chosen refusal to make any marks at all. . . . We are the most dangerous species of life on the planet, and every other species, even the earth itself, has cause to fear our power to exterminate. But we are also the only species which, when it chooses to do so, will go to great effort to save what it might destroy . . . in the decades to come, it will not be only the buffalo and the trumpeter swan who need sanctuaries. Our own species is going to need them too. It needs them now.[44]

This flourish of publicity over Echo Park did not sit well with public opinion in the Upper Basin states. To most residents there, it seemed odd that an all but unknown national monument had suddenly become a powerful symbol of wilderness. Proponents of the dam sought to counter the image of a precious preserve. They asserted that the National Park Service had agreed to the dam as early as the 1930s, when the boundaries of Dinosaur had been enlarged around the Green and Yampa Rivers. Utah Senator Arthur Watkins contended that the necessary "power withdrawals" had been established by the Federal Power Commission in the 1920s.[45] These arguments clinched the case in favor of the dam in the minds of people throughout the Upper Basin. And, lingering doubts vanished once the battle reached Congress in 1954 and southern California's strident opposition to the bill became evident. Nonresidents of the Upper Basin felt sure that those who protested the Echo Park Dam mere "mainly pawns of . . . California groups seeking to hog Colorado River water."[46] On the strength of these arguments, members of Congress from the Upper Basin refused to surrender the controversial dam.

Late in 1955 they changed their minds. By October of that year, the CRSP bill had been effectively blocked in the 1954 and 1955 sessions of Congress, not only because of pressure from preservationists but from other groups as well. Farm states in other regions of the country opposed the legislation, charging that agricultural surpluses depressed commodity prices and that additional cultivated land would exacerbate the problem.[47] Some members of Congress denounced the bill for its costly irrigation projects and its questionable cost accounting.[48] California water and power interests also continued to oppose the legislation. Upper Basin lawmakers finally concluded that the bill stood no chance of passage without being changed, and the key interest group in the coalition of opponents was the preservationists. At a crucial meeting in Denver in early November, Upper Basin senators, led by Democrat Clinton P. Anderson of New Mexico, resolved to eliminate Echo Park Dam from the bill in hopes of winning approval for the rest of the CRSP project (including Glen Canyon Dam) in the following session.[49] Conservation groups, representing several million Americans, at once withdrew their opposition and swung their support to the bill, clearing the way for its passage. President Dwight Eisenhower signed the bill on 11 April 1956, with a provision that prohibited dams in any area of the national park system along the Colorado River.[50]

The successful effort to preserve Echo Park and Dinosaur National Monument proved to be a milestone in the history of the American environmental movement. By blocking the dam, the coalition of organizations resoundingly reaffirmed the sanctity of the park system and strengthened the mandate, that parks be conserved "unimpaired," contained in the National Park Service Act of 1916. While it can hardly be said that defeat of Echo Park Dam ended all threats to parks, it became substantially more difficult to offer serious proposals for dams, mines, or other intrusive structures inside park boundaries. Even in the more famous conflict over the Grand Canyon in the 1960s, the Bureau of

Reclamation did not propose to locate its dams inside the national monument or national park, but just outside their boundaries. With the Bureau thus claiming that the two dams posed no direct intrusion, the Grand Canyon controversy revolved around how the dams might alter the "living Colorado."[51] Since the 1960s threats to parks have come increasingly from sources such as oil and gas drilling companies and power plants outside the parks, rather than from intrusive developments within their boundaries. The successful outcome of the Echo Park battle fortified the barriers of national parks and monuments against such threats.

In another way, the outcome of the controversy enhanced the power of many conservation groups, who discovered a remarkable degree of strength when they united in a common cause. Some seventy-eight organizations joined the Echo Park campaign, far outnumbering the handful that aided John Muir with the Hetch Hetchy battle.[52] Numbers alone did not guarantee victory, of course. The single-interest groups had never before worked together on this scale, and they spent much time formulating strategy and coordinating efforts. Yet no one could deny the impressiveness of their coalition. Irving Brant, historian and former adviser to the Franklin D. Roosevelt administration on conservation issues, told President Harry Truman early in the controversy that it was the first time he could recall so many groups "having been stirred up to a joint campaign." Later, Brant told Adlai Stevenson that "there is no issue on which conservationists are more deeply stirred, or which has done more to galvanize them into a nationwide organization."[53] Harold Bradley recalled that the controversy unified "the conservation forces into a single, very formidable fighting force," while *New York Times* columnist John Oakes, who had done much to publicize the threatened monument in his columns, wrote that it demonstrated "what strength the conservation movement has at last achieved in the United States." Almost exactly thirty years after the end of the controversy, Oakes recalled in an interview that Echo Park had been "of vital importance in demonstrating for the future that these kinds of battles could be won."[54]

The controversy had special significance for those organizations most interested in parks and wilderness—the Sierra Club, the National Parks Association, and the Wilderness Society. Small by today's standards, with only a few thousand members each, these groups could not exert much pressure on Congress by themselves, and they had to count heavily on larger bodies like the National Wildlife Federation and Izaak Walton League. But if the larger groups provided essential political clout, the preservationists had the most to gain, for the threat to Echo Park provided them with an excellent opportunity to laud "primeval parks" and to incorporate their agenda into the broader conservation movement. The Wilderness Society gained national attention from the battle and took advantage of the triumph to resume its campaign for a national wilderness system. Echo Park offered the society a major opportunity to publicize the kind of pressures on public lands that its leaders contended would soon eliminate remaining parcels of wilderness unless they were permanently protected by Congress.

Howard Zahniser, executive secretary of the Wilderness Society, had recognized at the outset that upholding the integrity of the park system must precede establishment of a national wilderness system. Zahniser was encouraged immeasurably by the campaign's success. To him, the outpouring of public support demonstrated something that he already deeply believed: that the last remaining wilderness offered a type of outdoor experience unavailable elsewhere and that an important segment of the public was willing to help protect such lands. Zahniser, whose role in the Echo Park controversy has been for too long overshadowed by Brower's, took satisfaction that the public had spoken decisively. On the very same day that President Eisenhower signed the CRSP legislation, Zahniser sent a letter to key members of Congress asking them to sponsor a bill for a national wilderness system.

While Zahniser found encouragement, David Brower and the Sierra Club gained a new public image and an important new place in the pantheon of conservation organizations. Echo Park proved to be a key episode in the club's history, bringing an end to its traditional focus on the Sierra Nevada and launching the club from its California base into the national conservation arena. Brower developed his own national reputation from the Echo Park battle. In his first major campaign as executive director of the club, he became renowned as a bold champion of wilderness and a fierce combatant against those who threatened it—in this case, the Bureau of Reclamation. Everyone involved with the campaign considered him invaluable and understood him to be a new voice for conservation that was bound to be heard from again. More than any of his counterparts, Brower knew instinctively how to grab public attention for threatened preserves. He knew the power of pictures and films: the film *Wilderness River Trail* and the book *This Is Dinosaur*, two highly successful tools of the campaign, had been his ideas. He knew how to dramatize a sense of place, and Echo Park and the canyons of Dinosaur lent themselves to rhetorical flourish about the merits of wilderness. Brower spearheaded new tactics of wilderness propaganda. He developed filmmaking and book publishing skills that became essential weapons in the campaigns of the 1960s to establish Point Reyes National Seashore and North Cascades and Redwoods National Parks and in the battle over dams in the Grand Canyon.

Perhaps the greatest impact of the controversy was in the American West. In regional terms, the elimination of Echo Park Dam from CRSP brought to an end the first great clash between water developers and preservationists and established a major theme of environmental conflict in the West for the next two generations. For westerners involved in the battle, the memories of the great conflict remained for years to come. Residents of small towns near Dinosaur Monument, like Vernal, Utah, remained bitter after the loss of the dam, and such memories linger even today.[56] One reason they persist is that the passions of the battle have hardly disappeared in many parts of the West where resentment against "environmentalists" remains strong. The loss of Echo Park Dam caused rural westerners to be aware of a new force in their lives, a force that seemed to many of them to be ignorant of water and power needs in their region, but which nevertheless seemed able to determine their fate. Local people sensed it might be a force not easily turned back, and in their frustration some of them found a new dimension to the old colonialism that held sway in the region's past. Conservationists from the urban East and the West Coast had now become a force to be reckoned with. Echo Park clearly had thrust guardians of the public lands into prominence.

From the vantage point of those who won, Echo Park not only reversed the verdict of Hetch Hetchy, but proved to be an important victory at the dawn of a new era in the West. To the opponents of the dam in Dinosaur, the Bureau of Reclamation, with its concept of managed rivers and natural resources, had become associated with loggers, miners, and ranchers in viewing nature's bounty as a resource to be exploited. By defeating Echo Park Dam, conservation groups turned back the development-minded interests in a decisive way. The battle taught them much about the economic, political, and technical aspects of water projects in the West and instilled in them a skepticism of the Bureau of Reclamation. According to the bureau's opponents, the Echo Park affair exposed the bureau's "credibility gap," and they would carefully scrutinize bureau projects in the years ahead. By successfully challenging the bureau's tight grip on information about the Colorado River and water "development," a core of wilderness activists emerged from the conflict with an understanding of evaporation rates, the legal complexities of the Colorado River Compact, and the economics of large water and power project This body of knowledge carried over into subsequent wilderness and water controversies.[57]

Still, while gaining an understanding of the bureau proved helpful in future clashes, the bureau in many ways set the terms of the Echo Park debate and for subsequent development of the Colorado River. Despite their eventual triumph in eliminating the Echo Park Dam from the Colorado River Storage

Project, conservationists proved unable and unwilling to block approval of Glen Canyon Dam, and, much to their regret, some priceless scenery disappeared beneath its reservoir. The fate of Glen Canyon added to their bitterness toward the bureau.[58] Further controversy ensued when they accused the bureau of reneging on its agreement to protect Rainbow Bridge from Lake Powell. In the mid-1960s conservationists challenged the bureau's efforts to construct two dams in the Grand Canyon. All of this testified to the bureau's substantial power in the American West in the decades following World War II. The Echo Park controversy gave birth to an icy relationship between the bureau and wilderness champions. By the 1960s and 1970s their contention had grown into a virtual cold war.

Despite disappointments over Glen Canyon and Rainbow Bridge, the battle for Echo Park opened an exciting new region of wild and scenic wonders to many of those involved in the campaign. The controversy brought the Colorado River and Plateau into the consciousness of the Sierra Club, National Parks Association, Wilderness Society, and other groups. Here was a region largely unknown to most wilderness enthusiasts who were more familiar with forested and alpine areas like the Adirondacks, Lake of the Woods, Grand Tetons, and High Sierra. Dinosaur Monument and Glen Canyon offered a new kind of wilderness terrain, one dominated by canyons, mesas, and a vast array of geologic features carved in the earth's surface by millions of years of river flows. The Dinosaur controversy—and the battles it ignited at Glen Canyon and Grand Canyon—brought the canyons and rivers of the Southwest into the spotlight, and the region thereafter became a focal point of environmental activism in the American West.

The Echo Park affair represents the classic tale of environmental controversy in the American West. Indeed, the battle initiated a pattern of conflict between the kind of managed use of natural resources that became woven into the fabric of the West in the first part of the twentieth century and the postwar interest in wilderness and environmentalism. The struggle has played itself out across the West in the decades since the Echo Park battle. From the energy boom of the 1970s to the Sagebrush Rebellion to the spotted owl, westerners have faced the same difficult and sometimes agonizing conflict between the traditional brand of conservation, first espoused by Gifford Pinchot and Theodore Roosevelt to ensure a constant supply of water, timber, and rangelands, and the postwar interest in preserving the aesthetic qualities of public lands.

In a letter in 1984, Wallace Stegner wrote that Echo Park was "in many ways . . . a key squabble, the interests involved are the classical and unavoidable interests, the warring points of view, the endlessly repetitive exploitation v.

preservation points of view."⁵⁹ In that sense, Echo Park is a story that lies at the birth of the modern American West. It is a touchstone for a pattern of conflict that has shaped the region's history in the last half century.

NOTES

1. Roderick Nash, *Wilderness and the American Mind*, 3rd ed. (New Haven: Yale University Press, 1982), 209–19. The National Parks Association was later renamed the National Parks and Conservation Association.

2. Norris Hundley, Jr., *The Great Thirst: Californians and Water, 1770s–1990s* (Berkeley: University of California Press, 1992), 220–30.

3. Ibid., 209–13.

4. Salt Lake Tribune, 9 April 1946.

5. Ibid., 4 April 1954.

6. Ibid., 16 October 1949.

7. Samuel P. Hays, "Three Decades of Environmental Politics: The Historical Context," in *Government and Environmental Politics: Essays on Historical Developments Since World War Two*, ed. Michael J. Lacey (Washington, D.C.: Woodrow Wilson Center Press, and Baltimore: Johns Hopkins University Press, 1991), 22.

8. Samuel P. Hays, *Beauty, Health, and Permanence: Environmental Politics in the United States, 1955–1985* (New York: Cambridge University Press, 1987), 2–5.

9. "Annual Report of the National Park Service," in *Annual Report of the Secretary of the Interior for Fiscal Year Ended 30 June 1947* (Washington, D.C.: Department of the Interior, 1947), 344.

10. William P. Wharton, "The National Primeval Parks," *National Parks Bulletin* 13 (February 1937): 3–5; Alfred Runte, *National Parks: The American Experience*, 2nd ed. (Lincoln: University of Nebraska Press, 1987), 132.

11. The notion of protecting samples of the "original" landscape is sprinkled throughout many preservationists' writings in the immediate postwar era. See Aldo Leopold, "Wilderness," in *A Sand County Almanac, And Sketches Here and There* (New York: Oxford University Press, 1968), 188–92.

12. News release, 6 July 1947, folder 28, box 29, Wilderness Society Papers [hereafter Wilderness Society Papers], Western History Department, Denver Public Library, Denver, CO (hereafter DPL).

13. Weldon F. Heald, "The Squeeze Is On the National Parks," *National Parks* 24 (January–March 1950): 3–4.

14. Fred Packard, "Grand Canyon Park and Dinosaur National Monument in Danger," *National Parks* 23 (October–December 1949): 11–13; "Glacier View Dam," *National Parks* 22 (October–December 1948): 3–4; Waldo Leland and Frank Setzler to Morris L. Cooke, 31 March 1950, box 18, Records of the President's Water Resources Policy Commission, Harry S. Truman Library, Independence, Missouri [hereafter Truman Library].

15. Fred Packard, "Grand Canyon and Dinosaur," 12; Devereux Butcher, "Stop the

Dinosaur Power Grab," *National Parks* 24 (April–June 1950): 61–65; Fred Packard to Phillip Sirotkin, 26 August 1955, box 34, Wilderness Society Papers.

16. Quoted in Barry Mackintosh, *The National Parks: Shaping the System* (Washington, D.C.: Department of the Interior, 1985), 18.

17. The remark appears in the minutes of the 1956 annual meeting of the Wilderness Society Council, box 18, Wilderness Society Papers.

18. For an alternate interpretation of Newton Drury's role in the early stages of the controversy, see Susan Rhoades Neel, "Newton Drury and the Echo Park Dam Controversy," *Forest & Conservation History* 38 (April 1994): 56–66.

19. Mackintosh, *National Parks*, 52. See also Susan Rhoades Neel, "Irreconcilable Differences: Reclamation, Preservation, and the Origins of the Echo Park Dam Controversy" (Ph.D. diss., University of California, Los Angeles, 1990), 122–30.

20. Olaus Murie to Oscar Chapman, 20 January 1951, National Park Service Records (hereafter NPS Records), 1951–1953, Dinosaur National Monument, Jensen, Utah; Robert W. Righter, *Crucible for Conservation: The Creation of Grand Teton National Park* (Boulder: Colorado Associated University Press, 1982), 89–92, 127; Butcher, "Dinosaur Power Grab," 61–65.

21. Newton Drury to Oscar Chapman, 30 December 1949, NPS Records, 1945–1950.

22. Fred Packard to Phillip Sirotkin, 16 August 1955, box 34, Wilderness Society Papers; Arthur Carhart to Bernard DeVoto, 11 March 1954, box 120, Arthur Carhart Papers, CPL; Press Release, Secretary of the Interior, 27 June 1950, box 14, Bryant Stringham Papers Department of Special Collections, Utah State University, Logan, Utah.

23. Carhart to Chapman, 30 June 1950, Central Classified Files, Bureau of Reclamation [microfiche copies of these records in author's possession].

24. Drury's two quotes appear in an oral history, "Parks and Redwoods, 1919–1971," conducted from 1960 to 1970 by Amelia Fry and Susan R. Schrepfer, vol. 2, 1972, p. 52, Regional Oral History Office, Bancroft Library, University of California, Berkeley (hereafter Bancroft Library).

25. Letters from numerous conservationists protesting Drury's removal are in folder labeled "National Park Service, 1948–1953," box 54, Official File of Harry S. Truman, Truman Library; *Denver Post*, 28 February 1951.

26. *Saturday Evening Post*, 22 July 1950, p. 42.

27. Howard Zahniser to John Saylor, 29 July 1955, folder 9, box 34, Wilderness Society Papers.

28. U.S. Grant III, "The Dinosaur Dam Sites Are Not Needed," *Living Wilderness* 15 (Autumn 1950): 17–24.

29. On the preservationist position on Glen Canyon Dam see Mark W.T. Harvey, "Echo Park, Glen Canyon, and the Postwar Wilderness Movement," *Pacific Historical Review* 60 (February 1991): 43–67.

30. *Washington Post*, 6 June 1954; *New York Times*, 15 May 1954; *San Francisco Chronicle*, 7 July 1954. See also Joseph C. Bradley, "This, Your Land, Is In Danger," *(Madison) Wisconsin State Journal*, 20 September 1953. The Council of Conservationists published a pamphlet listing nearly a dozen newspaper editorials against the dam. The pamphlet is found in

box 114, Harold Bradley Papers (hereafter Bradley Papers), Sierra Club Records, Bancroft Library.

31. "The Dinosaur Case," *Garden Club Bulletin* (March 1953), 85.

32. Richard Bradley, interview with author, 17 June 1988, Colorado Springs, CO.

33. Ibid.

34. Harold Bradley, "Danger to Dinosaur," *Pacific Discovery* 7 (January–February 1954): 4; news item, *Sierra Club Bulletin* 38 (February 1953): 11–12.

35. Bradley, "Danger to Dinosaur," 8; Ruth Aiken to Harold Bradley, 23 June 1953, box 114, Bradley Papers.

36. David Brower, *For Earth's Sake: The Life and Times of David Brower* (Salt Lake City: Gibbs Smith, 1990), 306.

37. Richard Bradley interview, 17 June 1988; news item, *Sierra Club Bulletin* 39 (February 1954): 2; David R. Brower, "Environmental Activist, Publicist, and Prophet," an oral history conducted 1974–1978 by Susan R. Schrepfer, p. 113, Regional Oral History Office, Bancroft Library.

38. Michael P. Cohen, *History of the Sierra Club, 1892–1970* (San Francisco: Sierra Club Books, 1988), 156.

39. Wallace Stegner, ed., *This is Dinosaur: Echo Park Country and Its Magic Rivers* (New York: Alfred A. Knopf, 1955); ibid., 2nd ed. (Boulder, CO: Roberts Rinehart, 1985).

40. Sierra Club press release, 31 March 1955, folder 4, box 34, Wilderness Society Papers.

41. Stegner, *This Is Dinosaur*, 1st ed., v.

42. Ibid., 2nd ed., 85.

43. Ibid., 1st ed., brochure insert.

44. Ibid., 1st ed., 17.

45. Senator Watkin's argument can be found in the *Congressional Record*, 84th Cong., 1st sess., 28 March 1955, vol. 101, pt. 3:3806–19. For other arguments see *Denver Post*, 9 March 1954; and *Deseret News*, 17 April 1954.

46. *Casper* (Wyoming) *Tribune-Herald*, 29 August 1954. See also *Salt Lake Tribune*, 11 June 1954.

47. The Colorado River Storage Project Act was first introduced in Congress in 1950 by Utah Senator Arthur Watkins. Congress did not debate the bill fully until 1954.

48. As an example, see remarks by Illinois Senator Paul Douglas, *Congressional Record*, 84th Cong., 1st sess., 19 April 1955, vol. 101, pt. 4:4634–41; see also the statement of the Engineers Joint Council, Senate Committee on Interior and Insular Affairs, *Hearings on S. 1555 on Colorado River Storage Project*, 83rd Cong., 2d sess., 1954, 681–85.

49. "Actions of Conference on Upper Colorado River Legislation," copy in box 356, Joseph C. O'Mahoney Papers, American Heritage Center, University of Wyoming, Laramie; Howard Zahniser to Olaus Murie, 4 November 1955, folder 11, box 79, Wilderness Society Papers.

50. William Dawson and Wayne Aspinall to Edgar Wayburn, 26 January 1956, box 114, Bradley Papers; Horace Albright, Ira Gabrielson, and Howard Zahniser to Usher Burdick, 23 January 1956, box 8. Usher Burdick Papers, Special Collections, University of North Dakota, Grand Forks.

51. Runte, *National Parks*, 189–96; François Leydet, *Time and the River Flowing: Grand Canyon* (San Francisco, Sierra Club Books, 1964).
52. Nash, *Wilderness and the American Mind*, 212.
53. Irving Brant to Harry S. Truman, 16 February 1951. Brant to Adlai Stevenson, 7 September 1957, box 21, Irving Brant Papers, Manuscript Division, Library of Congress, Washington, D.C.
54. Harold Bradley to Fred Smith, 12 November 1957, box 114, Bradley Papers; *New York Times*, 1 January 1956; John Oakes, telephone interview with author, 27 February 1986.
55. "Echo Park Controversy Resolved," *Living Wilderness* 20 (Winter–Spring 1955–56): 23–43.
56. Roy D. Webb, *If We Had a Boat: Green River Explorers, Adventurers, and Runners* (Salt Lake City: University of Utah Press, 1986), 136.
57. David Brower, interview with author, 24 July 1985, San Francisco, California.
58. Harvey, "Echo Park and Glen Canyon," 44–45; Eliot Porter, *The Place No One Knew: Glen Canyon on the Colorado* (San Francisco: Sierra Club Books, 1963).
59. Wallace Stegner to author, 14 July 1984 (in author's possession).

CHAPTER ELEVEN

"WE CAN WAIT. WE SHOULD WAIT."
Eugene's Nuclear Power Controversy, 1968–1970

DANIEL POPE

In November 1968 voters in Eugene, Oregon, overwhelmingly approved a measure allowing their municipal electric utility to sell bonds to fund construction of a nuclear power plant. Eighteen months later, on 26 May 1970, citizens again went to the polls; this time, however, they voted to impose a four-year moratorium on their utility's nuclear construction plans. Why this rapid shift? What had changed in the intervening months to convert general approval to rejection of nuclear power plans? The answer lies in an examination of national, regional, and local developments.

Eugene's aborted affair with nuclear energy came near the end of a long period of American energy abundance. Domestic crude petroleum at about $3 a barrel, imports at about $1.25 a barrel, seemingly inexhaustible supplies of coal, and convenient natural-gas resources had fueled the nation's post–World War II boom. The spread of electricity to rural America had been followed by the rapid intensification of household energy use. Between 1961 and 1970, while the nation's average residential electricity rates declined by about 15 percent, the average home's monthly usage grew about 83 percent.[1]

The late 1960s saw signs of the end of the era of energy abundance. Domestic oil and gas reserves both began to decline in 1968. Energy consumption grew more rapidly than real gross national product in the 1960s, a pattern not repeated since 1970. Those years also saw the birth of the modern environmentalist movement, culminating in the massive Earth Day celebrations in April 1970; the movement called attention to the ecological costs of an energy-intensive way of life.

Meanwhile, nuclear energy showed signs of realizing the long-standing hopes of its proponents. By the end of 1970 there were nineteen operating power reactors, with 1.9 percent of the nation's electrical utility generating capability, a fivefold increase over 1965.[2] Commercial nuclear power had gotten off to a rocky start in the 1950s and early '60s, but many had come to see it as the best source of safe, cheap electricity. Some were predicting that nuclear reactors would generate half of the nation's electric power by the end of the century.

National policy makers perceived only dimly the end of an era of energy abundance. In the Pacific Northwest, however, power planners knew that major changes were imminent. Construction of the Bonneville and Grand Coulee dams during the Great Depression had helped launch an era of rapid growth. Blessed with the nation's greatest hydropower potential in the Columbia River and its tributaries, and with heavy federal investments in dams and transmission facilities, the Pacific Northwest had the nation's lowest electricity rates. In Eugene, Oregon, customers of the city's Water and Electric Board paid less than one cent per kilowatt-hour and used about two-and-one-half times as much electricity as the national average. Regional planners had already determined that future growth would have to rest upon thermal generation, since major hydropower sites were practically all occupied. The Bonneville Power Administration (BPA), the agency charged with transmitting and marketing federally generated electricity, had convened a Joint Power Planning Council in 1966, with representatives from 180 public and investor-owned utilities. The council's 1968 Hydro-Thermal Power Plan was, to put it mildly, ambitious. It envisioned twenty large nuclear plants in the Pacific Northwest by 1990, along with investments to increase the capacity of the hydro system and transmission facilities. These would be needed, BPA and the utilities contended, to meet an anticipated tripling of demand by 1990. Already, in 1967, Portland General Electric Company had undertaken to build a nuclear facility, the Trojan plant on the Columbia near Rainier, Oregon.

The city of Eugene had, like much of the region, thrived on low-cost hydropower. From 20,838 in 1940, the community had reached a population of 76,346 by 1970. The seat of Lane County, Eugene was both a major wood-products manufacturing center and home of the University of Oregon, which had grown apace with the city. The presence of the university meant that Eugene's population was generally well educated, though not particularly affluent. Since 1911 the Eugene Water and Electric Board (EWEB) had supplied the city's power needs. Like several of the other larger Northwest municipal systems and public utility districts, EWEB had built some dams, but it purchased most of its energy from Bonneville.

A five-member elected board governed the utility; they served four-year terms and received no compensation. Obtaining office in nonpartisan elections, EWEB members had rarely provoked public controversy. They came in most instances from local retail and construction businesses and almost invariably sought to run the utility on "sound business principles." The board minutes from 1968 to 1970 reveal that all motions received unanimous votes. EWEB's operations were in the hands of General Manager Byron Price, an experienced utility administrator who enthusiastically shared the board's penchant for expansion.

By the mid-1960s EWEB had begun to plan for nuclear energy. A 1966 staff report indicated "that EWEB would need extensive new generating facilities before the end of 1977." Board members discussed how to finance a nuclear plant in the upper Willamette River Valley and how to market a share of the plant's output.[3] In 1967 the board took further steps. It joined in lobbying to pass a bill permitting public and private utilities in Oregon to cooperate in thermal power facilities. After its enactment, the board considered whether to take a share of the capacity of the Trojan plant. It also explored "possible sponsorship of a large plant in this part of Oregon." A resolution endorsing "active planning for future power sources" passed—unanimously, of course.[4]

The utility promoted demand growth at the same time it was forecasting such growth. Three weeks after the board endorsed nuclear planning, it approved a "loadbuilding program" of cash incentives to convert homes to electric heat.[5] A few months later, Ehrman Giustina, board member and local mill owner, presented a personal statement at a board meeting. He urged EWEB to get more involved in recruiting businesses to Eugene. Giustina cited a consultant's review that noted: "A comprehensive area development plan and action program to attract new industry would enhance the probability of EWEB's growth projections being realized, and of the community remaining prosperous." In conclusion, Giustina proposed that EWEB mandate one of its top managers to drop day-to-day tasks in favor of industrial promotion. Fellow board member John Tiffany, who owned several local drug stores, endorsed the statement and thought other members would also concur.[6] Although the proposal did not generate a formal motion, it indicated how EWEB worked to create the demand that it subsequently used to justify supply growth.

EWEB took steps to ready itself and its customers for the nuclear path. The board and staff members took "a most interesting and informative trip," in Giustina's words, to Boston, New York, and Washington, D.C., in October 1967. They met with architect-engineering firms, investment bankers, and members of Oregon's congressional delegation.[7] The next month a consulting firm held a two-day "management familiarization seminar on nuclear power."[8] In January 1968 EWEB appropriated $7,430 to fund a lecture-demonstration program on "This Atomic World" for area secondary schools. That spring EWEB joined the Atomic Industrial Forum, a national trade association.⁹

In the summer of 1968 EWEB decided to present its nuclear visions to the electorate. For the utility to issue revenue bonds for a plant, Eugene voters would have to approve a city charter amendment. In August the board submitted a proposed amendment and the city council placed it on the ballot for the November election. The measure sought permission to borrow $225 million to finance construction of a thermal generating facility somewhere in Lane County (which is almost as large as the state of Connecticut). Although the wording did not commit EWEB to a nuclear reactor, this was generally understood to be the utility's intention.

The election campaign was distinctly low-key. EWEB staff devised a public information program to win support, but the measure provoked almost no organized opposition. The *Eugene Register-Guard*, the city's sole daily newspaper, favored the bond authorization but expressed "definite moral reservations."¹⁰ A mail-in survey in October brought only 801 returns, but 82.5 percent were favorable, 15.5 percent undecided, and only 2 percent opposed. The actual vote was almost as lopsided. The tally was 23,261 to 6,081 (79.3 percent in favor). The measure won two-thirds or more of the votes in every Eugene precinct.¹¹ The support for EWEB's bond issue contrasted with voters' rejection of two tax levies on the same ballot.

Two days after the election, when the EWEB board convened, members happily reported widespread appreciation of their campaign. The board voted to hire advisers to prepare a schedule for the project, which was expected to take eight years to complete. They discussed expanding office space to house the new venture and appropriated funds for a cooperative program with the United States Geological Survey to monitor water conditions in the Willamette River below Eugene. At the same time, EWEB agreed to join with the state's privately owned utilities to fund a chair in nuclear engineering at Oregon State University in Corvallis. General Manager Price, endorsing the proposal for the endowed chair, noted that a training program would benefit the utility's nuclear efforts. Only one small cloud loomed on the horizon. Charles O. Porter, a former U.S. congressman and now a liberal gadfly lawyer in Eugene, had sent a letter asking how much EWEB had spent on the ballot measure campaign and questioning EWEB's authority to dispense these funds. Staff reported the campaign had cost $6,710.38, and the board referred Porter's letter to legal counsel for a reply.¹²

EWEB's drive for a nuclear power plant reflected, of course, its understanding of the city's energy needs and its consensus in favor of business-led urban growth. Meanwhile, regional and national forces helped prod EWEB toward nuclear power. The Joint Power Planning Council (JPPC) had spelled out the need for massive thermal plant construction; although formally a body representing local utilities, the JPPC was in fact a creation of the Bonneville Power Administration (BPA) and was largely dependent upon its technical expertise and demand forecasts. Bonneville transmission lines carried the majority of the region's electricity, and thus BPA assumed a formidable role in electrical system planning.

Indeed, it was clear from the Hydro-Thermal Power Plan that new plants would be built to supply the region as a whole. During the planning process in 1969, EWEB indicated that it expected to maintain a 55 percent interest in its facility, with the two large private utilities in the state owning most of the remainder.[13] The electricity itself would be fed into the Bonneville transmission grid. When the EWEB plant began to send energy to BPA, Bonneville would then grant credit on EWEB's monthly bill for the value of that power. This "net billing" procedure proved during the 1970s to be one of the elements that undermined the plans of the Washington Public Power Supply System to build and run five nuclear plants for the region. As of the late 1960s, however, net billing looked like a way to tie EWEB commitments to regional needs and to share the risks of a major undertaking.[14]

Other external forces entered the EWEB story as well. The planning process for the reactor necessitated links to a web of engineering, legal, and financial firms. The Boston engineering company, Stone and Webster, drew up specifications for the nuclear steam supply system (the heart of the generating plant) and its fuel supply and prepared plans to solicit proposals from qualified manufacturers. Stone and Webster also handled preliminary site selection research. Meanwhile, EWEB sought the services of municipal finance specialists to handle its bond sales. In August 1969 it hired a leading New York law firm—Wood, King, Dawson, Love and Sabatine—to serve as bond counsel. It also engaged an underwriting syndicate led by New York investment bankers Blyth & Co. Half a dozen other large New York investment bankers joined as syndicate managers.[15]

The easy approval of EWEB's nuclear plan disguised a paradoxical situation. The legitimacy of EWEB's growth strategy depended on its claim to serve local needs and goals. However, to implement its strategy, the board had to rely on external planning, technical, legal, and financial institutions. This reliance left it vulnerable to later charges that community interests were being subordinated or neglected by these cosmopolitan organizations. When the time came to defend its project to Eugene's voters, the utility found it difficult to reassert its democratic roots.

As a university town, Eugene was home to many well-educated and articulate people with an orientation toward civic activism. The traditions and institutions of direct democracy, legacies of Oregon's Progressive years, set the stage for protest and gave it a tactical focus. Finally, environmentalist sentiments in the region, soon to be dubbed "Ecotopia," help account for the opposition which EWEB's nuclear plans soon engendered.

Jane Novick in some respects typifies those who opposed EWEB's nuclear plant. For her, community involvement had ranged from the PTA through the American Civil Liberties Union and the Lane County Democratic Party, along with roles in numerous candidates' campaigns. These activities were "what I've done all my life," she recalled in 1986. Novick had paid little attention to the bond measure, in part because the November 1968 election featured far more dramatic contests, in part because she lived just outside the city limits and hence was ineligible to vote on the issue, although EWEB served her home.[16] (It was an anomaly of the opposition to EWEB's plans that several of the leading foes lived outside city limits. During the moratorium campaign, the utility and its supporters never raised this issue or labeled the opponents as outsiders; Novick, in retrospect, wondered why and thought it might have been an effective tactic.)

Novick participated in a League of Women Voters' book discussion group; and in the spring of 1969 she was asked to substitute for another member and review *The Careless Atom*, a critique of nuclear power by science journalist Sheldon Novick (no relation to Jane). She had considered herself generally well informed on environmental matters before this assignment but was impressed by the force of the author's case against the rapid development of nuclear energy. Over the following months and years, Jane Novick continued to study the issues and became a formidable expert in her own right on nuclear power. Willingness to research the issue was characteristic of Eugene's nuclear plant opponents; activists had high regard for the role of knowledge in public debate. They also knew that supporters of the EWEB plant would pounce on any factual mistakes in the antinuclear case. As Chris Attneave, another leader, noted, "We were . . . paranoid about being caught in errors."[17]

Attneave was another experienced civic reformer drawn to question the wisdom of the EWEB plant in 1969. For her, the initial impetus to doubt was awareness that evaporation from a nuclear plant's cooling tower or cooling pond would exacerbate the Willamette Valley's already severe fog problems. Attneave soon concluded that fog was only one of several potential problems.

Joseph Holaday described himself as "more of a lay person" than Novick or Attneave, both of whom were married to University of Oregon professors. A former high school teacher who owned a nursery in Eugene, Holaday had a long-standing interest in the environment. Jane Novick was a customer at his nursery and discussed the EWEB proposal with him; Holaday, already "doubtful" about nuclear energy, proved a receptive listener.[18]

These three—middle-class, well educated, reformist, politically liberal—personified the movement that eventually foiled EWEB's nuclear plans. Richard S. Lewis, describing opponents of a breeder reactor planned for Pennsylvania, could have been writing about Eugene's protesters. These were

> people who vote carefully at elections, pay their taxes, perform useful work in the community, help each other in times of stress, offer their children as much higher education as the children will take, and are concerned about the future of their community, nation, and humanity.[19]

At first, however, a different constituency spoke out. These protesters were farmers in the Willamette Valley north of Eugene. Although EWEB never reached a decision on the site for its plant, initial discussion had focused on a nearby valley location. Agricultural interests were not placated by EWEB's project to investigate the benefits of warm-water irrigation using discharge from the reactor's cooling circuits. In February 1969, 115 farmers crowded into a protest meeting north of Eugene. Later, however, as the utility shifted its siting studies to areas along the coast, about sixty miles west of the Willamette Valley, the farmers' opposition vanished. The activists who pursued the cause through the moratorium vote in May 1970 were free of any implication of material self-interest, but they lacked the motivation that a direct stake in the outcome can provide.[20]

During the spring and summer of 1969 a small band of critics pressed EWEB to subject its nuclear plans to public examination. The utility, confident of its path, hesitated to open the issue for lay debate. It did not take long for an adversarial relationship between EWEB and those questioning its plans to develop. Seventeen years later, when asked about relations with EWEB members, Novick still snapped, "We were treated like dogs." Novick commented, "They loathed us with such a passion that you could see their teeth grinding in their jaws; they were *livid* with rage when they'd see us walk into the room."[21]

On 7 July 1969 Eugene's city council met with the EWEB board to discuss whether to call a public hearing on its nuclear project. Citizens in attendance were not allowed to testify, and when EWEB refused to schedule a hearing, some walked out. In the hallway, the Eugene Future Power Committee was born; sixteen men and women signed a notebook which Joe Holaday passed around indicating that they would keep working together on the issue. Three of the signers were farmers opposing a valley site; the others had more general objections to the EWEB plans.

A day later, EWEB reversed itself and acceded to demands for an open hearing, although the utility announced that citizen statements could take no more than ten minutes and had to be prepared in writing in advance of the meeting. Thirty-four people planned to testify, most of them opponents of EWEB's plans and "ready for a fight." The hearing went on for four-and-one-half hours and reconvened the next morning. EWEB chose not to respond to the questions which protesters raised in their statements.[22]

That summer the Eugene Future Power Committee took on an organizational structure. Bylaws, a mailing list, and officers (Holiday as president, Novick as secretary) were chosen. Charles Brooks, an administrator for a psychological consulting firm, suggested the group's name; he was chosen as treasurer. The activists sought to expand their numbers and began compiling a mailing list.

Later that summer EWEB rebutted its critics. For the evening of 9 September it scheduled a presentation by seven experts who worked for nuclear power industry firms, Bonneville, and Oregon State University's Agriculture School. If the utility hoped that the presentation would assuage opponents' concerns, their plans certainly backfired. An initial crowd of 150 in the city council chambers dwindled by two-thirds as the EWEB panel spent four full hours on its case for nuclear power. When it was over, the authorities responded to written questions from the audience. The answers, however, were brief, and the tone, in many cases, curt. In a statement following the meeting, the Future Power Committee said, "We regret that the panel appeared to be promotional rather than informational."[23]

The EWEB Board initially revealed no fears that opponents could derail the project. In August it voted to engage a bond counsel and an investment banking syndicate to market its securities. At the same meeting, an agreement with Oregon State University Extension Service to study plans for warm-water irrigation was reported.[24] Site selection work was also continuing apace. However, by their meeting of 27 October 1969 the board apparently felt a need to show that its nuclear ambitions had community backing. Although most EWEB meetings had very few visitors, that night when board president Tiffany asked guests if they wanted to be heard, eight men offered their "unsolicited expressions of confidence" in the plant. One stated that he was attending "primarily as a newsman" but added that "it was unfortunate that local publicity seemed to reflect a minority opinion" of plant foes.[25] Two weeks later the board noted support messages for a nuclear plant from governmental agencies in the coastal city of Florence and from the chamber of commerce in the remote community of Oakridge.[26]

Meanwhile, a new development that autumn strengthened the hand of the foes of the nuclear project. The previous spring, the *Eugene Register-Guard* hired Gene Bryerton, a young investigative reporter who had been at the *Saint Louis Post-Dispatch*. Bryerton had already been interested in the debate over nuclear energy and immediately concentrated on that issue. By the fall, he had prepared a fourteen-part investigative series, "The Nuclear Dilemma," which the *Register-Guard* published between 26 September and 9 October. The paper compiled Bryerton's articles into a brochure, and the material formed the basis of a book of the same name, which Ballantine Books published as a mass-market paperback in 1970. Bryerton's study was not a work of advocacy. Nevertheless, it gave careful attention to the critics of nuclear energy and presented their concerns with evident sympathy.

Since the *Register-Guard* had generally aligned its editorial stance with business elements favoring the city's economic development, foes of EWEB's plans were quite surprised at the prominent display of Bryerton's guarded evaluation of nuclear energy. However, they counted their blessings, and in later interviews they agreed that the *Register-Guard*'s position had helped their cause considerably. Former Congressman Charles Porter had jousted with the newspaper throughout his career, but granted that Alton Baker, Jr., head of the family that owned the paper, had a "first-amendment streak" of integrity which he sometimes displayed.[27]

At least one source indicates that the opponents of the utility's plans had thought about a petition campaign to overturn it within weeks of passage of the bond measure.[28] Yet the Eugene Future Power Committee did not decide until late in 1969 to launch a drive to place an initiative measure on the city's ballot. This strategy entailed some tactical decisions. The group strove to project an image of moderation and common sense; therefore, they decided they should not be seen as unalterably opposed to all nuclear energy. They would not challenge EWEB's involvement with the Trojan nuclear plant. Nor would they call for a permanent halt to all EWEB nuclear projects. Rather, they drafted a moratorium proposal. It would amend the Eugene city charter:

> by providing that funds may not be expended by Eugene Water and Electric Board prior to January 1, 1974 for design and construction of nuclear electric generating plant except for investment in Portland General Electric Trojan nuclear facility, collection of safety data, and environmental studies at possible sites.[29]

In other words, as Jane Novick recalled, "We said, 'We are not against nuclear power. . . . We need time to look it over.' "³⁰ Novick and other EFPC activists maintained that this accurately reflected their views at the time; their doubts about nuclear power had not crystallized into implacable opposition. The Future Power Committee devised a slogan to fit its limited objective. Fred Keppel, who designed irrigation equipment and was a registered Republican, proposed, "We can wait. We should wait." This became the watchword of the nuclear plant's foes.

As the Future Power Committee prepared for its petition drive, it gained the support of other civic and environmental organizations. The local League of Women Voters endorsed the measure in November and joined with the Future Power Committee when it announced the petition drive at a press conference on 19 January 1970.³¹ Also joining the effort at that time were the Oregon Environmental Council and the Eugene Sierra Club.³²

Putting the measure on the ballot was a major task. The law required 4,528 valid signatures of registered Eugene voters to bring the measure to a vote. This equaled over 15 percent of the vote total on the bond issue of 1968 and nearly three-quarters of the number of votes against that proposal. To get the moratorium on the May 1970 special election ballot, the petitioners had to gather the signatures by late-February. The EFPC noted at its press conference that it was mailing petitions to 350 residents who had indicated an interest in the cause.

However, signature collection lagged for several weeks. At a 5 February meeting, Jane Novick remarked that the petitions were "not being returned to us very quickly." The Future Power Committee decided to focus its energies on a rally and petition drive on Saturday, 14 February, Valentine's Day.³³ Scores of volunteers fanned out to public places to gather signatures; by late-February, the names were gathered and the city council certified the initiative as Ballot Measure 52.

Meanwhile, EWEB and its friends had been rallying to defend the project. In December 1969 the defenders had created CODE, Citizens for the Orderly Development of Electricity, chaired by Wayne Shields, a local contractor and principal owner of the region's largest shopping center. CODE's backers included the secretary-treasurer of the Lane County Building Trades Council, a physics professor at the university, and several business leaders and public officials.³⁴ During the campaign, CODE provided speakers at debates on the moratorium and operated an information center downtown. A second organization opposing the moratorium, Scientists and Citizens for Atomic Power, was headed by the chairman of the University of Oregon physics department.

In public presentations he stressed the support of eminent scientists for nuclear power.³⁵ EWEB also received petitions and letters from around the county expressing interest in siting the nuclear plant in their area of the county.[36] Although most of the visible support for the plant came from business groups, including chambers of commerce, realtors, and home builders, the Lane County Labor Council also endorsed the project, despite internal dissent on the issue.[37]

Proponents of nuclear power in the county had to contend with the problem of siting the plant while they combatted local opposition. Engineering consultants Stone and Webster had investigated at least seven locations by the fall of 1969, and in October added four more possibilities to the list. At a November meeting, EWEB heard Stone and Webster report that five sites were still being studied, four in or near the Willamette River Valley and one on the Siuslaw River near the coastal town of Florence.[38] Reedsport, a community south of Florence, also soon became a potential location.

In reality, the project's siting options were quite limited. Farmers in the rich agricultural area north of Eugene had, as noted above, rallied against a plant in their area, despite the blandishments of warm-water irrigation. Beyond this, Governor Tom McCall had named a task force to study questions of thermal plant siting, led by L.E. Wilkinson, his nuclear power coordinator.[39] That group paralleled the work of a legislative group in the state's House of Representatives, chaired by George Wingard, a maverick Republican with a strong environmentalist record. Four days after Stone and Webster had reported that valley sites were under consideration, the Wingard group announced its opposition to any nuclear plants in the Willamette Valley, noting that the plants might exacerbate the area's fogging problems and that population densities were too high to tolerate a nuclear facility.[40] Wilkinson, speaking to the House Task Force in February 1970, agreed that "strong favor should be placed on nuclear plant sites outside the Willamette Valley." Although he denied several days later that he had meant to rule out a valley plant entirely, it was clear that statewide pressures would make such a location a very difficult choice.[41] By March, the board had to admit that site selection was probably at least two months away; and that meant a final determination would not come until just before the 26 May initiative vote. In April EWEB reported to the Governor's Task Force that sites near Florence were their first choice, with four possible locations in that area.[42]

Although the uncertain progress of site selection may have won endorsements for the nuclear plant from around the county, as localities vied for the economic advantages of a quarter-billion-dollar investment, it could hardly have won EWEB support in Eugene itself. Without a definite site, it was difficult to portray the plant as technologically a sure thing and the project as secure and well managed.

Despite the supportive news coverage in the *Register Guard*, the momentum of the successful petition campaign, and the utility's problems in siting the project, the Eugene Future Power Committee faced an uphill struggle. To reverse a four-to-one majority for the bond issue required changing many people's minds in the face of strong opposition from the local business community and from EWEB itself. The committee lacked substantial financial resources, professional staff, or prominent spokespeople.[43]

The committee did, however, have a core group of eager volunteers, a talkative and independent bunch who made meetings "very, very chaotic," despite chairman Holaday's efforts to adhere to parliamentary procedure.[44] They had studied the topic of nuclear power with care and managed to present themselves as well informed and thoughtful. Additionally, they had the backing of many scientists at the University of Oregon. For instance, Jane Novick's husband Aaron, a biochemist who had worked at Los Alamos on the Manhattan Project, took issue with EWEB's forecasts of rapid growth and worked out more modest projections. Years later, Keith Parks, an EWEB official who became general manager in 1974, readily conceded that "Aaron Novick was right on the button."[45] When the pro-nuclear Scientists and Citizens group claimed scientific backing for its stand, the EFPC was able to release a poll of science faculty at the university indicating that seventy-one of eighty-eight respondents supported the moratorium.[46]

During the spring campaign, the Future Power Committee showed itself to be remarkably adept at finding opportunities to present its case for the moratorium. It sought out speaking opportunities before virtually every civic and community group in Eugene and managed to debate nuclear power proponents in several encounters around town. When a popular local radio station broadcast an editorial favoring the nuclear plans, the committee exercised its right of reply and got free air time to present its position. In the last days of the campaign, the EFPC publicized a phone number to call for responses to what Joe Holaday called "the often biased information presently being given" by EWEB.[47]

Opponents of the plant also sought legitimacy by presenting scientists who opposed nuclear energy development. The Oregon Environmental Foundation, a group backing the moratorium, sponsored a series of lectures by John Gofman, a nationally known expert on radiation, by University of Minnesota biochemist Charles Huver, and by Richard Rosa, a researcher on alternative energy generation with Avco Everett Research Co.

In order to make contact with those who failed to attend civic organizations or educational forums, the EFPC in the last weeks of the campaign devised a "survey." Canvassers knocked on doors with a set of questions about attitudes on nuclear energy. The poll was meant more to persuade, to "sow seeds" of doubt about the nuclear project, than to predict the election's outcome. Future Power Committee members recalled it as an effective tactic.[48] In questioning the wisdom of EWEB's project, the committee proved adept at linking the concerns of the emerging national opposition to nuclear power to the distinctive situation facing Eugene. They stuck closely to their self-assigned task, persuading Eugeneans that it was not the time for their utility to build such a large nuclear plant in their county.

Sheldon Novick, the author whose book had first ignited Jane Novick's concern, had outlined a trail of dangers ranging from accidents and natural disasters to radiation perils from reactor operation to radioactive waste disposal. Richard Curtis and Elizabeth Hogan, in *The Perils of the Peaceful Atom*, rehearsed a similar litany of abuses. Both books had criticized the undemocratic and secretive nature of nuclear power planning and had raised doubts about the economic and energy demand forecasts which underlay the case for the atom.

Although the Future Power Committee raised questions about health and safety dangers in the nuclear fuel cycle, it concentrated its fire on the specifics of EWEB's plans. It actively challenged EWEB's predictions of a doubling of demand between 1970 and 1976 and pointed out that the proposed plant would be supplying more energy to aluminum plants elsewhere in the region than to Eugene residents. It also emphasized the uncertain economics of nuclear energy, noting that cost overruns were probable, that the plant would provide few permanent jobs for local workers, and that a reactor built outside the city would not enhance Eugene's tax base. Clearly, opponents of the nuclear plans were not relying on scientific arguments alone to carry the vote.[49]

EWEB justified its nuclear plans with a similar mixture of technical and economic appeals. In its public statements it contended that operating nuclear plants released virtually no radiation. The utility pointed out that a coastal plant using ocean water for cooling would not contribute to fogging problems. Waste disposal methods, according to EWEB, assured safety as well. However, like the opposition, the utility led off its case (in a brochure mailed to all ratepayers days before the vote) by noting the effect of delay on the economy, construction costs, power supplies, and local taxes.[50]

Both EWEB and the Future Power Committee sought to depict themselves and their cases as "rational, professional and factual."[51] Yet both were careful to emphasize local concerns and conditions. Technical expertise about the health and safety aspects of nuclear energy served to persuade voters but also to establish the legitimacy and respectability of each side in the debate.

Neither supporters nor foes of Ballot Measure 52 felt sure of the outcome before election day, and indeed the contest was very close. Citizens gave a majority to the moratorium in only twenty-three of the city's forty-seven pre-educated voter

cincts, but the margins in these were enough to yield a 12,060 to 11,209 victory for the Future Power Committee (51.8 percent to 48.2 percent).

The closeness of the vote contrasted with the overwhelming approval of the bond measure a year and a half earlier. An examination of the May 1970 vote on the moratorium ballot measure and the bond vote of 1968 gives an idea of the bases of opposition to the EWEB nuclear plant. The election totals are available on a precinct basis for each election. They do not allow one to draw inferences about individuals' voting behavior, but they do permit one to identify the voting characteristics of different neighborhoods of the city on the two nuclear measures. They also make possible an evaluation of the moratorium vote in relationship to two other measures on the May 1970 ballot. Ballot Measure 51 polled the citizenry on an emotionally charged symbolic issue. In 1964 John Alltucker of Eugene Sand and Gravel had donated materials and labor to erect a large concrete cross on city land on Skinner's Butte, overlooking downtown Eugene. The cross was illuminated at night during the Christmas season. After the Oregon Supreme Court ruled that the cross on public property violated the First Amendment, Eugeneans were called upon to vote on a measure that declared the cross to be a war memorial rather than a religious symbol. Approval of the proposal by an almost three-to-one margin (17,185 to 6,009) indicated a desire to keep the cross in place. The vote can be seen as a test of what political analysts were soon to call a "social issue." Following the nuclear moratorium measure on the ballot was a different kind of question. Ballot Measure 53 would require collective bargaining for Eugene's civil servants. Although the measure received little attention in the *Register Guard*, a close vote expressed divisions in the community over labor union rights. The measure won by the slim margin of 10,766 to 10,694 (50.2 percent to 49.8 percent).

A precinct-level correlation of the votes on the measures on the May ballot reveals, not unexpectedly, that the areas which voted in favor of the moratorium were also relatively "liberal" on the other issues. They were less likely to give overwhelming majorities to retaining the cross as a war memorial. That measure carried all but two of the forty-seven precincts. The correlation between "yes" votes on the moratorium and on the cross was .85 (significant at the .001 level). Neighborhoods supporting the moratorium were generally those which approved of collective bargaining rights for city employees. Here, the correlation was +.46 (also significant at the .001 level).

More surprising is the positive precinct-level correlation between opposition to EWEB's nuclear plant in 1970 and strong support for the bond issue in 1968. The correlation here is +.34 (significant at the .01 level). Those areas which had voted most strongly to facilitate EWEB's nuclear plans were also likely to vote to derail the project.⁵² Of the ten precincts backing the moratorium most strongly, eight had given above-average support to the bond measure.

The positive correlation between precincts supporting the bond issue and those backing the moratorium shows a likelihood that many voters who backed nuclear development in 1968 were wary enough to favor delay in 1970. This speaks well for the effectiveness of the Eugene Future Power Committee's campaign. The record of other referenda on nuclear energy is at best a mixed one for the foes of these plants; although voters have sometimes favored limits on waste disposal methods, they have been reluctant to delay or reject proposed plants and have been virtually unanimous in rejecting measures to halt construction or operation of plants which were already under way. Only in the 1989 case of California's troubled Rancho Seco plant did voters in the Sacramento Municipal Utility District shut down an operating reactor. At least before the Three Mile Island accident in 1979, public opinion polls showed a consistent majority supporting nuclear power in general and a somewhat smaller edge in favor of the idea of a nuclear plant nearby. In Oregon, four surveys between 1970 and 1978 showed that most respondents favored the Trojan Plant.

As has been noted, the Future Power Committee, despite (or perhaps because of) its amateurism, campaigned creatively and energetically. Although its expenditures totaled only a few thousand dollars, the EFPC drew upon countless hours of volunteer time from community activists. It took advantage of the openness of the small city's media, which filled their columns and air time with accounts of the press conferences, forums, and debates that the moratorium advocates produced.

EWEB's tactics also probably benefited the antinuclear forces. The board and general manager Price at first dismissed the protesters as annoyances and then treated them with scarcely disguised hostility. "These people were a real nuisance," noted Keith Parks in retrospect.⁵⁴ Novick and Attneave both indicated years later that EWEB's animosity had only stimulated greater efforts. Parks also remarked about the technical experts whom EWEB used to promote its position: they were "totally out of touch with the reality of the people" and the scientists were "scary to the average John Doe."⁵⁵

Certainly, the Future Power Committee's success depended, too, on the nature of the community. Described a decade later by the *Wall Street Journal* as the "last refuge for the terminally hip," Eugene was fertile ground for a movement that questioned the wisdom of a large-scale, high-technology project with perhaps inescapable connotations of warfare and destruction. Parks was not far off the mark when he stated: "In this community, there's opposition to anything."⁵⁶

The student and ex-student community almost surely tipped the balance in the moratorium vote. More than a third of the city's vote margin came from the precinct that included the University of Oregon campus. Moratorium support in precincts adjacent to campus or where university housing developments were located were more than enough to overcome the negative vote elsewhere. Yet this was not a student movement. Future Power Committee activists were drawn heavily, although not entirely, from the university community, but very few students took part in the campaign. The voting age was still twenty-one, and on campus, issues concerning the war in Vietnam and racism understandably overshadowed a local controversy on which most undergraduates could not vote.

Moreover, Eugene voters were not always unorthodox. After all, on the same day they approved the moratorium, they voted by a three-to-one margin to make the cross on Skinner's Butte a war memorial and only barely endorsed collective bargaining for city employees. Since the EWEB nuclear project had won approval easily in November 1968, voters were clearly not consistently hostile to development plans.

Any explanation of the moratorium victory can only be tentative. However, one factor that seems to have been significant was the ability of the Eugene Future Power Committee to enunciate positions that sounded civic-minded and responsible rather than obstructionist and extreme. The decision to press for a moratorium rather than a ban bespoke an open-mindedness about the technological issues. The choice not to interfere with EWEB's commitment to the Trojan plant suggested a spirit of compromise. The committee's campaign emphasis on EWEB's plan to sell the majority of the plant's output to utilities elsewhere may have appealed to a parochial and "selfish type of attitude," as Parks called it, but the committee capitalized on EWEB's reliance on outside expertise and regional commitments to raise doubts about whether the utility had the community's well-being at heart.[57] Ironically, as has been noted, because so many of the committee's leaders lived outside the city limits (although generally speaking within the EWEB service area), they were themselves open to accusations of carpetbagging.

The Future Power Committee's success in seizing the high ground of civic responsibility may help explain the positive correlation of precinct support for the bond issue and later for the moratorium. Voters who backed the plant in 1968 and the moratorium in 1970 were in a sense not really changing their minds. In each instance, they were voting for a measure that looked fiscally prudent, was backed by sober-minded and serious community leaders, and appealed to their sense of community concern.

The May moratorium vote left utility planners uncertain. The board remained convinced that its project would soon be needed in the region and in Eugene itself. This unprecedented measure tied their hands for four years, but they were unwilling to foreclose the nuclear option entirely. The Future Power Committee was also not certain what it had wrought. For one thing, it was faced with a lawsuit by a local businessman, Andy Maxon, seeking to overturn the measure. It was not until 1971 that Maxon's legal challenge was thrown out. For another, EFPC was genuinely interested in solving energy problems; although it focused on alternative sources and conservation, it was not unalterably opposed to all nuclear energy. Jane Novick's victory statement called for cooperation with EWEB in energy planning; the adversarial relationship of the campaign prevented later cooperation, but the hope was not insincere.

In the longer run, however, the moratorium killed off EWEB's nuclear construction plans. By 1974 Keith Parks had become general manager. He could read the evidence of declining demand growth and recognized the political difficulties of the nuclear path. In late-1974 the utility abandoned its option on the Big Creek site near Florence. When WPPSS and BPA offered Northwest Utilities a chance to invest in shares of WPPSS plants 4 and 5, EWEB declined. Perhaps the final word on the moratorium campaign should belong to Parks. A loyal EWEB staff member who had felt that the Future Power Committee was "sabotaging something that was good," Parks nevertheless concluded in retrospect: "They did a great favor for this community. They saved its butt."[58]

NOTES

1. U.S. Federal Power Commission, *Status of Privately-Owned Electric Utilities in the United States December 1971* (Washington D.C., 1972), xxxiii.

2. U.S. Bureau of the Census, *Statistic Abstract of the United States: 1982–1983* (Washington, D.C., 1982), 589 (table 1007).

3. Eugene Water and Electric Board, "Record of Proceedings," vol. 9, 9 May 1966, 80–81 [henceforth cited as EWEB, "Record"], Eugene Water and Electric Board, Eugene, Ore.

4. Ibid., 26 July 1967, 187.

5. Ibid., 14 August 1967, 190.

6. Ibid., 13 May 1968, 255–56.

7. Ibid., 9 October 1967, 209.

8. Ibid., 6 November 1967, 209–10.

9. Ibid., 8 January 1968, 228, and 25 March 1968, 258.

10. *Eugene Register-Guard*, 3 November 1968, 10A.

11. State of Oregon Abstract of Votes, Lane County City Measures, 5 November 1968, Lane County Board of Elections Division, Eugene, Oregon. Figures for 1970 ballot measures discussed below come from the 26 May 1970 Abstract.

12. EWEB, "Record," 7 November 1968, 331–32.
13. Ibid., 28 May 1969, 376.
14. For a fuller description of net billing and its implications, see Kai N. Lee and Donna Lee Klemka with Marion E. Marts, *Electric Power and the Future of the Pacific Northwest* (Seattle: University of Washington Press, 1980), 75–86.
15. EWEB, "Record," 25 August 1969, 396.
16. Jane Novick, interview by author, 23 September 1986.
17. Chris Attneave, telephone interview by author, 26 September 1986.
18. Joseph Holaday, interview by author, 22 September 1986.
19. Richard S. Lewis, *The Nuclear-Power Rebellion* (New York: Viking, 1972), 9.
20. This statement is not strictly true. Keith Parks, the general manager of EWEB after Byron Price, suggested in 1986 that University of Oregon scientists' opposition to nuclear power was related to research funds which oil companies donated to the university, but I have found no evidence for this claim.
21. "Second Thoughts on Future Power," in *The Grass Roots Primer*, ed. James Robertson and John Lewallen(San Francisco: Sierra Club Books, 1975), 89; Novick, interview.
22. *Eugene Register-Guard*, 8 July 1969, lA, describes the initial Council EWEB meeting. For a full description of the hearing, see Rhoda Love, "Growing Concern Culminates in Formation of EFPC," 32–38, a typescript of a section of a longer, unpublished work on the EFPC prepared in 1971 by a member (copy in possession of Jane Novick and used with her permission).
23. Love, "Growing Concerns," 40–42.
24. EWEB, "Record," 25 August 1969, 395–96.
25. Ibid., 27 October 1969, 400–1.
26. Ibid., 10 November 1969, 408.
27. Charles O. Porter, interview by author, October 1986.
28. Lynton Hayes, "Nuclear Energy as a Stimulant to Economic Growth," typescript draft in files of Joseph Holaday.
29. This passage is from the statement of purpose on the initiative petition (copy of petition in files of Jane Novick, who lent it to the author).
30. Novick, interview.
31. *Portland Oregonian*, 22 November 1969, 12.
32. *Eugene Register-Guard*, 19 January 1970, lA.
33. *Eugene Register-Guard*, 6 February 1970, 9B.
34. *Portland Oregonian*, 12 December 1969, 29; *Eugene Register-Guard*, 6 February 1970.
35. *Eugene Register-Guard*, 29 April 1970, lB, and 13 May 1970, 4B.
36. EWEB, "Record," 12 January 1970, 419.
37. *Eugene Register-Guard*, 21 May 1970, lA.
39. *Portland Oregonian*, 1 November 1969, 11.
40. Ibid., 29 November 1969, 15.
41. *Eugene Register-Guard*, 2 February 1970, lA, and 11 February 1970, lB.
42. Ibid., 16 March 1970, 12A, and 16 April 1970, lA.
43. The EFPC raised money essentially by passing the hat among activists. Signers of an

"We Can Wait. We Should Wait." 219

advertisement supporting the moratorium, which appeared in the *Register-Guard*, contributed small amounts. After the election, a rummage sale earned funds to pay legal expenses and other debts. Its expenditures totaled $2,763.33, according to a memo, "To All Members and Well Wishers of the Eugene Future Power Committee," located in EFPC files and lent to the author by Jane Novick. The committee estimated that pro-nuclear outlays reached about $20,000.

44. Attneave, interview.
45. Keith Parks, interview by author, 16 October 1986.
46. *Eugene Register-Guard*, 11 May 1970, lB.
47. Ibid., 22 May 1970, 12A.
48. Novick, interview.
49. "Yes 52: We Can Wait . . . We Should Wait" (n.d.) is the leaflet which the Future Power Committee distributed most extensively. A copy is in the files of Jane Novick. See also the statement of Joseph Holaday in the *Eugene Register-Guard*, 21 May 1971, 10B.
50. Eugene Water and Electric Board, "Nuclear Electric Power for the Eugene Area" (n.d.), pamphlet in Jane Novick files. See also John Tiffany's statement in the *Eugene Register-Guard*, 21 May 1970, 10B.
51. The quoted terms are from the EWEB brochure quoted above (see note 50).
52. It cannot be assumed that individuals who favored the bond measure were especially likely to support the moratorium; precinct-level vote correlations do not permit inferences about individual voter behavior. To do so would be to commit the "ecological fallacy."
53. Although the interlude between the two votes did see the publication of some popular antinuclear tracts and the Bryerton series in the *Register-Guard*, this literature alone would not seem to explain the shift from overwhelming support for nuclear energy in 1968 to defeat for nuclear energy in 1970.
54. William L. Rankin, Stanley M. Nealey, and Barbara Desow Melber, "An Overview of National Attitudes Toward Nuclear Energy: A Longitudinal Analysis," in *Public Reaction to Nuclear Power: Are There Critical Masses?*, American Association for the Advancement of Science Selected Symposium 93, ed. William R. Freudenburg and Eugene A. Rosa (Boulder, CO: Westview Press, 1984), 41–67. On the Oregon polls, see Stanley M. Nealey, Barbara D. Melber, and William L. Rankin, *Public Opinion and Nuclear Energy* (Lexington, Ma: Lexington Books, 1983), 41. These two books provide comprehensive reports and interpretations of public opinion data.
55. Parks, interview.
56. *Wall Street Journal*, 13 October 1981, 35; Parks, interview. For an earlier view of Eugene's outdoors, environmentalist way of life, see Philip Hamburger, "Notes for a Gazetteer: XLII-Eugene, Ore.," *New Yorker*, 1 June 1963, pp. 95–99.
57. Parks, interview.
58. Ibid.

CHAPTER TWELVE

THE COMMUNITY VALUE OF WATER
Implications for the Rural Poor in the Southwest

F. LEE BROWN AND HELEN M. INGRAM

"Water flows uphill to money" is a popular adage in the arid Southwest.[1] It succinctly captures the well understood economic and political reality that water is, indeed, controlled by wealth and power. Philip L. Fradkin has remarked, "When it comes to distributing water in the West, it has been the politically strong and aggressive who get it. To be tenacious and knowledgeable helps."[2] Such persons almost invariably have been members of the dominant Anglo culture.

The corollary proposition to the adage, namely that water runs from the poor and powerless, is less well understood and virtually unstudied. This article examines the relationship between water and poverty in an arid region where throughout human history control over scarce water resources has caused conflict. Two questions are addressed: (1) How is water important to the rural poor in the Southwest—principally Indians and Hispanics? (2) What realistic, water-based strategies offer the best hope for advancing the interests of the rural poor?

The response to these questions consists of three interrelated propositions. First, there is an extremely strong bond between water and the cultural values of the rural poor. For this reason, perceived threats to their water interests evoke from the poor a vigorous and emotional reaction. Successfully gaining or preserving control over water contributes to the improvement of their general welfare by affirming and fulfilling cultural values.

Second, effectively participating in water policy in order to control water will strengthen the capacity of the poor to advance their general interest within the larger society of which they are a part. While water may not be a unique vehicle to build capacity, water provides a major opportunity for the rural poor because it so strongly motivates participation. Further, fundamental changes are occurring in western water institutions and policies which provide openings not before available whereby the rural poor can insert their interests into water resources policy making. By collectively acting to promote and preserve water values, the poor have an opportunity to build capability to improve their condition in areas beyond simply realizing water values.

Third, the economic condition of poor, rural people can be improved by water-based strategies that offer economic benefits as well as being culturally compatible. Solid evidence suggests that the water development strategy most acceptable and desirable to rural communities involves irrigated agriculture. While there is no guarantee that agricultural strategies that build upon community desires and values will succeed, strategies that ignore these values are certainly doomed. The obstacles to the development of a successful and profitable irrigation economy are substantial; none are insurmountable, however, given time, will, and judicious assistance commensurate with the challenge.

During the development period that has characterized western water affairs over the last hundred years, the rural poor of the region neither extensively benefited from, nor effectively participated in, decisions involving river compacts, dam or reclamation project construction, or most other major water events. In a number of important respects developmental activities have actually been injurious to their interests.

Historically, the development of the surface streams of the region, particularly the Colorado, was made possible by the existence of a powerful, politically skilled coalition which could translate its development aims into federal dollars by making use of a reclamation ethic having substantial national support.[3] While debate was often acrimonious and contention frequent, there was a pervasive desire throughout the region to harness the rivers with dams, generate cheap hydroelectric power, and construct aqueducts to transport water to irrigation projects and population centers.

Neither the coalition nor developmental activity it fostered represented the full range of regional constituencies. One conspicuous exception may be collectively termed the rural poor and includes both Indians and Hispanics. For Indians a clear symbol of their absence from the coalition is contained in the language of the 1922 Colorado River Compact, one of the elder water institutions in the region. Despite the fact that Indian "reserved rights" had been given judicial recognition by the United States Supreme Court during the first decade of the century,[4] the Indian tribes of the Colorado were not included in the 1922 negotiations. The resulting Compact, which divided the flow of the Colorado between its upper and lower basins and opened the door to construction of Hoover Dam and the development of the river, side-stepped potential Indian rights with the language, "Nothing in this compact shall be construed as affecting the obligations of the United States of America to Indian tribes."[5] Though the negotiators were aware of potential tribal claims, the Indians were not included in this most basic developmental decision on the Colorado.

The Hispanics of the upper Rio Grande in southern Colorado and northern New Mexico, have, in large measure, also not been party to the developmental coalition. As a consequence, many water storage facilities and water delivery systems built in the upper Rio Grande region have been mainly for the benefit of extraregional interests.[6] Many of the Hispanic community irrigation systems (*acequias*) suffer from inadequate earthen and brush diversion works which wash out with spring floods and from inadequate water flow in the latter months of the growing season because there is insufficient storage capacity to hold some of the spring runoff for the drier months ahead.

In those locations in which development has proceeded without the participation of the rural poor, they not only failed to share in benefits, but their interests have frequently even been damaged. For example, it is politically more difficult for tribes to implement their court-granted "paper rights" and obtain "wet water" when all of the water supply in a basin is already being used. An even more extreme example has occurred in the upper Rio Grande in which two of the oldest water-using societies in the nation, Pueblos and Hispanics, have been pitted against one another in the wake of developmental decisions on the Rio Grande. There are few exceptions to the general rule of the water-development period which ignored or further impoverished the poor. Even the largest Indian water development, the Navajo Irrigation Project, followed a tortuous funding path which lagged far behind its politically negotiated, non-Indian counterpart, the San Juan–Chama diversion project.[7]

OPPORTUNITY AND CHALLENGE IN THE MANAGEMENT ERA

Now, just as some leaders of the rural poor have facility with the rules of the development game, western water institutions and policies are fundamentally changing. In the words of former Arizona governor Bruce Babbitt, "the old institutions are crumbling."[8] This change is most pronounced in the Southwest where surface-water supplies are essentially fully appropriated and many ground-water aquifers are being mined, yet economies and population continue to grow. Coupled with declining federal developmental funding, these factors have created strong pressures that are shifting the region's water institutions from their almost exclusive focus on development. Increasingly, the functions performed by these institutions are more accurately described as *water management* in which conservation, reallocation, and quality preservation and improvement assume greater importance relative to the traditional development activity.

As the region begins the transition into the management era,[9] an important opportunity exists to include the rural poor in discussion processes that will reshape the region's water institutions, policies, and events. Effective participation is essential not only from the standpoint of the rural poor themselves but also in the interest of the larger society within which those excluded become destabilizing forces. The consequence of tribes having been left out of past water development decisions is the increasingly costly and time consuming litigation and negotiation over Indian water rights that currently clog the courts and political arena. Everyone suffers: business and governmental leaders fear the extensive delays caused by protracted dispute; Indian leaders fear yet another assault on a valuable natural resource they regard as part of their heritage and legal right.

The inequity that many rural poor perceive as having taken place in the development period has left strong feelings of bitterness and suspicion. They are no longer willing to remain neglected constituencies. In the upper Rio Grande, for example, recent decades have seen a proposed dam and conservancy district defeated by Hispanics,[10] Indian water rights asserted in a protracted and emotional court battle,[11] and resistance to market sales of water rights by Hispanics. These clashes are just part of a general struggle over water which is taking place between the rural poor and municipal and industrial interests in the region. Less developed, traditional cultures and communities whose capacity to act effectively in modern circumstances is just being forged are confronting the often aggressive, more affluent communities whose values and objectives are secure.

Despite the strong tensions surrounding water issues, which wax and wane with the advent and resolution of particular disputes, there is a growing opportunity for both the political and economic interests of the rural poor to be effectively inserted into water-decision processes. That assessment rests on four factors: (1) the strength and extent of their claim on water; (2) the increased possibility during a period of fundamental institutional change; (3) the democratization of the water decision process that is occurring coupled with the arrival of a new generation of water leaders; and (4) most importantly, the growing skill and determination with which their interests are asserted.

Whether the opportunity will be successfully pursued remains to be determined. The battles over water have been and will be arduous. The responsibility for articulating and asserting the water-related values and interests of the rural poor in the Southwest ultimately rests with those communities and their leaders. External assistance, judiciously directed, can be useful, and societal receptivity to these values is also essential.

THE RURAL POOR IN THE SOUTHWEST

Just who are the rural poor in this region? First, we must give some context to the meaning and occurrence of poverty. Low income is clearly an element in the poverty condition, but, as many authorities have noted, it is not relative economic status alone that signals poverty. Instead, poverty also connotes an inability on the part of people to exercise substantial control over their own lives and to cope effectively with outside pressures. Poor people often lack the level of education necessary to perform successfully in complex modern society. Very frequently, poor people also suffer substantial social and political barriers when their poverty condition is combined with an accompanying status as racial or religious minorities. For these reasons, an identification of rural poverty should be based on criteria including powerlessness as well as economic status.

Poverty in the Southwest, as measured by economic statistics,[12] tends to be accompanied by membership in ethnic minorities, including Indians and Hispanics. These poor minorities make up a large proportion of the population in reservation areas and in the Rio Grande Valley of southern Colorado and northern New Mexico. These groups are also of interest because of substantial, unrealized water claims, as is the case of Indian tribes, or because of their large existing water use.

THE CONTROL AND USE OF WATER BY THE RURAL POOR

The water rights system of the Southwest is highly complex. Moreover, it treats Indian tribes differently from rural Hispanics. A brief historical review explains how the system evolved. For hunters and gatherers[13] who moved from water source to water source in the region for thousands of years, water-use customs were relatively simple. Subsequent agricultural settlement, particularly during Spanish colonialization (1540–1821), gave rise to the need for a system of enforceable rights to the use of water. Formally, the Spanish colonial system honored grants of land made on parchments issued under authority of the king, and the right to reasonable water use was attendant to the land. In practice, disputes over water use were seldom resolved by courts and officials on the basis of written title alone but also took account of prior use, need, third-party rights, intent, governmental priorities, municipal and Pueblo preferences, and notions of equity and common good.[14] Both Spanish and the later Mexican (1822–46) reign saw water allocated in a process balancing formal title claims with people's needs and expectations.[15]

Rights created under Spanish and Mexican rule were recognized by the United States in a protocol accompanying the 1848 Treaty of Guadalupe Hidalgo. Congress did not prescribe a water-rights system for the region, leaving the territorial legislatures and courts free to declare a public interest in water while rushing to make rights private through the rule of prior appropriation (first-in-time, first-in-right).[16] Federal rights in navigable water were announced only occasionally, and the legal doctrine of reserved water rights for Indians was not pronounced by the U.S. Supreme Court until 1908,[17] long after most Indian reservations had been created. Similar rights for reserved federal lands, such as National Forests and National Parks, were recognized even later, in 1963.[18]

Several strata of water rights thus were laid down in the settlement of the Southwest and persist today: (1) pre-1848 rights of use officially granted or awarded under Spanish and Mexican rule; (2) post-1848 appropriative rights perfected under territorial and state law; (3) sundry rights asserted by states over their public waters; (4) federal navigational and reserved rights; and (5) Indian reserved rights. Hispanics hold both pre-1848 and post-1848 rights. Indians on reservations variously hold or claim pre-1848 rights (e.g., Pueblo or aboriginal) and reserved rights. As mentioned, many of the Indian rights have not been quantified or adjudicated, leaving considerable uncertainty as to their extent.[19]

While water use differs considerably among different groups of the rural poor, there are some constants. Agriculture is everywhere a primary consumptive user, though the type of agriculture varies from cotton in southern Arizona to alfalfa in northern New Mexico. The significance to the rural poor of the agriculture for which water is used invariably is more than simple economic benefit. Agriculture is an important part of lifestyle, even though it may be a part-time activity. The agriculture enterprise attaches indigenous people to a place, and provides a link to the past. Even the part-time farmer gains a sense of security and independence from the predominantly Anglo world in which he or she may be employed. The relationship of water to human values that transcend economic returns is a subject to which we now turn.

In the Southwest, as in arid regions generally, water is a special resource. Because water is essential to the quality of life and also to a secure future,

Westerners place a value on water which transcends its material worth. While water is certainly a means to economic improvement, its role in establishing or maintaining a feeling of community well-being results in a strong symbolic and emotional attachment. This community value of water is particularly strong among many Indians and rural Hispanics. Their continued proximity to the natural environment and the insecurity of their cultures in modern circumstances make the presence of water crucial to their community well-being. This special attitude is sufficiently prevalent in all communities, however, that it compels examination as part of any effort by the rural poor, or others, to build a viable water strategy.

What evidence supports the assertion that water is a "special" resource, and what function does it serve other than contributing to material gain? If water is perceived in symbolic terms, what values does it symbolize? What is the relationship between the community and commodity values of water? Finally, what are the implications for the rural poor of these values and the water environment they help engender?

A SPECIAL RESOURCE

Many economists view as misguided the sentiment that water is a "different" resource with a special meaning linked to survival and security that transcends its commodity value—and their testimony that such attitudes do in fact exist is particularly compelling. Kenneth Boulding has written that public emotions and feelings associated with water are so strong that what he regards as rational solutions to water problems may never be possible.[20] In an article centered around the "Water is Different Syndrome," Maurice Kelso argued that water is bound up in a number of myths such as "water is priceless," and that water is important to the establishment of strong democratic institutions, or agrarian fundamentalism, all of which set it apart from ordinary market mechanisms.[21] In their study of six irrigation communities Maass and Anderson concluded that water is "a special product" and that farmers generally believed it "should be removed from ordinary market transactions so that farmers can control conflict, maintain popular influence and control, and realize equity and social justice."[22]

Articles on water issues in regional newspapers further describe this emotional and symbolic significance. Phrases such as "life-giving," "precious," or stories emphasizing the importance of water to posterity or survival appeared in about one-fifth of the sample of newspapers. The type of treatment found and classified as symbolic can be illustrated in a few examples:[23]

Drought in the West is as certain as death and taxes . . . water is fundamental.

Water is our lifeblood in this state.

We are allocating scarcity.

In this water poor desert, water is life itself, if you know where to find it.

A Westerner's priorities are, in order: (1) water; (2) gold; (3) women. You may tamper with the latter two but not the first.

This substantial emotional and symbolic meaning we term the community value of water, and it is toward a more complete description of the term that we now turn.

WATER AND COMMUNITY

Long caught up in the web of human relationships and social dependencies, water is closely tied to social organization. After all, collective decision making first developed to provide basic goods and services that individuals could not provide, or not provide so well, for themselves. Among such services are fire and police protection, and water supply. Further, decisions about water have commonly been made on the basis of community rather than individual interest. Spanish colonial water law, from which much of our present Western water law is derived, went to great lengths to protect the public interest and place it above parties seeking particular water rights.

The water retained by the Spanish-American town was held in trust for the benefit of the entire community. It was not the property of the inhabitants of that town, either individually or collectively, but rather was the property of the collective body itself.[24]

The *acequia* system of the Hispanics in northern New Mexico is one example of the importance of water to community. As the community developed, the farmers built, maintained, and operated the water system together. Raymond Otis created fiction from the long-standing Hispanic obligation to share in water construction, allocation, and management. According to Otis, a man can stand the loss of a wife, a field, an arm, but not the ditch. After a bad storm, every man in the *Little Valley* grabs a shovel and restores his segment of the ditch before he even thinks of damages to his house or anything else. The *mayordomo* (ditch boss), a job that is passed around, is always obeyed, and the irrigation schedule is faithfully kept.[25]

The same sort of reciprocity or mutual trust that is basic to community life that Otis wrote about among the Hispanics also emerged among Anglo settlers. Mormon pioneers established communities on the basis of irrigation. Ralph Moody wrote of resolving water conflicts through reasonable give-and-take among homesteaders. Moody and his near neighbors at the tail of the irrigation ditch found themselves in pitched battles with farmers at the ditch head during a drought when water seldom reached the tail until people got together and agreed to share shortages.[26]

The saguaro harvest ceremony central to Tohono O'odham community life was regarded as responsible for bringing the rain. The O'odham were similar to many other native peoples living in arid lands who believed that droughts were not merely acts of nature but the vengeance of the gods brought upon them by the breach of ceremony or obligation. Florence Cranell Means created a story of a three-year drought among the Hopi during which many suffered. It is only when one by one the surviving Hopi publicly confess their sins that the gods relent and a long, gentle, life-giving rain falls.[27]

Water was central to Indian culture not just for its life-giving properties, but also for its role in myth. "Because the Pueblo Indians have been a people without a written language, historical fact can often change to myth; but since myth is their only window to history, the dimensions of myth itself form an autonomous reality."[28] The mythical role of water is strongly reflected in Frank Waters's *The Man Who Killed the Deer*. A Pueblo father tells his son who is about to be initiated into manhood:

> You will be taught the whole history of our people, of our tribe. How they had their last arising from the deep turquoise lake of life at the center of the world, the blue lake in whose depths gleams a tiny star, our Dawn Lake. How they emerged from a great cave whose lips dripped with water to congeal into perpetual flakes of ice white as eagle-down. You will understand, then, son, why those of our clan are called the Deep Water people: why our kiva, this kiva, is called the eagle-down kiva; the meaning of our masks, our dances, our songs. You will see this cave. You will see this lake—our Dawn Lake.[29]

Fairness, like water, is fundamentally important to social arrangements, and so it is not at all surprising that communities have been preoccupied with the fairness of water distribution. In Spanish colonial law governing water allocation such relatively objective criteria as title and prior appropriation were tempered by a number of other, more subjective issues related to fairness. Judges attended to the 1720 Santa Fe guideline, "divide the waters always verifying the greatest need . . . and giving to each one that which he needs."[30] Rights of third parties that might be damaged by new upstream water uses were protected. Above all, the rights of the collective community were considered. "There is no question that in Spanish and Mexican judicial systems the rights of the corporate community weighed more heavily than those of the individual."[31]

The rules of fairness in water allocation are honed to their sharpest edge in irrigation communities. In their historical study of six such communities in the United States and Spain, Maass and Anderson found that equity far outweighed efficiency as a decision rule for water allocation. In the cases they studied, unequal treatment of individuals in the same situation or category is avoided. Rotation is the common practice and on the whole is faithfully honored.[32] As Oliver LaFarge notes in *The Mother Ditch/La Acequia Madre*, a book about irrigation in northern New Mexico, a good neighbor "closes their gate on time."[33] In Valencia the ordinances state a generous general principle that all farmers have an obligation to give aid to those who have the greater need.[34]

Fairness is not the same thing as equality in irrigation communities. Water is distributed by rule of seniority (first in right, first in time) or proportionally by size of holding. As a rule, irrigators are more concerned with fairness in process than equality in results. The fairness of procedures is judged according to whether or not the procedures prevent control from being imposed from outside the community, or from arbitrary actions from the community's officers. Above all, community members must be satisfied by their own participation in determining procedures.[35]

THE COMMUNITY VALUE OF WATER: PARTICIPATION AND LOCAL CONTROL

The attitude that the people themselves should have a strong voice in water decisions is found among local water users throughout the world. Maass and Anderson found, in the six irrigation communities they studied, a variety of institutions and procedures that were used to resolve conflicts over water allocation. The maintenance of local participation and control was the overriding concern in settling disputes, and when institutions or procedures threatened these overriding values, they were replaced with others.[36]

The proportion and content of articles in western newspapers dealing with participation in water resources attests to its importance as a regional value. Nearly 10 percent of the articles about water referred to participation, and of these more than a quarter treated participation in symbolic and emotional terms. People react strongly if they believe their participation has been denied. In one case, the apportionment of funds by the state legislature for a common water system to service two small New Mexico villages, Canjilon and Cebolla, was vigorously protested by Canjilon because its residents had not been notified of pending legislation. One resident termed the measure "a political maneuver which would rob Canjilon of water which rightfully belongs there."[37]

Scholarly literature attests to the importance of community involvement in the development of successful irrigation projects worldwide.[38] The strength of indigenous irrigation systems is that they involve the farmers in allocation of water, maintenance of the system, and establishing the rules for resolving disputes. The staying power of the *acequia* system in northern New Mexico lies to a large extent in local autonomy in management according to local customs and traditions. Individuals are bound to the community through their participation in water matters.

> Individuals identified dams and ditches by names of ancestors who through the centuries, helped maintain them—thus proclaiming their inheritance through their communities. This is not done in a romanticized fashion; rather, it serves as an important community focus for personal identity.[39]

Prevailing western dialogue indicates that if you have water, you have a chance; if not, you are done for. Western newspapers make very clear the opportunity value of water. New Mexico State Engineer Steve Reynolds described water "as simply the limiting factor."[40] According to former Utah Governor Scott Matheson, "we have little time left to take charge of the small amount of water that gives us life."[41] The unquestioned reason given for the slow death of the little New Mexico town of Colonias was loss of water when the Pecos River changed channels and left the town high and dry.

> Before the Pecos began to wander, Colonias was a prosperous farming village on the main wagon road from Santa Rosa to Las Vegas. Only when it was deprived of water did the town begin to die.[42]

Water represents opportunity less because the water itself has economic value than because control over it signals social organization. A community that cannot hold on to its water probably does not have what it takes to be very much. Strong communities are able to hold on to their water and put it to work. In their thorough study of the relationship of water to social organization, Robert and Eva Hunt observe, "What does emerge from this material is a quite consistent picture of a strong relationship between irrigation and power."[43]

Indian peoples in the Southwest and Hispanics in the Upper Rio Grande Valley feel that they have been denied the community values associated with water, including equity, participation, and opportunity. For Indians, failure to secure a water supply is but one more example of the dominant culture's failure to afford equity to Indian peoples. Consider the words of Wendell Chino, president of the Mescalero Apaches:

> In the nineteenth century, direct force was used to take over Indian lands. The danger today is more subtle. Under the cloak of legal strategy and executive department policy decisions, a real threat is stalking all Indian leaders. These decisions regarding legal strategy and policy are designed to dry up the water from what little land is left to us. We must be alert. We must protect our rights. The consequences of losing our water would be as serious as those following the loss of our lands in the nineteenth century.[44]

Broken promises mark the trail of Indian water rights controversies. Nowhere is the gap between promise and reality more poignantly portrayed than with the Fort Belknap tribe in Montana, which won the water-rights lawsuit enunciating the *Winters* doctrine, which "reserved" water rights for Indians sufficient to "practice the arts of civilization." Today, the water use of Fort Belknap Indians is about what it was at the time of the historic court decision, while the diversions of others have increased greatly because of federally financed reclamation projects not available to the Indians. The quality of life of the Indians has remained among the poorest in the nation.[45]

The Salt River Indians near Phoenix feel twice deprived of their rightful voice in water matters. First, they object to the Kent Decree of 1910, issued by a federal territorial court before Arizona became a state, which they believe unfairly limits their rights. Second, they object to the removal of their lands from the service area of the Salt River Project Water Users Association, which they regard as representing only Anglo development interests.[46]

Hispanics in the upper Rio Grande and the Tohono O'odham in southern Arizona whom we have interviewed made it very clear that money was no substitute for the opportunity value of water. As one Indian said: "Money is just spent and the people are left with nothing. With water, there is something in the future."[47] A Hispanic replied when asked how he felt about people who sell water, "It is clear they have forgotten how their grandfathers made their living."[48]

Frank Waters writes about the association of control over water and community survival. The elders of the tribe count the slipping away of the old, good ways to the loss of Dawn Lake, now under the control of the United States Forest Service and open to Hispanic grazers and Anglo picnickers. In frustration an Indian agent shouts at the Chief:

> Let us forget this thing. Why should you cause trouble to the Government? Hasn't it given you a hospital, school, a new ditch so you can have water for your fields, a threshing machine?
> The old man answers patiently:
> They schools, but mebbe soon no children. They hospital, but mebbe soon people they dead, no sick. They new acequia and aquina, but mebbe soon earth she die too. This Dawn Lake our church. From it come all the good things we get. The mountains, our land, Indian land. The Government promised. We no forget. This we say.[49]

In Anglo society, too, water is symbolic of opportunity. Farmers in Arizona have continued to support the Central Arizona Project despite mounting evidence that project water will cost them more than they can afford to pay. From their viewpoint, it is inconceivable that gaining control over water could be a mistake, regardless of the cost.[50]

THE COMMUNITY VALUE OF WATER: CARING FOR THE RESOURCES

Westerners who live close to the land are aware of water's importance in the natural environment. But even native westerners living in urban areas hold a respect for water. From water running down the street due to an overirrigated lawn to massive fountains heralding the latest residential land development, waste offends people who care for an important resource.

The perception that water is being wasted does not mean that people make a rational choice to plug leaks and fix leaky faucets, but rather that people have a deep feeling that waste or excess is incompatible with and even irreverent toward such a valuable and fundamental part of the environment. Spanish water law clearly recognized that a water right did not include freedom to waste. Even in the case of spring or well water originating on private property, the owner could not deprive his neighbor of its use simply by wasting that which he did not need.[51]

One non-Indian observer of the disputes over tribal claims to water, while concerned about the potential consequences for non-Indians, nevertheless privately expressed the opinion that tribes would probably make better managers of the water than the current system because of their stronger sense of stewardship. Indeed, it is likely that some of the existing excess in western water use would not have occurred had the traditional practice of leaving it in the stream until it is needed not been replaced by a "beneficial use" and "priority" system that hastened the capture of water.

COMMUNITY VALUE, COMMODITY VALUE, AND MATERIAL IMPROVEMENT

The hard-nosed analyst may be skeptical of the utility of poor rural people pursuing the community value of water. If material improvement is sought, surely water should be treated like a material commodity instead of pursuing a mirage like "community value." Moreover, isn't it this same noneconomic viewpoint that "water is different" which fueled the "develop at any cost" passion which stands so much in need of correction in this time of budget austerity? Community values are basically nonmaterial and symbolic—the human stuff about which novelists write, or the basic belief systems to which politicians appeal. It is certainly arguable that if economic development is really the aim of the poor rural communities, water should be treated like any other commodity that can be used or sold depending on what brings the highest economic returns.

But the fact is that many, perhaps most, people do not feel about water the way the purely rational analyst suggests that they should. Case studies of Hispanics in the upper Rio Grande and the Tohono O'odham Nation in Arizona indicate that although economic development is valued, the community value of water is overriding. Further, this attitude is shared to some extent by westerners in the dominant culture. It was the strong belief that "water is different" coupled to the symbolic engine of "making the desert bloom" that drove water development policy for virtually a hundred years.

In economically developed communities the emotional relationship to water has become obscured because the smooth running of institutions that secure water supply is taken for granted; the availability of water is as easy as turning on a faucet. Even in economically developed communities, however, when control over water is threatened, the strength of the emotional tie reemerges. When the United States District Court struck down New Mexico's anti-export statute allowing El Paso access to groundwater supplies in New Mexico, citizens of the state reacted angrily. The *Albuquerque Journal*'s editorial was, if anything, understated.

Quite simply New Mexico's border has been breached. Theoretically, outside municipalities or states—or anybody—can now apply to claim every unappropriated drop of New Mexico water. It is ironic because New Mexico has carefully guarded and conserved its water resources, only to lose a lawsuit to a city and state that takes far less care of the resource.[52]

Since the community value of water is a common thread among a variety of cultures, both developed and undeveloped, it is reasonable to suggest that the community and commodity value of water need not be incompatible, but may instead work together. Indeed, both the stewardship component of water's community value and the marketplace value of the water commodity signal the need for increased care in its use under conditions of scarcity.

The lessons from the literature on international development projects are also relevant.[53] They teach that the process of development occurs in stages. In the earliest, difficult but essential, stage the goal of efficiency takes a backseat to other objectives, such as engaging the participation and commitment of community leaders and members. Once community capacity to act effectively has been developed, then attention can shift to efficiency. Because of the close emotional ties of water to community, some international development experts have recommended rural water projects as a means for building community capacity.[54]

The community value of water should be regarded as the broad context within which economic values may be pursued. Unless community values of water are satisfied, debilitating attitudes of hopelessness persist among the rural poor, fostered by perceptions of injustice, lack of efficacy, and loss of opportunity. Rather than being inimical to economic progress, their fulfillment seems to be a necessary condition for that progress. Moreover, the acquisition of secure control of water through a community's own initiative and effort may supply the key ingredients of participation and belief in the possibilities of the future that are essential to sustained economic improvement.

THE CHALLENGE FOR THE RURAL POOR

Whether or not the rural poor can fashion a water resource strategy which is responsive to community values and also provides economic benefits will be determined in particular cases. While in the abstract community and economic values are compatible, in the real world of events the strong assertion of the commodity view of water by the dominant Anglo culture presents a formidable challenge to particular poor communities. By way of conclusion, the following lessons can be drawn from our discussion:

(1) Water ownership is unlikely to produce a collective windfall affluence for either tribes or Hispanic communities. Market prices for rights will increase and some individual rightholders may profit handsomely, but in most instances the holdings are either too small or the community population too large to support an "oil sheik" strategy, even if the political process would tolerate such an approach. There may be local areas in which the development of major commercial or industrial enterprises such as ski resorts or power plants dramatically increase the price of water rights, but no general strategy should be built on the assumption that sizable economic gains can be achieved simply by holding rights and waiting. A second, and more important, reason for discarding this strategy flows from the notion of community value. Unless the cultural and community base is secure, any strategy for economic gain through the sale or lease of water rights is likely to be met with hostility.

(2) Low-valued or economically marginal uses of water will be put under increasing pressure practically everywhere, even though dramatic increases in market prices will occur only in some localities. For tribes, this means that it will be increasingly difficult to gain control of water that is already fully appropriated, particularly if the proposed uses of that water are low-valued economically. Resistance against Indian control over water may be softened if Indians agree to lease back their water to existing or other users, at least for a period of transition. Yet for different reasons tribal leasing itself is sometimes opposed. Nevertheless, in a water-scarce environment such as the Southwest, it is likely that some enterprising municipalities or businesses will be interested in leasing Indian water.

Hispanics in the upper Rio Grande are seeing water rights in their communities being bid away acre by acre. Although the process is not fast-paced or substantial at this point, it can be expected to increase over time. Particularly vulnerable, since they produce relatively few dollars per acre-foot of water applied, are the alfalfa and forage crop uses of water which predominate in most Hispanic communities. In the continuing shift of water from irrigated agriculture described above, it is the water rights associated with these crops that are most likely to be the target of prospective buyers.

The production of crops of economically higher value is more sustainable. Again, it is a question of degree. Relatively little irrigation water can go a long way toward meeting municipal and industrial needs. Despite the extent of regional growth, quantitatively there does not have to be wholesale reduction in irrigated agriculture to meet the water needs of this growth.[55] In the marketplace, other things equal, water from lower-valued crops can be obtained more cheaply than water from such crops as vegetables and fruits.

(3) The protection provided to existing and prospective Indian and Hispanic water users by legal institutions is threatened. Fundamental institutional and policy change is occurring in western water practices. The economic scrutiny recently applied to Indian water claims is illustrative. Similarly, the Rio Grande Compact that divides the United States's share of that river among Colorado, New Mexico, and Texas has effectively insulated Hispanic water rights in the upper basin region from water demand in Santa Fe, Albuquerque, or even El Paso.[56] Yet with the challenges to compacts that are occurring as a result of the U.S. Supreme Court *Sporhase* decision, this protection should no longer be considered absolutely secure.[57] Old rules and habits cannot be considered as fixed. Instead, the newly emerging institutions, policies, and patterns must be discerned and learned. It is an "old saw" that such times breed opportunity for the wary and danger for the unwary.

(4) Hispanics and Indians must be actively involved in water politics if their water strategies are to be successful. With both the marketplace and the judiciary promoting the commodity value and view of water, Indians and Hispanics must assert their community values politically through elective and agency processes. Further, political and governmental arenas will cast the new rules for the water management era to come.

There may appear to be disadvantages to a political strategy. The rural poor were not effectively included in the political arena when coalitions were formed to promote previous water development. Why should they expect to have any greater success at this juncture? Further, water politics is currently in disrepute because appropriations for "hometown" water projects are commonly perceived to be one of the most flagrant abuses of pork-barrel, redistributive politics.

Nevertheless, the political strategy offers the best hope. Chances of succeeding can be improved by acquisition of relevant education, training, and experience. The art of succeeding in politics can be learned just as the arts of succeeding in the marketplace or the courtroom. Indian success in the latter arena over recent decades has come at least in part from learning just how the judicial process works. Moreover, neither Indians nor Hispanics are novices in water politics, although the general posture until recently has been more reactive than assertive.

The disreputable image of water politics stems in part from an absence of economic accountability. Projects have been built which were economically unjustified and did not enlarge the material pie, but only redivided it. In a former age of affluence such political decisions may have been tolerated, but in the present era of austerity this is less likely to be the case. Therefore, the "involvement in water politics" by Indian and rural Hispanics cannot simply mean pursuit of the "water business as usual." The commodity value of water will play a strong future role, and this value must be accommodated by the poor for their water strategy to be successful.

ACKNOWLEDGMENTS

The authors thank Gilbert Bonem, Wade Martin, Thomas R. McGuire, Stephen P. Mumme, Luis Torres, Mary Wallace, and Gary Weatherford for their contributions to this article and the larger study from which it is drawn. This article is a revised version of two chapters from the forthcoming *Water and Poverty in the Southwest* (University of Arizona Press, fall 1987).

NOTES

1. Although subject to various geographical definitions, as used in this article the Southwest loosely refers to the Four Corners states of Colorado, Utah, Arizona, and New Mexico.
2. Philip L. Fradkin, *A River No More: The Colorado River and the West* (Tucson: University of Arizona Press, 1986), 155. Much the same point is made even more strongly in Donald Worster, *River of Empire: Water, Aridity, and the Growth of the American West* (New York: Pantheon Books, 1985).
3. See, generally, Norris Hundley, *Water and the West* (Berkeley: University of California Press, 1975), and *Dividing the Waters* (Berkeley: University of California Press, 1966).
4. *Winters v United States*, 207 US 564 (1908).
5. Article VII, Colorado River Compact, 24 November 1922, Santa Fe, NM.
6. Many of the facilities which have been built, e.g., El Vado and Abiquiu Dams, were designed as either flood protection or storage facilities to serve municipal and agricultural interests lower in the basin.
7. "While the San Juan–Chama Division and the Navajo Indian Irrigation Project were authorized together, congressional appropriations for NIIP lagged far behind. By 1970, the SJD was 65 percent complete while NIIP was only 17 percent constructed." Charles DuMars, Helen Ingram, Ronald Little, Bahe Billy and Phil Reno, "The Navajo Irrigation Project: A Study of Legal, Political and Cultural Conflict," in *Water and Agriculture in the Western U.S.: Conservation, Reallocation, and Markets*, ed. Gary D. Weatherford (Boulder, CO: Westview Press, 1982), 110.
8. Speech to the Western Governors Association, 22 May 1984, Palm Springs, California.
9. This theme of regional transition from water development to water management is not new (see, for example, Weatherford, *Water and Agriculture in the Western U.S.*) Yet it is not

well understood outside of the West, nor is its fundamental character always fathomed in the West itself.

10. The dam was Indian Camp Dam, which was accompanied by a proposed Taos Conservancy District. See John Nichols, "Land and Water Problems in Northern New Mexico," paper presented at the University of New Mexico School of Law, 16 October 1975.

11. *New Mexico v Aamodt*, 537 F. 2d, 1102 (10th Cir 1976).

12. County aggregate data were examined using two-income criteria: county per-capita income, and percentage of persons living below the poverty level in the county (i.e., counties where 20 percent or more of the population was below the poverty level were considered "poor").

13. For a description of Indian movement along New Spain's northern frontier from the sixteenth through the nineteenth centuries, see Albert Schroeder, "Shifting Survival in the Spanish Southwest," in *New Spain's Northern Frontier*, ed. David Weber (Albuquerque: University of New Mexico Press, 1979), 237–55.

14. See, generally, Michael C. Meyer, *Water in the Hispanic Southwest: A Social and Legal History, 1550–1850* (Tucson: University of Arizona Press, 1984); also Charles T. DuMars, Marilyn O'Leary, and Albert E. Utton, *Pueblo Indian Water Rights: Struggle for a Precious Resource* (Tucson: University of Arizona Press, 1984).

15. Ibid.

16. See, generally, Wells A. Hutchins, *Water Rights in the Nineteen Western States, Vol. I* (Washington: U.S. Government Printing Office, 1971), 159–75. The territorial legislature of New Mexico recognized prior appropriation in 1851; Colorado in 1861, Arizona in 1864, and Utah in 1881.

17. *Winters v United States*, 207 US 564 (1908).

18. *Arizona v California*, 373 US 546 (1963).

19. For a survey of recent Indian water disputes, see John A. Folk-Williams, *What Indian Water Means to the West* (Santa Fe: Western Network, 1982).

20. Kenneth Boulding, "The Implications of Improved Water Allocation Policy," in *Western Water Resources: Coming Problems and the Policy Alternatives*, Marvin R. Duncan, ed. (Boulder, CO: Westview Press, 1980), 300.

21. Maurice Kelso, "The Water is Different Syndrome, or What is Wrong with the Water Industry?" paper presented to the American Water Resources Association, San Francisco, 9 November 1967.

22. Arthur Maass and Raymond Anderson, *And the Desert Shall Rejoice: Conflict, Growth, and Justice in Arid Environments* (Cambridge: MIT Press, 1978), 5.

23. Helen Ingram and Stephen P. Mumrne, "Public Perceptions of Water Issues in the Four Corners States as Indicated Through a Survey of Regional Newspapers: A Preliminary Report," paper presented at the Western Social Science Association's 25th Annual Conference, Albuquerque, NM, 27–30 April 1983.

24. Meyer, *Water in the Hispanic Southwest*, 157.

25. Raymond Otis, *Little Valley* (1937; reprint, Albuquerque: University of New Mexico Press, 1980).

26. Ralph Moody, *Little Britches: Father and I Were Ranchers* (New York: W.W. Norton, 1950).
27. Florence Cranell Means, *The Rains Will Come* (Boston: Houghton Mifflin, 1954).
28. DuMars, O'Leary, Utton, *Pueblo Indian Water Rights*, 8.
29. Frank Waters, *The Man Who Killed the Deer* (1941; reprint, New York: Pocket Books, 1973), 80.
30. Meyer, *Water in the Hispanic Southwest*, 151.
31. Ibid., 157.
32. Maass and Anderson, And the Desert Shall Rejoice.
33. Oliver LaFarge, *The Mother Ditch, La Acequia Madre* (1954; reprint, Santa Fe: Sunstone Press, 1983), 50.
34. Maass and Anderson, *And the Desert Shall Rejoice*, 27.
35. Ibid., 396.
36. Ibid., 370.
37. *Rio Grande* (Española, New Mexico) *Sun*, 28 February 1980.
38. E. Walter Coward, *Irrigation and Agricultural Development in Asia: Perspective From the Social Sciences* (Ithaca: Cornell University Press, 1980).
39. Sue Ellen Jacobs, "Top-Down Planning: Analysis of Obstacles to Community Development in an Economically Poor Region of the Southwestern United States," *Human Organization* 37 (fall 1978): 252.
40. *Albuquerque Journal*, 25 February 1980.
41. *Salt Lake Tribune*, 16 December 1977.
42. Steve Winston, "A Slow Death on the Pecos," *Albuquerque Journal*, 25 February 1980.
43. Robert C. Hunt and Eva Hunt, "Canal Irrigation and Local Social Organization," *Current Anthropology* (September 1976): 394–97.
44. American Indian Lawyer Training Program, *Indian Water Policy in a Changing Environment* (Oakland, CA: American Indian Lawyer Training Program, Inc., 1982), 56.
45. Norris Hundley, Jr., "The 'Winters' Decision and Indian Water Rights: A Mystery Reexamined," *Western Historical Quarterly* 13 (January 1982): 17–42.
46. Folk-Williams, *What Indian Water Means to the West*, 38.
47. C. Vandemoer and R. Peters, "Indigenous Response to Water in an Arid Environment: A Papago Case Study," *John Muir Institute for Environmental Studies*, 28 February 1984, 44–45.
48. Hispanic field interviews conducted by the author.
49. Waters, *Man Who Killed the Deer*, 24.
50. Helen M. Ingram, William E. Martin, and Nancy K. Laney, "A Willingness to Play: Analysis of Water Resources Development in Arizona," in Weatherford, *Water and Agriculture in the Western U.S.*
51. Novesima Recopilacion, Libro III, Titulo 28, lev 31, as quoted in Meyer, *Water in the Hispanic Southwest*, 21.
52. *Albuquerque Journal*, 21 January 1983.
53. See, for example, David Korten, "Community Organization and Rural Development:

A Learning Process Approach," *Public Administration Review* (September/October 1980): 480–510.

54. Steve Cox and Sheldon Annis, "Community Participation in Rural Water Supply Grassroots Development" *Journal of the Inter-American Foundation* 6 (1982).

55. For a more extended discussion of this point together with a review of some quantitative evidence, see B. Delworth Gardner, "The Untried Market Approach to Water Allocation" in *New Courses for the Colorado River*, ed. Gary D. Weatherford and F. Lee Brown (Albuquerque: University of New Mexico Press, 1986).

56. The terms of the Compact make it practically very difficult to transfer a right across a Compact accounting point, one of which occurs at Otowi Bridge just above Santa Fe.

57. See *Sporhase v Nebraska*, 81 US 613 (1982). In this decision the Supreme Court declared: (1) water to be a commodity subject to interstate commerce clause scrutiny; and (2) the public ownership of water asserted by most western state statutes and constitutions to be a "legal fiction."

CHAPTER THIRTEEN

MEXICAN AMERICAN WOMEN GRASSROOTS COMMUNITY ACTIVISTS
"Mothers of East Los Angeles"

MARY PARDO

The relatively few studies of Chicana political activism show a bias in the way political activism is conceptualized by social scientists, who often use a narrow definition confined to electoral politics.[1] Most feminist research uses an expanded definition that moves across the boundaries between public, electoral politics and private, family politics; but feminist research generally focuses on women mobilized around gender-specific issues.[2] For some feminists, adherence to "tradition" constitutes conservatism and submission to patriarchy. Both approaches exclude the contributions of working-class women, particularly those of Afro-American women and Latinas, thus failing to capture the full dynamic of social change.[3]

The following case study of Mexican-American women activists in "Mothers of East Los Angeles" (MELA) contributes another dimension to the conception of grassroots politics. It illustrates how these Mexican-American women transform "traditional" networks and resources based on family and culture into political assets to defend the quality of urban life. Far from unique, these patterns of activism are repeated in Latin America and elsewhere. Here, as in other times and places, the women's activism arises out of seemingly "traditional" roles, addresses wider social and political issues, and capitalizes on informal associations sanctioned by the community.[4] Religion, commonly viewed as a conservative force, is intertwined with politics.[5] Often, women speak of their communities and their activism as extensions of their family and household responsibility. The central role of women in grassroots struggles around quality of life, in the Third World and in the United States, challenges conventional assumptions about the powerlessness of women and static definitions of culture and tradition.

In general, the women in MELA are longtime residents of East Los Angeles; some are bilingual and native born, others Mexican-born and Spanish dominant. All the core activists are bilingual and have lived in the community over thirty years. All have been active in parish-sponsored groups and activities; some have had experience working in community-based groups arising from schools, neighborhood watch associations, and labor support groups. To gain an appreciation of the group and the core activists, I used ethnographic field methods. I interviewed six women, using a life history approach focused on their first community activities, current activism, household and family responsibilities, and perceptions of community issues.[6] Also, from December 1987 through October 1989, I attended hearings on the two currently pending projects of contention—a proposed state prison and a toxic waste incinerator—and participated in community and organizational meetings and demonstrations. The following discussion briefly chronicles an intense and significant five-year segment of community history from which emerged MELA and the women's transformation of "traditional" resources and experiences into political assets for community mobilization.[7]

Political science theory often guides the political strategies used by local government to select the sites for undesirable projects. In 1984 the state of California commissioned a public relations firm to assess the political difficulties facing the construction of energy-producing waste incinerators. The report provided a "personality profile" of those residents most likely to organize effective opposition to projects:

> Middle and upper socioeconomic strata possess better resources to effectuate their opposition. Middle and higher socioeconomic strata neighborhoods should not fall within the one-mile and five-mile radii of the proposed site. Conversely, older people, people with a high school education or less are least likely to oppose a facility.[8]

The state accordingly placed the plant in Commerce, a predominantly Mexican-American, low-income community. This pattern holds throughout the state and the country: three out of five Afro-Americans and Latinos live near toxic waste sites, and three of the five largest hazardous waste landfills are in communities with at least 80 percent minority populations.[9]

Similarly, in March 1985, when the state sought a site for the first state prison in Los Angeles County, Governor Deukmejian resolved to place the 1,700-

inmate institution in East Los Angeles, within a mile of the long-established Boyle Heights neighborhood and within two miles of thirty-four schools. Furthermore, violating convention, the state bid on the expensive parcel of industrially zoned land without compiling an environmental impact report or providing a public community hearing. According to James Vigil, Jr., a field representative for Assemblywoman Gloria Molina, shortly after the state announced the site selection, Molina's office began informing the community and gauging residents' sentiments about it through direct mailings and calls to leaders of organizations and business groups.

In spring 1986, after much pressure from the fifty-sixth assembly district office and the community, the Department of Corrections agreed to hold a public information meeting, which was attended by over seven hundred Boyle Heights residents. From that moment on, Vigil observed, "the tables turned, the community mobilized, and the residents began calling the political representatives and requesting their presence at hearings and meetings."[10] By summer 1986 the community was well aware of the prison site proposal. Over two thousand people, carrying placards proclaiming "No Prison in ELA," marched from Resurrection Church in Boyle Heights to the Third Street bridge linking East Los Angeles with the rapidly expanding downtown Los Angeles.[11] This march marked the beginning of one of the largest grassroots coalitions to emerge from the Latino community in the last decade.

Prominent among the coalition's groups is "Mothers of East Los Angeles," a loosely knit group of over four hundred Mexican-American women.[12] MELA initially coalesced to oppose the state prison construction but has since organized opposition to several other projects detrimental to the quality of life in the central city.[13] Its second large target is a toxic waste incinerator proposed for Vernon, a small city adjacent to East Los Angeles. This incinerator would worsen the already debilitating air quality of the entire county and set a precedent dangerous for other communities throughout California.[14] When MELA took up the fight against the toxic waste incinerator, it became more than a single-issue group and began working with environmental groups around the state.[15] As a result of the community struggle, AB58 (Roybal-Allard), which provides all Californians with the minimum protection of an environmental impact report before the construction of hazardous waste incinerators, was signed into law. But the law's effectiveness relies on a watchful community network. Since its emergence, MELA has become centrally important to just such a network of grassroots activists, including a select number of Catholic priests and two Mexican-American political representatives. Furthermore, the group's very formation, and its continued spirit and activism, fly in the face of the conventional political-science beliefs regarding political participation.

Predictions by the "experts" attribute the low formal political participation (i.e., voting) of Mexican American people in the United States to a set of cultural "retardants," including primary kinship systems, fatalism, religious traditionalism, traditional cultural values, and mother country attachment.[16] The core activists in MELA may appear to fit this description, as well as the state-commissioned profile of residents least likely to oppose toxic waste incinerator projects. All the women live in a low-income community. Furthermore, they identify themselves as active and committed participants in the Catholic Church; they claim an ethnic identity—Mexican American; their ages range from forty to sixty; and they have attained at most high school educations. However, these women fail to conform to the predicted political apathy. Instead, they have transformed social identity—ethnic identity, class identity, and gender identity—into an impetus as well a basis for activism. And, in transforming their existing social networks into grassroots political networks, they have also transformed themselves.

TRANSFORMATION AS A DOMINANT THEME

From the life histories of the group's core activists and from my own field notes, I have selected excerpts that tell two representative stories. One is a narrative of the events that led to community mobilization in East Los Angeles. The other is a story of transformation, the process of creating new and better relationships that empower people to unite and achieve common goals."

First, women have transformed organizing experiences and social networks arising from gender-related responsibilities into political resources.[18] When I asked the women about the first community, not necessarily "political," involvement they could recall, they discussed experiences that predated the formation of MELA. Juana Gutierrez explained:

> Well, it didn't start with the prison, you know. It started when my kids went to school. I started by joining the Parents Club and we worked on different problems here in the area. Like the people who come to the parks to sell drugs to the kids. I got the neighbors to have meetings. I would go knock at the doors, house to house. And I told them that we should stick together with the Neighborhood Watch for the community and for the kids![19]

Erlinda Robles similarly recalled:

> I wanted my kids to go to Catholic school and from the time my oldest one went there, I was there every day. I used to take my two little ones with me and I helped one way or another. I used to question things they did. And the other mothers would just watch me. Later, they would ask me, "Why do you do that? They are going to take it out on your kids." I'd say, "They better not." And before you knew it, we had a big group of mothers that were very involved.[20]

Part of a mother's "traditional" responsibility includes overseeing her child's progress in school, interacting with school staff, and supporting school activities. In these processes, women meet other mothers and begin developing a network of acquaintanceships and friendships based on mutual concern for the welfare of their children.

Although the women in MELA carried the greatest burden of participating in school activities, Erlinda Robles also spoke of strategies they used to draw men into the enterprise and into the networks:[21]

> At the beginning, the priests used to say who the president of the mothers guild would be; they used to pick 'em. But, we wanted elections, so we got elections. Then we wanted the fathers to be involved, and the nuns suggested that a father should be president and a mother would be secretary or be involved there [at the school site].[22]

Of course, this comment piqued my curiosity, so I asked how the mothers agreed on the nuns' suggestion. The answer was simple and instructive:

> At the time we thought it was a "natural" way to get the fathers involved because they weren't involved; it was just the mothers. Everybody [the women] agreed on them [the fathers] being president because they worked all day and they couldn't be involved in a lot of daily activities like food sales and whatever. During the week, a steering committee of mothers planned the group's activities. But now that I think about it, a woman could have done the job just as well![23]

So women got men into the group by giving them a position they could manage. The men may have held the title of "president," but they were not making day-to-day decisions about work, nor were they dictating the direction of the group.

Erlinda Robles laughed as she recalled an occasion when the president insisted, against the wishes of the women, on scheduling a parents' group fund-raiser—a breakfast—on Mother's Day. On that morning, only the president and his wife were present to prepare breakfast. This should alert researchers against measuring power and influence by looking solely at who holds titles.

Each of the cofounders had a history of working with groups arising out of the responsibilities usually assumed by "mothers"—the education of children and the safety of the surrounding community. From these groups, they gained valuable experiences and networks that facilitated the formation of MELA." Juana Gutierrez explained how pre-existing networks progressively expanded community support:

> You know nobody knew about the plan to build a prison in this community until Assemblywoman Gloria Molina told me. Martha Molina called me and said, "You know what is happening in your area? The governor wants to put a prison in Boyle Heights!" So, I called a Neighborhood Watch meeting at my house and we got fifteen people together. Then, Father John started informing his people at the Church and that is when the group of two to three hundred started showing up for every march on the bridge.[24]

MELA effectively linked up pre-existing networks into a viable grassroots coalition.

Second, the process of activism also transformed previously "invisible" women, making them not only visible but the center of public attention. From a conventional perspective, political activism assumes a kind of gender neutrality. This means that anyone can participate, but men are the expected key actors. In accordance with this pattern, in winter 1986 an informal group of concerned businessmen in the community began lobbying and testifying against the prison at hearings in Sacramento. Working in conjunction with Assemblywoman Molina, they made many trips to Sacramento at their own expense. Residents who did not have the income to travel were unable to join them. Finally, Molina, commonly recognized as a forceful advocate for Latinas and the community, asked Frank Villalobos, an urban planner in the group, why there were no women coming up to speak in Sacramento against the prison. As he phrased it, "I was getting some heat from her because no women were going up there."[25]

In response to this comment, Veronica Gutierrez, a law student who lived in the community, agreed to accompany him on the next trip to Sacramento.[26] He also mentioned the comment to Father John Moretta at Resurrection

Catholic Parish. Meanwhile, representatives of the business sector of the community and of the 56th assembly district office were continuing to compile arguments and supportive data against the East Los Angeles prison site. Frank Villalobos stated one of the pressing problems:

> We felt that the Senators whom we prepared all this for didn't even acknowledge that we existed. They kept calling it the "downtown" site, and they argued that there was no opposition in the community. So, I told Father Moretta, what we have to do is demonstrate that there is a link (proximity) between the Boyle Heights community and the prison.[27]

The next juncture illustrates how perceptions of gender-specific behavior set in motion a sequence of events that brought women into the political limelight. Father Moretta decided to ask all the women to meet after mass. He told them about the prison site and called for their support. When I asked him about his rationale for selecting the women, he replied:

> I felt so strongly about the issue, and I knew in my heart what a terrible offense this was to the people. So, I was afraid that once we got into a demonstration situation we had to be very careful. I thought the women would be cooler and calmer than the men. The bottom line is that the men came anyway. The first times out the majority were women. Then they began to invite their husbands and their children, but originally it was just women.[28]

Father Moretta also named the group. Quite moved by a film, *The Official Story*, about the courageous Argentine women who demonstrated for the return of their children who disappeared during a repressive right-wing military dictatorship, he transformed the name "Las Madres de la Plaza de Mayo" into "Mothers of East Los Angeles."[29]

However, Aurora Castillo, one of the cofounders of the group, modified my emphasis on the predominance of women:

> Of course the fathers work. We also have many, many grandmothers. And all this IS with the support of the fathers. They make the placards and the posters; they do the security and carry the signs; and they come to the marches when they can.[30]

Although women played a key role in the mobilization, they emphasized the group's broad base of active supporters as well as the other organizations in the

"Coalition Against the Prison." Their intent was to counter any notion that MELA was composed exclusively of women or mothers and to stress the "inclusiveness" of the group. All the women who assumed lead roles in the group had long histories of volunteer work in the Boyle Heights community; but formation of the group brought them out of the "private" margins and into "public" light.

Third, the women in MELA have transformed the definition of "mother" to include militant political opposition to state-proposed projects they see as adverse to the quality of life in the community. Explaining how she discovered the issue, Aurora Castillo said:

yup

> You know if one of your children's safety is jeopardized, the mother turns into a lioness. That's why Father John got the mothers. We have to have a well-organized, strong group of mothers to protect the community and oppose things that are detrimental to us. You know the governor is in the wrong and the mothers are in the right. After all, the mothers have to be right. Mothers are for the children's interest, not for self-interest; the governor is for his own political interest.[31]

The women also have expanded the boundaries of "motherhood" to include social and political community activism and redefined the word to include women who are not biological "mothers." At one meeting a young Latina expressed her solidarity with the group and, almost apologetically, qualified herself as a "resident," not a "mother," of East Los Angeles. Erlinda Robles replied:

> When you are fighting for a better life for children and "doing" for them, isn't that what mothers do? So we're all mothers. You don't have to have children to be a "mother."[32]

Good for her!

At critical points, grassroots community activism requires attending many meetings, phone calling, and door-to-door communications—all very labor-intensive work. In order to keep harmony in the "domestic" sphere, the core activists must creatively integrate family members into their community activities. I asked Erlinda Robles how her husband felt about her activism, and she replied quite openly:

> My husband doesn't like getting involved, but he takes me because he knows I like it. Sometimes we would have two or three meetings a week. And my husband would say, "Why are you doing so much? It is really

getting out of hand." But he is very supportive. Once he gets there, he enjoys it and he starts in arguing too! See, it's just that he is not used to it. He couldn't believe things happened the way that they do. He was in the Navy twenty years and they brainwashed him that none of the politicians could do wrong. So he has come a long way. Now he comes home and parks the car out front and asks me, "Well, where are we going tonight?"[33]

When women explain their activism, they link family and community as one entity. Juana Gutierrez, a woman with extensive experience working on community and neighborhood issues, stated:

Yo como madre de familia, y como residente del Este de Los Angeles, seguiré luchando sin descanso por que se nos respete. Y yo lo hago con bastante cariño hacia mi comunidad. Digo "mi comunidad," porque me siento parte de ella, quiero a mi raza como parte de mi familia, y si Dios me permite seguiré luchando contra todos los gobernadores que quieran abusar de nosotros. (As a mother and a resident of East L.A., I shall continue fighting tirelessly, so we will be respected. And I will do this with much affection for my community. I say "my community" because I am part of it. I love my "raza" [race] as part of my family; and if God allows, I will keep on fighting against all the governors that want to take advantage of us.)[34]

Like the other activists, she has expanded her responsibilities and legitimated militant opposition to abuse of the community by representatives of the state.

Working-class women activists seldom opt to separate themselves from men and their families. In this particular struggle for community quality of life, they are fighting for the family unit and thus are not competitive with men.[35] Of course, this fact does not preclude different alignments in other contexts and situations.[36]

Fourth, the story of MELA also shows the transformation of class and ethnic identity. Aurora Castillo told of an incident that illustrated her growing knowledge of the relationship of East Los Angeles to other communities and the basis necessary for coalition building:

And do you know we have been approached by other groups? [She lowers her voice in emphasis.] You know that Pacific Palisades group asked for our backing. But what they did, they sent their powerful lobbyist that they pay thousands of dollars to get our support against the drilling in

Pacific Palisades. So what we did was tell them to send their grassroots people, not their lobbyist. We're suspicious. We don't want to talk to a high salaried lobbyist; we are humble people. We did our own lobbying. In one week we went to Sacramento twice.[37]

The contrast between the often tedious and labor-intensive work of mobilizing people at the "grassroots" level and the paid work of a "high salaried lobbyist" represents a point of pride and integrity, not a deficiency or a source of shame. If the two groups were to construct a coalition, they must communicate on equal terms.

The women of MELA combine a willingness to assert opposition with a critical assessment of their own weaknesses. At one community meeting, for example, representatives of several oil companies attempted to gain support for placement of an oil pipeline through the center of East Los Angeles. The exchange between the women in the audience and the oil representative was heated, as women alternated asking questions about the chosen route for the pipeline:

"Is it going through Cielito Lindo [Reagan's ranch]?" The oil representative answered, "No." Another woman stood up and asked, "Why not place it along the coastline?" Without thinking of the implications, the representative responded, "Oh, no! If it burst, it would endanger the marine life." The woman retorted, "You value the marine life more than human beings?" His face reddened with anger and the hearing disintegrated into angry chanting.[38]

The proposal was quickly defeated. But Aurora Castillo acknowledged that it was not solely their opposition that brought about the defeat:

We won because the Westside was opposed to it, so we united with them. You know there are a lot of attorneys who live there and they also questioned the representative. Believe me, no way is justice blind. . . . We just don't want all this garbage thrown at us because we are low-income and Mexican-American. We are lucky now that we have good representatives, which we didn't have before.[39]

Throughout their life histories, the women refer to the disruptive effects of land use decisions made in the 1950s. As longtime residents, all but one share the experience of losing a home and relocating to make way for a freeway. Juana Gutierrez refers to the community response at that time:

> Una de las cosas que me caen muy mal es la injusticia y en nuestra comunidad hemos visto mucho de eso. Sobre todo antes, porque creo que nuestra gente estaba mas dormida, nos atrevíamos menos. En los cincuentas hicieron los freeways y así, sin más, nos dieron la noticia de que nos teníamos que mudar. Y eso pasó dos veces. La gente se conformaba porque lo ordeno el gobierno. Recuerdo que yo me enojaba y quería que los demás me secundaran, pero nadia queria hacer nada. (One of the things that really upsets me is the injustice that we see so much in our community. Above everything else, I believe that our people were less aware; we were less challenging. In the 1950s—they made the freeways and just like that they gave us a notice that we had to move. That happened twice. The people accepted it because the government ordered it. I remember that I was angry and wanted the others to back me but nobody else wanted to do anything.)[40]

The freeways that cut through communities and disrupted neighborhoods are now a concrete reminder of shared injustice, of the vulnerability of the community in the 1950s. The community's social and political history thus informs perceptions of its current predicament; however, today's activists emphasize not the powerlessness of the community but the change in status and progression toward political empowerment.

Fifth, the core activists typically tell stories illustrating personal change and a new sense of entitlement to speak for the community. They have transformed the unspoken sentiments of individuals into a collective community voice. Lucy Ramos related her initial apprehensions:

> I was afraid to get involved. I didn't know what was going to come out of this and I hesitated at first. Right after we started, Father John came up to me and told me, "I want you to be a spokesperson." I said, "Oh no, I don't know what I am going to say." I was nervous. I am surprised I didn't have a nervous breakdown then. Every time we used to get in front of the TV cameras and even interviews like this, I used to sit there and I could feel myself shaking. But as time went on, I started getting used to it. And this is what I have noticed with a lot of them. They were afraid to speak up and say anything. Now, with this prison issue, a lot of them have come out and come forward and given their opinions. Everybody used to be real "quietlike."[41]

She also related a situation that brought all her fears to a climax, which she confronted and resolved as follows:

When I first started working with the coalition, Channel 13 called me up and said they wanted to interview me and I said OK. Then I started getting nervous. So I called Father John and told him, "You better get over here right away." He said, "Don't worry, don't worry, you can handle it by yourself." Then Channel 13 called me back and said they were going to interview another person, someone I had never heard of, and asked if it was OK if he came to my house. And I said OK again. Then I began thinking, what if this guy is for the prison? What am I going to do? And I was so nervous and I thought, I know what I am going to do!

Since the meeting was taking place in her home, she reasoned that she was entitled to order any troublemakers out of her domain:

If this man tells me anything, I am just going to chase him out of my house. That is what I am going to do! All these thoughts were going through my head. Then Channel 13 walk into my house followed by six men I had never met. And I thought, Oh, my God, what did I get myself into? I kept saying to myself, if they get smart with me I am throwing them ALL out.[42]

At this point her tone expressed a sense of resolve. In fact, the situation turned out to be neither confrontational nor threatening, as the "other men" were also members of the coalition. This woman confronted an anxiety-laden situation by relying on her sense of control within her home and family—a quite "traditional" source of authority for women—and transforming that control into the courage to express a political position before a potential audience all over one of the largest metropolitan areas in the nation.

People living in Third World countries as well as in minority communities in the United States face an increasingly degraded environment.[43] Recognizing the threat to the well-being of their families, residents have mobilized at the neighborhood level to fight for "quality of life" issues. The common notion that environmental well-being is of concern solely to white middle-class and upper-class residents ignores the specific way working-class neighborhoods suffer from the fallout of the city "growth machine" geared for profit.[44]

In Los Angeles, the culmination of postwar urban renewal policies, the growing Pacific Rim trade surplus and investment, and low-wage international labor migration from Third World countries are creating potentially volatile conditions. Literally palatial financial buildings swallow up the space previously occupied by modest, low-cost housing. Increasing density and development not matched by investment in social programs, services, and infrastructure erode the quality of life, beginning in the core of the city.⁴⁵ Latinos, the majority of whom live close to the center of the city, must confront the distilled social consequences of development focused solely on profit. The Mexican-American community in East Los Angeles, much like other minority working-class communities, has been a repository for prisons instead of new schools, hazardous industries instead of safe work sites, and one of the largest concentrations of freeway interchanges in the country, which transports much wealth past the community. And the concerns of residents in East Los Angeles may provide lessons for other minority as well as middle-class communities. Increasing environmental pollution resulting from inadequate waste disposal plans and an out-of-control "need" for penal institutions to contain the casualties created by the growing bipolar distribution of wages may not be limited to the Southwest.⁴⁶ These conditions set the stage for new conflicts and new opportunities, to transform old relationships into coalitions that can challenge state agendas and create new community visions.⁴⁷

Mexican-American women living east of downtown Los Angeles exemplify the tendency of women to enter into environmental struggles in defense of their community. Women have a rich historical legacy of community activism, partly reconstructed over the last two decades in social histories of women who contested other "quality-of-life issues," from the price of bread to "Demon Rum" (often representing domestic violence).⁴⁸

But something new is also happening. The issues "traditionally" addressed by women—health, housing, sanitation, and the urban environment—have moved to center stage as capitalist urbanization progresses. Environmental issues now fuel the fires of many political campaigns and drive citizens beyond the rather restricted, perfunctory political act of voting. Instances of political mobilization at the grassroots level, where women often play a central role, allow us to "see" abstract concepts like participatory democracy and social change as dynamic processes.

The existence and activities of MELA attest to the dynamic nature of participatory democracy, as well as to the dynamic nature of our gender, class, and ethnic identity. The story of MELA reveals, on the one hand, how individuals and groups can transform a seemingly "traditional" role such as "mother." On the other hand, it illustrates how such a role may also be a social agent drawing members of the community into the "political" arena. Studying women's contributions as well as men's will shed greater light on the network dynamics of grassroots movements.⁴⁹

The work MELA does to mobilize the community demonstrates that people's political involvement cannot be predicted by their cultural characteristics. These women have defied stereotypes of apathy and used ethnic, gender, and class identity as an impetus, a strength, a vehicle for political activism. They have expanded their—and our—understanding of the complexities of a political system, and they have reaffirmed the possibility of "doing something."

They also generously share the lessons they have learned. One of the women in MELA told me, as I hesitated to set up an interview with another woman I hadn't yet met in person:

> You know, nothing ventured nothing lost. You should have seen how timid we were the first time we went to a public hearing. Now, forget it, I walk right up and make myself heard and that's what you have to do.[50]

On 15 September 1989 another version of this paper was accepted for presentation at the 1990 International Sociological Association meetings to be held in Madrid, Spain, 9 July 1990.

NOTES

1. See Vicky Randall, *Women and Politics, An International Perspective* (Chicago: University of Chicago Press, 1987) for a review of the central themes and debates in the literature. For two of the few books on Chicanas, work, and family, see Vicki L. Ruiz, *Cannery Women, Cannery Lives: Mexican Women, Unionization, and the California Food Processing Industry, 1930–1950* (Albuquerque: University of New Mexico Press, 1987), and Patricia Zavella, *Women's Work & Chicano Families* (Ithaca, NY: Cornell University Press, 1987).

2. For recent exceptions to this approach, see Anne Witte Garland, *Women Activists: Challenging the Abuse of Power* (New York: The Feminist Press, 1988); Ann Bookman and Sandra Morgan, eds., *Women and the Politics of Empowerment* (Philadelphia: Temple University Press, 1987); Karen Sacks, *Caring by the Hour* (Chicago: University of Illinois Press, 1988). For a sociological analysis of community activism among Afro-American women see Cheryl Townsend Gilkes, *Holding back the Ocean with a Broom: The Black Woman* (Beverly Hills, CA: Sage Publications, 1980).

3. For two exceptions to this criticism, see Sara Evans, *Born for Liberty: A History of Women in America* (New York: The Free Press, 1989), and Bettina Aptheker, *Tapestries of Life: Women's Work, Women's Consciousness, and the Meaning of Daily Experience* (Amherst: University of Massachusetts Press, 1989). For a critique, see Maxine Baca Zinn, Lynn Weber Cannon, Elizabeth Higginbotham, and Bonnie Thornton Dill, "The Costs of Exclusionary Practices in Women's Studies," *Signs* 2 (Winter 1986).

4. For cases of grassroots activism among women in Latin America, see Sally W. Yudelman, *Hopeful Openings: A Study of Five Women's Development Organizations in Latin American and the Caribbean* (West Hartford, CN: Kumarian Press, 1987). For an excellent case analysis of how informal associations enlarge and empower women's world in Third World countries, see Kathryn S. March and Rachelle L. Taqqu, *Women's Informal Associations in Developing Countries: Catalysts for Change?* (Boulder, CO: Westview Press, 1986). Also, see

Carmen Feijoó, "Women in Neighbourhoods: From Local Issues to Gender Problems," *Canadian Woman Studies* 6 (fall 1984), for a concise overview of the patterns of activism.

5. The relationship between Catholicism and political activism is varied and not unitary. In some Mexican-American communities, grassroots activism relies on parish networks. See Isidro D. Ortiz, "Chicano Urban Politics and the Politics of Reform in the Seventies," *Western Political Quarterly* 37 (December 1984): 565–77. Also, see Joseph D. Sekul, "Communities Organized for Public Service: Citizen Power and Public Power in San Antonio," in *Latinos and the Political System*, ed. F. Chris Garcia (Notre Dame, IN: University of Notre Dame Press, 1988). Sekul tells how COPS members challenged prevailing patterns of power by working for the well-being of families and cites four former presidents who were Mexican-American women, but he makes no special point of gender.

6. I also interviewed other members of the Coalition Against the Prison and local political office representatives. For a general reference, see James P. Spradley, *The Ethnographic Interview* (New York: Holt, Rinehart and Winston, 1979). For a review essay focused on the relevancy of the method for examining the diversity of women's experiences, see Susan N.G. Geiger, "Women's Life Histories: Method and Content," *Signs* 11 (winter 1982): 334–51.

7. During the last five years, over three hundred newspaper articles have appeared on the issue. Frank Villalobos generously shared his extensive newspaper archives with me. See Leo C. Wolinsky, "L.A. Prison Bill 'Locked Up' in New Clash," *Los Angeles Times*, 16 July 1987, sec. 1, p. 3; Rudy Acuna, "The Fate of East L.A.: One Big Jail," *Los Angeles Herald Examiner*, 28 April 1989, A15; Carolina Serna, "Eastside Residents Oppose Prison," *La Gente UCLA Student Newspaper* 17 (October 1986), 5; Daniel M. Weintraub, "10,000 Fee Paid to Lawmaker Who Left Sickbed to Cast Vote," *Los Angeles Times*, 13 March 1988, sec. 1, p. 3.

8. Cerrell Associates, Inc., "Political Difficulties Facing Waste-to-Energy Conversion Plant Siting," report prepared for California Waste Management Board, State of California, Los Angeles, 1984, 43.

9. Jesus Sanchez, "The Environment: Whose Movement?" *California Tomorrow* 3 (fall 1988): 13. Also see Rudy Acuna, *A Community Under Siege* (Los Angeles: Chicano Studies Research Center Publications, UCLA, 1984). The book and its title capture the sentiments and the history of a community that bears an unfair burden of city projects deemed undesirable by all residents.

10. James Vigil, Jr., field representative for Assemblywoman Gloria Molina, 1984–1986, interview by author, Whittier, CA, 27 September 1989. Vigil stated that the Department of Corrections used a threefold strategy: political pressure in the legislature, the promise of jobs for residents, and contracts for local businesses.

11. Edward J. Boyer and Marita Hernandez, "Eastside Seethes over Prison Plan," *Los Angeles Times*, 13 August 1986, sec. 2, p. 1.

12. Martha Molina-Aviles, currently administrative assistant for Assemblywoman Lucille Roybal-Allard, 56th assembly district, and former field representative for Gloria Molina when she held this assembly seat, interview by author, Los Angeles, 5 June 1989. Molina-Aviles, who grew up in East Los Angeles, used her experiences and insights to help forge strong links among the women in MELA, other members of the coalition, and the assembly office.

13. MELA has also opposed the expansion of a county prison literally across the street from Wlliam Mead Housing Projects, home to two thousand Latinos, Asians, and Afro-Americans, and a chemical treatment plant for toxic wastes.

14. The first of its kind in a metropolitan area, it would burn 125,000 pounds per day of hazardous wastes. For an excellent article that links recent struggles against hazardous waste dumps and incinerators in minority communities and features women in MELA, see Dick Russell, "Environmental Racism: Minority Communities and Their Battle against Toxics," *Amicus Journal* 11 (spring 1989): 22–32.

15. Miguel G. Mendívil, field representative for Assemblywoman Lucille Roybal-Allard, 56th assembly district, interview by author, Los Angeles, 25 April 1989.

16. John Garcia and Rudolfo de la Garza, "Mobilizing the Mexican Immigrant: The Role of Mexican American Organizations," *Western Political Quarterly* 38 (December 1985): 551–64.

17. This concept is discussed in relation to Latino communities in David T. Abalos, *Latinos in the U.S.: The Sacred and the Political* (Indiana: University of Notre Dame Press, 1986). The notion of transformation of traditional culture in struggles against oppression is certainly not a new one. For a brief essay on a longer work, see Frantz Fanon, "Algeria Unveiled," in *The New Left Reader*, ed. Carl Oglesby (New York: Grove Press, 1969), 161–85.

18. Karen Sacks, Caring by the Hour.

19. Juana Gutierrez, interview by the author, Boyle Heights, East Los Angeles, 15 January 1988.

20. Erlinda Robles, interview by the author, Boyle Heights, Los Angeles, 14 September 1989.

21. Mina Davis Caulfield, "Imperialism, the Family, and Cultures of Resistance," *Socialist Revolution* 29 (1974): 67–85.

22. Erlinda Robles, interview.

23. Ibid.

24. Juana Gutierrez, interview.

25. Frank Villalobos, architect and urban planner, interview by author, Los Angeles, 2 May 1989.

26. The law student, Veronica Gutierrez, is the daughter of Juana Gutierrez, one of the cofounders of MELA. Martin Gutierrez, one of her sons, was a field representative for Assemblywoman Lucille Roybal-Allard and also central to community mobilization. Ricardo Gutierrez, Juana's husband, and almost all the other family members are community activists. They are a microcosm of the family networks that strengthened community mobilization and the Coalition Against the Prison. See Raymundo Reynoso, "Juana Beatrice Gutierrez: La incansable lucha de una activista comunitaria," *La Opinion*, 6 Agosto de 1989, Acceso, p. 1, and Louis Sahagun, "The Mothers of East L.A. Transform Themselves and Their Community," *Los Angeles Times*, 13 August 1989, sec. 2, p. 1.

27. Frank Villalobos, interview.

28. Father John Moretta, Resurrection Parish, interview by author, Boyle Heights, Los Angeles, 24 May 1989.

29. The Plaza de Mayo mothers organized spontaneously to demand the return of their

missing children, in open defiance of the Argentine military dictatorship. For a brief overview of the group and its relationship to other women's organizations in Argentina, and a synopsis of the criticism of the mothers that reveals ideological camps, see Gloria Bonder, "Women's Organizations in Argentina's Transition to Democracy" in *Women and Counter Power*, ed. Yolanda Cohen (New York: Black Rose Books, 1989): 65–85. There is no direct relationship between this group and MELA.

30. Aurora Castillo, interview by author, Boyle Heights, Los Angeles, 15 January 1988.

31. Aurora Castillo, interview.

32. Erlinda Robles, interview.

33. Ibid.

34. Reynoso, "Juana Beatriz Gutierrez," 1.

35. For historical examples, see Chris Marín, "La Asociación Hispano Americana de Madres Y Esposas: Tucson's Mexican-American Women in World War II," Renato Rosaldo Lecture Series 1, 1983–1984 (Tucson, AZ: Mexican-American Studies Center, University of Arizona, Tucson, 1985), and Judy Aulette and Trudy Mills, "Something Old, Something New: Auxiliary Work in the 1983–1986 Copper Strike," *Feminist Studies* 14 (summer 1988): 251–69.

36. Caulfield, "Imperialism, the Family and Cultures of Resistance."

37. Aurora Castillo, interview.

38. As reconstructed by Juana Gutierrez, Ricardo Gutierrez, and Aurora Castillo.

39. Aurora Castillo, interview.

40. Juana Gutierrez, interview.

41. Lucy Ramos, interview by author, Boyle Heights, Los Angeles, 3 May 1989.

42. Ibid.

43. For an overview of contemporary Third World struggles against environmental degradation, see Alan B. Durning, "Saving the Planet," *Progressive* 53 (April 1989): 35–59.

44. John Logan and Harvey Molotch, *Urban Fortunes* (Berkeley: University of California Press, 1988). Logan and Molotch use the term in reference to a coalition of business people, local politicians, and the media.

45. Mike Davis, "Chinatown, Part Two? The Internationalization of Downtown Los Angeles," *New Left Review* 164 (July/August 1987): 64–86.

46. Paul Ong, *The Widening Divide, Income Inequality and Poverty in Los Angeles* (Los Angeles: The Research Group on the Los Angeles Economy, 1989). This UCLA-based study documents the growing gap between "haves" and "have-nots" in the midst of the economic boom in Los Angeles. According to economists, the study mirrors a national trend in which rising employment levels are failing to lift the poor out of poverty or boost the middle class; see Jill Steward, "Two-Tiered Economy Feared as Dead End of Unskilled," *Los Angeles Times*, 25 June 1989, sec. 2, p. 1. At the same time, the California prison population will climb to more than twice its designed capacity by 1995. See Carl Ingram, "New Forecast Sees a Worse Jam in Prisons," *Los Angeles Times*, 27 June 1989, sec. 1, p. 23.

47. The point that urban land-use policies are the products of class struggle—both cause and consequence—is made by Don Parson, "The Development of Redevelopment: Public Housing and Urban Renewal in Los Angeles," *International Journal of Urban and Regional*

Research 6 (December 1982): 392–413. Parson provides an excellent discussion of the working-class struggle for housing in the 1930s, the counterinitiative of urban renewal in the 1950s, and the inner-city revolts of the 1960s.

48. Louise Tilly, "Paths of Proletarianization: Organization of Production, Sexual Division of Labor, and Women's Collective Action," *Signs* 7 (1981): 400–17; Alice Kessler-Harris, "Women's Social Mission," in *Women Have Always Worked* (Old Westbury, N.Y.: The Feminist Press, 1981), 102–35. For a literature review of women's activism during the Progressive era, see Marilyn Gittell and Teresa Shtob, "Changing Women's Roles in Political Volunteerism and Reform of the City," in *Women and the American City*, ed. Catharine Stimpson et al. (Chicago: University of Chicago Press, 1981): 64–75.

49. Karen Sacks, *Caring by the Hour*, argues that often the significance of women's contributions is not "seen" because they take place in networks.

50. Aurora Castillo, interview.

CHAPTER FOURTEEN

THE DROUGHT OF 1988, THE GLOBAL WARMING EXPERIMENT, AND ITS CHALLENGE TO IRRIGATION IN THE OLD DUST BOWL REGION

JOHN OPIE

The Great Drought of 1988 was most severe on the Northern Plains in the Dakotas and Montana. Drought was less severe in the old Dust Bowl heartland of southwest Kansas and the Oklahoma Texas panhandle. As the summer progressed, however, the awful combination of sun, wind, and lack of rain once again began to hammer the Dust Bowl. This region was best prepared to protect itself from lack of rain. Not only had lessons been learned the hard way from the 1930s, but local farmers, suppliers, bankers, and extension agents had additionally been tested by the droughts of the 1950s and 1970s.[1] Called the "Filthy Fifties," the 1950s drought did not last as long as the one in the 1930s, but it was often as severe. Almost twenty-one million acres had been seriously damaged and remained vulnerable to more harm. Resettlement and reversion to grassland no longer were acceptable alternatives. This left the search for more water as the only answer, and points of irrigation from groundwater were beginning to appear. Even after it rained in May 1955, and farmers could expect a good crop, sixteen million acres were still ready to blow. The mid-1970s would again bring severe drought. Orbiting earth-satellite scanning in the late-winter of 1977 clearly showed newly seeded West Texas farms blowing dust into Oklahoma, while neighboring New Mexico grassland held steady.[2] But now, geographer John Borchert could write that there is "a widespread belief that, though there will be future droughts, there need be no future dust bowl."[3] Most important, the south-central High Plains now enjoyed their climate substitute: widespread irrigation from large groundwater supplies.

Another drought cycle can be anticipated for the 1990s. On the Plains one newsman called the spring of 1988, "this eerie spring—a spring without thunder, a spring without rain. . . . Some rangeland barely turned green this spring. . . . Many fields of spring wheat, pathetically trying to head out on plants only four inches high, would be almost too short to harvest." The rain delay had a finality about it: "Rain—even in torrents, even tomorrow—would come too late."[4] Irrigation supplies from the great Ogallala aquifer were declining, but they no doubt could carry farmers through one more time. The next time around, however, would the new threat of a man-induced global warming —the CO_2 or greenhouse effect—overwhelm the substitute?

Once more, it is clear that the Plains cannot be understood in isolation. The jet stream—high-altitude climate-maker—would spend the spring and summer of 1988 flowing off-course far to the north. It, in turn, was responding to below-average sea-surface temperatures along the equator in the Pacific.[5] As a result, a vast stretch of the United States, from Montana through Georgia, and most of the East, suffered through record stretches of 90°-plus temperatures. Forty-five days and nights of pounding heat were contrasted with Noah's flood of forty days and forty nights. Francis Bretherton of the National Center for Atmospheric Research (NCAR) in Boulder had already frightened people in October 1987 when he spoke prophetically to *Time* magazine of the "greenhouse effect" consequences of a doubled concentration of CO_2. "Suppose it's August in New York City. The temperature is 95 degrees; the humidity is 95%. The heat wave started on July 4 and will continue through Labor Day," a span of fifty-five days.[6] In reality, the New York heat wave lasted for forty-four continuous days, started well before July 4, the temperature and humidity in the 90s, and continued through August 18.

Across the Plains, the first to be hard-hit were ranchers; their cattle demanded daily water and access to grazing. By 1988 their second or third year of drought had been reached. As early as the middle of May, Texas ranchers were burning thorns off prickly pear cactus with gas torches. In Minnesota a Soil Conservation Service official reported, "There are drifts of dirt like snowdrifts in the ditches and lots of dirt in the air and lots of dirt in the houses." In a modern twist, one farmer near Minton, South Dakota, pulled out his snowblower. "It worked. I created a mini-dust storm blowing the soil back onto my wheat field."[7] Farmers used the 1930s phrase "black blizzards" to describe murky conditions that reached from the old Bonanza country of the Red River Valley of the North in the eastern Dakotas, down into eastern Colorado and southward beyond the Texas panhandle. In the southeast, an Atlanta water bureau official summed up farmers' feelings nationally: "There's a double drought, a drought of reality and a drought of anticipation." This reflected the fears of Plains farmers while they waited uneasily to see if 1988 was the beginning of the next Dust Bowl, scheduled for the early 1990s.[8] La Verne G.

Ausman, long-time Wisconsin farmer, and USDA official in charge of the federal response to the drought, spoke of the sense of impotence and despair brought on by day after day of no rain. "There isn't anything more disheartening. You're totally helpless. You just watch your crops wither away."9

By mid-August the U.S. Department of Agriculture made front-page news when it predicted a 37 percent plunge in the nation's corn crop, a 23 percent drop in soybeans, a 13 percent drop in wheat, and a 31 percent overall drop in all grains. In early August President Reagan belatedly signed a $3.9 billion disaster-relief measure, the costliest ever: "This bill isn't as good as rain, but it will tide you [farmers] over until normal [sic] weather" returns. The president's comments mimicked the old hope for better weather that farmers had laid to rest in the 1930s. One Plains insurance agent said. "Farmers aren't buying anything but crop insurance." A *New York Times* editorial on 26 August 1988 proposed that crop insurance be required of any farmer who benefited from federal income supports. In draconian language, it concluded, "Perhaps the only way to persuade individual farmers to take charge of their fate will be to deny them special relief the next time disaster strikes." The editorial may have signaled exasperation with federal farm support during endlessly cycling crises and the end of the nation's loyalty to the myth of the independent farmer. Ultimately, farmers consumed $20 billion to protect nin million acres during the 1988 drought.10

Some people profited from the drought. At the Chicago Board of Trade, where much of the nation's supply of wheat, corn, and soybeans is traded in the volatile futures market, the volume of trading soared to 170 percent of normal as big grain users—food processing companies—tried to lock in prices from grain elevator operators. At the beginning of May wheat futures hovered around $3.30 a bushel and reached $4.20 in early July. Corn stood at $2.30 in early May and neared $3.60 in early July. One market journalist claimed that "bullish traders—gleefully detail the damage that the drought has already wrought. . . . One would think that the entire corn crop will consist of nothing but cocktail-party-size baby ears. . . . Predictably, the price action has been galvanic."11 Another bullish soybean trader said with more passion than sense, "It will never rain again in your or my lifetime. Climatic changes are in the process of transforming Illinois and Iowa into the Gobi Desert. . . . Since we're repeating the weather conditions of the Dust Bowl years, it stands to reason that the corn crop is doomed to come in with yields 65 percent or so of normal like it did in 1936."12

Some Midwestern and Plains farmers, having already speculated on the right climate for their stunted fields, and lost, decided they could only survive by gambling a second time on the futures markets. Critics called it a highly volatile "de facto insurance policy on at least a portion of their crops to lock in prices before the harvest."[13] Prices soared in June, but scattered midwestern showers in July caused prices to collapse to their daily permitted limits. Wheat highs of $4.20 in early July were down to $3.70 by the end of the month, while corn futures fell from $3.60 to $2.75.[14] Shortages of durum wheat for pasta, corn oil for margarine, soybeans for mayonnaise, soybean meal for chickens, and feed corn for cattle pushed prices up 5 to 7 percent, the highest since 1980. A fifteen-ounce box of Cheerios that cost $1.98 in October 1987 sold for $2.14 in June 1988.[15] A one-pound loaf of white bread rose 10 percent between April 1987 and June 1988. A 6 percent increase in food prices would cost a typical American family of four more than $300 a year. Nevertheless, food costs still accounted for only 17 percent of the American consumer's pocketbook, less than half that of Europeans. Despite the swings, it was clear that consumer food prices would not soar because low consumer prices stood as a sacred tenet of the U.S. Department of Agriculture.

With worldwide drought damage in Canada, China, and the United States, the globe had only a fifty-four-day supply of stockpiled grain. This was down from eighty-nine days earlier in 1988, and below the sixty-day global minimum set by the United Nations Food and Agricultural Organization.[16] This was the lowest since the food crisis of 1972 and 1973, when wheat prices doubled and corn prices tripled. The drought laid waste Canada's farm belt, which is equally industrialized, and ravaged China's central and southern grain belts.[17] In this fourth drought since 1980, Canada's prairie grain crop fell to about 33 million tons, down a third from 1987 levels. "We've had people up here who've walked away from their crops." In China, 65 million acres of China's grain, out of a total of 196 million acres of farm land, had been hit by a heat wave. Because of drought, two-thirds of China's peanut and soybean crops were lost for the year. Chinese meteorologists believed the drought was global, the result of an increase in sunspot activity or the greenhouse effect. There were fears that if the United States, which produced a fifth of the world's grain, abandoned export subsidies because of short supplies, it would bring more famine to some poorer nations. The Export Enhancement Program had spent $2.1 billion to encourage foreign nations to buy American wheat, and build their need for it, at $25 to $33 a ton less than in the United States. In December 1987 the Soviet Union's dependence was assured when it bought wheat for $42 a metric ton less than in the domestic United States, a subsidy of more than $1 a bushel.[18]

The Green Revolution had been widely expected to overcome global food scarcity by the early 1960s. Historic importing nations, like India, China, Brazil and Mexico, became major exporters.[19] But if a worldwide drought hit for several consecutive years, the United Nations Food and Agriculture Organization in Rome concluded that it would collapse food surpluses and threaten widespread hunger on a global basis. The Green Revolution was running out of steam. A USDA report in early August noted that the 1988 American grain harvest was the smallest since 1970 and even smaller than the Soviet harvest. Lester Brown of the environmentally oriented Worldwatch Institute argued that even if the fifty-four million fallow acres in the United States were put back into production, the cumulative effects of loss of farmland to erosion and development, depletion of water for irrigation, and higher costs for farm equipment globally, meant less agricultural resiliency. Between 1965 and 1983, for example, India more than tripled its wheat harvest, but had hardly raised its output since.[20] Both conservatives and environmentalists agreed, however, that another global drought year in 1989 would be catastrophic.

The drought that drove half a million people off the land in the 1930s would not have the same effect in the 1980s.[21] As many as sixty-four million acres of land had been abandoned fifty years earlier, compared to only fourteen million acres in 1988. The dry spell was severe, even more severe for some Plains farmers than it was in the 1930s, but when a young farmer could raise three times more wheat per acre than his father or grandfather, or five times more corn, he had a greater capacity to recover in the next wetter season. Hence his borrowing power with the local banker was better and it could tide him over. In 1984, following the bad drought of 1983, harvests had been nearly double. Farmers were also protected by government subsidies and drought relief born out of the sufferings of the 1930s. After the 1930s experience, over fourteen million acres on the High Plains—grasslands, fallow lands, conservation reserves—had been set aside as unsuited for farming. In the 1980s, at considerable taxpayer cost (but willingly offered to support the farmer as a national symbol), government agencies prevented wild swings in prices, supplies, and production.[22] Up to two years' wheat and corn could be prudently set aside to stabilize supplies and prices.

Not the least, better farming resources meant that the 1988 winter wheat crop came through virtually unscathed compared to the minuscule wheat harvests of the 1930s. Genetically superior seed varieties were now available, together with better fertilizer and pesticide know-how. Following the creation of the USDA's Soil Conservation Service in the 1930s, and the widespread creation of soil conservation districts across the High Plains, farmers learned to contour and terrace the land, practice minimum tillage, plant grain in strips, and graze cattle more carefully. In addition, a combination of higher wheat prices (due to the drought), normal subsidies and emergency drought relief (due to the federal government), and crop insurance (farmer good sense) pulled farmers through.

Soil erosion in 1988 was severe and dust storms reappeared to cause heavy environmental damage. Plains farmers lost this year's crops and permanently damaged their land. One North Dakota state conservationist observed, "Landowners aren't only concerned about losing their crops this year; they're also concerned about losing their soil, because that's their livelihood." Damage was measured by a formula: an acre that lost about fifteen tons of topsoil, or three times more than the land can regenerate in a year, was a damaged acre. In such a case, nutrient loss—nitrogen, phosphates, potassium—runs $576 per inch of topsoil. Kansas State University researchers say that wheat yields drop 5.3 percent for every inch of topsoil lost to erosion, with corn more and grain sorghum a bit less.[23] In 1988, 1.5 million Kansas acres were vulnerable, 200,000 in the Oklahoma panhandle, and 3.8 million in West Texas, or four times normal.[24] Across the entire Plains, almost 12 million acres were erosion-damaged and a total of 19.5 million acres lacked cover and were at risk, ready to blow.

Drought in the 1980s and the 1990s would have a more global effect than in the 1930s because of the intensive high-yield farming that now prevails on the Plains. The high yields of modern industrial farming, with its reliance upon a small number of basic Plains crops—wheat, milo, alfalfa, and corn—matched to specific chemicals and irrigation methods, would be extremely vulnerable to lack of groundwater to overcome any future drought. The discovery and application of unused but more adaptable crops could take a decade. If the greenhouse effect took hold tomorrow, or a simple decade-long climate anomaly appeared in the 1990s (the opposite of the decade of extraordinary rain from 1876 to 1885), a long spell of hotter, drier conditions would find farmers once again vulnerable to extreme dryness.[25] Climate still did matter.

Amaranth is such a heat and drought-tolerant grain. But the young experimenter, Keith Allen, north of Sublette, Kansas, had learned that alternative crops require public acceptance and the creation of markets before they become profitable. The Land Institute in Salina, Kansas, is exploring means to improve the cattle forage abilities of natural grasses, such as Illinois bundleflower and leymus. The Institute also urged a move away from today's industrialized monoculture, no matter how productive, to the more flexible "diversity" field—polyculture—planting which would mix grain crops with legumes, sunflowers, and perennial grasses. In this way, farmers could mimic the historic prairie, which survived under all weather conditions. But it was unlikely that this alternative would offer the high-yield productivity upon which the agricultural economy depends. Plant physiologists are also pushing to identify why plants fail at times of extreme stress, such as heat or drought. Can proteins be inserted into crops to provide protection? Agricultural research, long a stepchild in the nation's federally funded research and development programs, has grown little since 1980.²⁶

Plains farming in 1988 was once more between a rock and a hard place. An eighty-eight-year-old Montana rancher, who had weathered the 1930s, told a reporter, "Back then, things cost less and everybody had a little money to get by. That's all you needed was a little income. Our expenses today are ten times higher, maybe twenty times higher. So even if you make a little money, you're still going to be short. You can't pay for the equipment. This drought is going to put a lot of ranches out of business."²⁷ Unlike traders in the Chicago pit, who profited handsomely, and unlike America's consumers, whose pocketbooks were not rifled, drought-stricken farmers bore the brunt of the collapse. Reduction of wheat storage from a two-year supply to a one-year supply, and of soybeans from a year to six months drove commodity prices to their highest prices in a decade, at a time when Plains farmers had little to sell. Farmers had looked forward to the beginnings of a recovery from a six-year-long debt-ridden agricultural recession. It had already forced the largest number of farm and bank failures since the 1930s. One in every six farmers in one Minnesota county had fallen behind on his debt payments. Commodity prices stayed high, but the ability of farmers to cover the inevitable lean years by taking advantage of a fat year or two slipped away with the drought. Echoing the mid-1930s, one farmer said, "High prices don't help if you go into town with an empty truck." By June 1988 farmland prices, upon which loan equity is often based, stopped climbing and started slipping. As the drought continued into early 1989, one Kansan from Hays wrote the *Wall Street Journal*: "Our little patch of wheat is dead. The evergreen shelter belt is more brown than green. The creek is barely running and most of the migrating water fowl are bypassing our country because the ponds are dry. The weather that scares the birds and animals has us worried too. But that's nothing new to Kansans. . . . [Your statement that] 'the name Kansas became a byword for the impossible and ridiculous' doesn't bother Kansans more. . . . Wall Street looks to a lot of us to be a hell of a lot less predictable than Kansas weather. . . . While Kansas weather may be ridiculous, Kansans are sublime."²⁸

One well-established regional alternative was to ignore the drought. Whoever irrigated his land could still proceed as if the Drought of 1988 had never come. Larry Hollis, a hands-on Wisconsin farmer with over three hundred acres in corn and hay who was also a Milwaukee stockbroker who watched farm production trends, observed in June that the driest Wisconsin spring in a century meant that his crops never came near their spring moisture requirements. By contrast, he noted, the irrigated High Plains were "in pretty decent shape" no matter the absence of rain.²⁹ Established irrigators on the High Plains could be envied for their ability to ignore the drought. They could schedule irrigation and not wait for rain. If, for a field of corn, more water was needed to support day forty through day fifty of the growth cycle, when the plants begin to grow substantially, then so be it at the flick of a pump switch. But without irrigation, said Hollis, "What we need is several days of nice gentle rain . . . a good soaking," of the kind provided by hitting a switch on the normally rainless Kansas Sandhills where the prosperous Gigot family, for example, operates over five hundred center-pivot units. "[Irrigators] can pick and choose."³⁰

The goal of irrigators was to tap the remaining two billion acre-feet (an acre-foot is approximately 326,000 gallons) of the groundwater of the Ogallala aquifer, which lay in two-hundred-foot-thick beds of sand and gravel fifty to one hundred feet below the surface. In 1950 the Ogallala had irrigated 3.5 million acres of farmland.³¹ In the 1970s irrigation peaked at 17 million acres and slipped slightly in the 1980s to 16 million acres. By May 1988 farmers' irrigation engines worked harder than ever before, and recorded approximately 50 percent more pumping time. Gary Baker, manager of Southwest Kansas Groundwater Management District No. 3, noted that farmers around Garden City had planted 10 percent more corn (wheat was a swing [rotation] crop for them) in 1988 than in 1987. In the drought, their rate of groundwater declined almost 3 feet in 1988 alone compared to a twenty-year average of 1.8 feet. This was all the more disappointing because local irrigators, using conservation methods, had in recent years reduced their annual decline to 1 foot or less.³² As an evil portent, dust storms swirled through all of Kansas west of Wichita; even when the wind settled from fifty miles per hour to thirty miles per hour, one soil scientist said, "It was dirty." SCS official Jeff Schmidt went out to check on crops damaged by sand and static electricity. He called for emergency tillage as serious soil erosion hit fields emptied of their crops.³³

Plains farming has its own distinctive history, associated mostly with the unending search for water. This is particularly true in the heart of the old Dust Bowl in southwest Kansas and the Oklahoma–Texas panhandle. Not until the post–World War II period did the combination of efficient deep-well pumps, low-cost energy to run gasoline or natural gas engines, inexpensive aluminum piping, center-pivot sprinklers, and other watering technologies allow farmers to ignore the lack of rain by tapping into what once was three billion acre-feet in the vast Ogallala aquifer.³⁴ But today the most enthusiastic boosters of High Plains farming admit that a mechanized "golden age" of irrigation, that has lasted approximately from 1960 to 1990, was only a temporary conquest because of declining water levels. On the central and southern High Plains, the long-term drought that would come with global warming from the CO_2 greenhouse effect would cut crop acreage as much as 25 percent.[35]

The gospel of efficiency took over on the High Plains. Beginning in the 1950s, the historic family farm became in many cases a private, heavily capitalized and mechanized industrial operation. Today, Plains irrigators, as they manage their equipment, are more similar to the nation's small businessmen than to their pioneer forebears.[36] Historian Donald Worster wryly wrote, "Agronomists promised [the Plains farmer] anew the tomorrow world of infinite abundance, when all the land would be contoured to the horizon, every drop of water captured and used, straight even rows of trees planted wherever they would grow—a landscape of engineering and efficiency."[37] By the late 1970s over half the market value of Kansas's farm production came from those counties with over twenty thousand acres irrigated.[38] Ogallala water was pumped onto the fields on a "use it or lose it" (prior appropriation) policy. By planting fence row to fence row at the highest efficiency, farmers could participate in the nation's ebullient postwar expansion. It was "get big or get out." Farm size in Kansas doubled, irrigation acreage tripled, and the number of farms decreased more than half between 1940 and 1983, from 159,000 to 76,000. Drought accelerated groundwater consumption to keep production up. In the drought year 1974, with rain seven inches below average, irrigators ran their pumps nearly twice as long as in the normal year 1973, and energy consumption rose 64 percent. Farmers recorded that their water levels declined as much as three feet a year. These high levels continued into 1975 and 1976 until the return of normal rainfall in 1978.[39] In Hugoton's Seward County, where "pick and choose" irrigators were plentiful, the 1982 National Resources Inventory recorded, probably at the peak of Plains irrigation, that 113,600 of the county's 263,800 cropland acres were irrigated. In Finney County, 244,400 acres, almost 30 percent of its farmland, was irrigated from 2,267 wells, largely with sprinklers. In Haskell County it was 263,900 acres, over 71 percent, from 988 wells, and 90 percent sprinkled.[40]

Irrigation on the High Plains is a very comforting technology that is inseparable from the irrigator's farm life. At the turn of a valve or flip of a switch the lucky farmer can flood his fields, sprinkle her crops, or drip water on his vegetables no matter how dry the weather is. Today's irrigator, on his typical 1,280 acres in the old Dust Bowl country of southwest Kansas and the panhandle country of Oklahoma and Texas, is now bound to a gushing steel umbilical, six to eight inches wide, plunging down 250 feet to the fresh water of the Ogallala (High Plains) aquifer. At the top end, another finite resource, natural gas, fires up the four cylinders of an International 605 or Minneapolis-Moline engine. The engine's shaft, rotating at 1,200 rpm, runs to a reduction gearbox that also shifts the torque from horizontal to vertical and powers a Johnson or Peerless impeller pump. The aquifer water rushes up the steel umbilical and into aluminum pipes at 800 to 1,200 gallons per minute. It flows, sprinkles, or drips onto nearby fields overloaded with wheat or corn or sorghum—human and animal food. Corn would take the most water during the season, as much as 156,000,000 gallons laid on a 160-acre quarter-section, wheat and sorghum half that amount, but alfalfa even more. A center-pivot sprinkler system, from drilling the well to watering the milo, would in 1990 cost a farmer $50,000 to $70,000 per 160 acres, depending on well depth and field needs. This is twice the cost back in 1976, and the typical irrigator, for security, would need eight units for his 1,280 acres.[41]

The High Plains will shortly experience as difficult times as it ever has. Everyone that this author interviewed between 1986 and 1990 in the old Dust Bowl region, from the local farmer and banker and equipment salesman, to the state or federal hydrologist and agronomist, agreed that intensive and profitable irrigation now faces the serious problem of rapid depletion of Ogallala groundwater.[42] Half the useable water has been spent; the current pace of consumption does not allow another thirty years, but instead maybe ten more years, or less if dire greenhouse desert predictions come true. All too soon, the remaining water will only be reached using prohibitively expensive deep pumps consuming costly fuels, a procedure that would require high-priced food, an unacceptable forecast in light of America's tradition of cheap food. The result will be a classic example of the limits set on the use of an environmental resource less by its complete depletion than by rising costs and diminishing returns.

At present, Ogallala farmers rest their hopes on better management skills to enhance the carrying capacity of the resource base, for example, a move toward sustainable development. When Plains farmers found by 1980 their groundwater levels declining two feet a year they learned to practice "water scheduling" to serve crops only at critical stages of growth, and "water harvesting" in tailwater ponds to return runoff onto the fields. New technologies ranged from center-pivot irrigation in the 1960s to in-ground moisture sensors and drip irrigation in the 1980s and genetically engineered crops that are less water-sensitive in the 1990s.

Irrigation farmers also joined new regional water management districts in Kansas and Texas, or formed their own independent associations, as in Oklahoma, to control the number and spacing of wells and pumps, to meter consumption, and to foster conservation and fight waste. The path-breaking Texas Groundwater Management District No. 1 opened its doors in Lubbock in 1952. Its mission was to promote controlled development of Ogallala water.⁴³ It has now shifted policies to protect the remaining supply for "beneficial use" only. This realization that, when the aquifer is set aside for farming, it serves a fundamental social good that benefits society is also stated in the legislative act that created the Oklahoma State Water Board in 1963.⁴⁴ But Oklahoma has no local districts, and local citizens' groups in the panhandle often find themselves in conflict with the state board, which they see as dominated by big oil. As to Kansas, a local vote in 1976 created the Kansas Groundwater Management District No. 3 headquartered in Garden City.⁴⁵ While there is a state groundwater engineer, groundwater policies and actions in Kansas belong to local boards of irrigators and local district officials. Water "mining" continues and a physical, technological, and economic limit will inevitably be reached. One radical new development bears watching. The Northwest Kansas Groundwater Management District Four, established in 1977, and overseeing more than 3,600 wells, set in 1990 a zero-depletion goal to be reached in as little as ten years. To quote one district official, "the declining levels meant zero-depletion anyway, so why not opt to reach the same goal earlier while retaining an acceptable quantity of water for future management options."⁴⁶ But the district is experiencing great difficulty in matching this draconian goal with the immediate needs by local farmers to irrigate for their survival.

Most High Plains farmers, heavily insulated by irrigation, efficient machinery, sophisticated plant science, and federal prices and credit supports, survived the Drought of 1988. As commodity prices were kept high, and grain surpluses the lowest in a decade, farmland prices began to rise by the fall of 1988. Hugoton, Kansas, grain farmer James Kramer, concluded that "we could see two or three years of a relatively stable farm economy, if the weather is fairly normal."⁴⁷ But some of the worst fears came true as the drought continued into 1989 on the central High Plains with little snow in the winter and belated rains in the spring.⁴⁸ In February farmers around Liberal, Kansas, did their regular tests of soil moisture. Instead of four feet moisture depth needed to plant their milo, they found two feet. Instead of forty-eight inches needed to plant their wheat, they found thirty inches of moisture. At Hayes in the middle of Kansas it was reported that gravediggers weren't hitting moisture at six feet. Those who planted wheat in the hope of fall rain and winter snow (neither appeared) were compelled to destroy their crops because the yields were so low it was too costly to harvest. In addition the crop had been unprotected by snow and left exposed to sandblasting winter winds. The result was "winter kill" of 20 percent of the crop, and by May, the USDA projected total loss on fourteen million acres out of fifty-five million planted in the fall. With no leaves, young wheat plants have to start all over again from the top of their root systems. But, farmer Larry Kern near Salina said, "The wheat's just not rooting down . . . to the subsoil because there's no moisture down there for it to get."

A year earlier, in the spring of 1988, the winter wheat crop had ripened before it was hit by intense drought in May and June, but in the spring of 1989 little rain or snow meant no ripening. In March the wheat plants were a shriveled brown instead of "greening up." Since bare land might blow, the Soil Conservation Service recommended "ghost crops" [no profit] of wheat for ground cover.[49] U.S. wheat stockpiles fell close to a seventeen-year low, only a third of the previous year. Kansas farmers abandoned one-fifth of their winter wheat acreage and the fields that were harvested dropped by ten bushels an acre. Kansas, the biggest wheat-producing state, one-quarter of the nation's entire output and a third of hard red winter wheat, faced the worst wheat harvest in thirty-two years, with its $1 billion wheat crop down more than a third. Heavy rains made up for deficits from late-April to mid-June, up to seventeen inches, but it was too late for wheat although just right for another try at milo.

In late December 1989 after six more months of drought, Kansas farmers once again worried over lost wheat crops and more severe damage from erosion. Little rain had fallen since September in wheat-growing regions. Lack of snow cover in southwest Kansas and the Oklahoma–Texas region, together with double-digit below-zero temperatures, meant failure. One central Kansas winter wheat farmer in 1989 harvested only seven bushels an acre instead of his usual thirty-five bushels, his net worth down to $12,000 after paying off the interest and part of the principal on last year's operating loan. A Kansas crop specialist worried, "Many farmers had operating loans last year that they couldn't repay because of crop failures. Now they are facing the threat of two in a row . . . we're looking at a rising number of distressed farms."[50]

High Plains drought reflected global climate events, including reports that it was worse because of human influences rather than only one of nature's capricious shifts. Not since 1936 had a dry April and May been followed by a very dry June and July, as took place in 1988. According to the standard measure, the Palmer Drought Index, the 1988 -drought was the fourth worst on record based on impact on cropland; the other truly bad years were 1934, 1936, and 1954.[51] Average global temperatures in the 1980s were the highest measured in the last 130 years, when reliable records were first kept. Seven of twelve years—1980, 1981, 1983, 1987, 1988, 1990 and 1991—had been the warmest on record. Temperatures had been rising very gradually and steadily for the last hundred years, but the sharper rise detected in the 1980s may be, according to some scientists, the beginning of much sharper increases yet to come over the next two decades. Since the nineteenth century, global temperatures had risen nearly 0.5°C, 28 percent of the warming (1.8°C) expected by the year 2030.[52] Federal climate program director Alan Hecht recently noted that the recent warming is unlikely to be part of a natural trend, since the earth is now in the later stages of a period between ice ages, meaning that the temperatures should be growing cooler with an approaching ice age.[53] What may temporarily protect our children and grandchildren from unbearable heat may be the world's oceans, which have heat-absorbing capacity more than forty times the atmosphere.

The cause could be a full-scale global greenhouse effect, evidence that industrial carbon dioxide, methane, nitrous oxide, and other gasses are trapping heat in the atmosphere.[54] Some greenhouse effect is natural; the earth would be frozen and lifeless without it. If rising man-made greenhouse gas emissions are unrestrained, and the climate is highly sensitive to a worst-case scenario, temperatures could rise at 0.8 degrees Celsius each decade, or sixteen times faster than the average rate of warming over the past industrialized hundred years. As it is, the world's oceans soak up heat and chemicals (half the carbon dioxide humans produce) and do much to slow and delay rising temperatures.

Geologist Dewey M. McLean of Virginia Polytechnic University testified at a congressional hearing, "If humans were not present on Earth, the climate would likely cool into a new ice age as it has done many times in the past." This may be a fine outcome, but "we may be moving through an entire geological epoch in a single century," an unacceptable pace, said John S. Hoffman of the global atmosphere program at EPA.[55] Geochemist Wallace Broecker at the Lamont–Doherty Geological Observatory spoke of a number of extreme climate "jumps" millions of years ago. He wondered whether the unusual temperature rise in the 1980s might "provoke the system into another mode of operation," as *Science* put it, "one not at all to the liking of humans and other living things." Human societies are remarkably resilient to climate changes that reach known extremes, but there is no history of extreme and repeated swings in temperature and moisture.[56] Stephen Schneider of the National Center for Atmospheric Research (NCAR) reminded *Science* readers that the planet Venus, with its dense CO_2 atmosphere and temperatures of 700K, and frigid Mars, with a thin CO_2 atmosphere, are both examples of "runaway greenhouse."[57]

Climate is one of the globe's most turbulent and unpredictable phenomena. Schneider, looking over NCAR's batteries of computers modeling tomorrow's trends from yesterday's data, concluded that any prediction more than six hours into the future is like sorcery.[58] Depending upon small changes, any given climate can have several "natural" behaviors or outcomes. Like most climatologists watching for greenhouse trends, he is concerned that "a kick from outside" [industrial gasses] can force climate to change states, from the unusual narrow-temperature-range steadiness of the last hundred years, to oscillating extremes in the twenty-first century. Then the oscillations could become the new steady pattern of irregular wide swings that would send humanity (and the rest of life on the globe) into a new existence. Climate extremes would rapidly consume its best information and resources to ensure ordinary survival. The long-term temperature trend is still upward. Schneider would probably agree with research meteorologist Edward Lorenz who said, "We might have trouble forecasting the temperature of [a cup of hot coffee] one minute in advance, but we should have little difficulty in forecasting it an hour ahead."[59] Nonlinear systems, such as climate, despite their short-term uncertainty, shape themselves toward identifiable long-term outcomes.

A single year's heat wave, as in the summer of 1988, made commodity prices first skyrocket and then wildly fluctuate. It tied up transport in shrunken rivers, decimated the world's food reserves, brought record smog levels to cities, terminated any last remaining hope for helpless populations in Africa's Sahel, intensified long-term water wars, and profoundly affected humanity's psychological security. Climate scientist Michael Oppenheimer of the Environmental Defense Fund noted that the new warmer world would be a place with no stability, only change.[60] Oppenheimer also noted that the world is already one degree Fahrenheit warmer than a century ago and, "within the lives of our grandchildren, it could become a blistering eight degrees hotter. . . . Remember that small temperature changes can remake the face of the earth: The planet was [only] eight degrees cooler during the glacial age."[61] With far greater extremes in regional climates, the unpleasantness of the spread of a Sahel-like desert onto the High Plains or the equivalent of a Bangeladesh monsoon in Chicago twenty years from now would have devastating effects upon population shifts, food resources, and a nation's well-being and security.[62] While a U.S. Weather Service official played down permanent impact ("This is a tough summer well within the normal range of variability"), the normally cautious James E. Hansen, a NASA research scientist, "shook up a lot of people" when he told a worried Senate subcommittee in June that he was 99 percent certain of a permanent man-induced greenhouse warming.[63] If this were so, the measures devised over the last fifty years to protect farmers against drought, such as irrigation, crop insurance, or emergency measures like the multibillion dollar federal drought bill of August 1988 would be like pitching money into a black hole.

The great food belt of middle America would likely move northward into central Canada, but with lower productivity because of poorer soils and a much shorter growing season. NASA and NOAA global climate modeling, despite its recognized coarseness, almost invariably placed the central and southern High Plains in harm's way if a CO_2 doubling took place, with hot, dry, cloudless weather. The High Plains would experience a perpetual Dust Bowl far more severe than anything experienced in the 1930s.[64] Soil moisture levels would drop 40 to 50 percent, levels which Gary Baker in Garden City, Kansas, reported destroyed wheat crops in the spring of 1989. Multiple studies correlated by Stephen Schneider indicated that the old Dust Bowl region (Arkansas–White–Red River basin) is the fourth most vulnerable to greenhouse impacts in the nation (after the Great Basin, Missouri basin, and California) because of extreme water consumption, extreme climate variability, and groundwater loss.[65] While the greenhouse effect would not be uniform across the globe, the United States would have specific "winners" and "losers," and the old Dust Bowl region is always listed as a "loser." James Hansen argues that while natural forces bring on droughts and other climate changes, the greenhouse effect acts on these natural forces to make such weather extremes even more likely.[66] Modern society has no comparable historic past experience with the temperature swings of a CO_2 doubling, other than distant ice age data.[67] Irrigation was an exceptionally successful response to Dust Bowl conditions, but how long can it support the High Plains under the pressure of both CO_2 doubling and aquifer depletion? Dryland crop yields could drop by 18 percent and irrigation yields reduced up to 21 percent. In compensation, using the example of Texas alone, thirty thousand more acres would have to be irrigated.[68]

Ironically, over geologic time the Plains region had contributed to global cooling. Over the last sixty-seven million years, the grasses increased the breakdown of soil minerals. When the vast inland seas covered the region, the resulting potassium, calcium, and magnesium ions were captured by marine organisms, which together with carbon dioxide (CO_2) formed the carbonate skeletons that made up layers of limestone. This systematic removal of carbon dioxide may have contributed to the overall global cooling during the so-called Tertiary period.[69]

Today the threat to the High Plains is desertification: "a self-accelerating process, feeding on itself, and as it advances, rehabilitation costs rise exponentially."[70] Michael H. Glantz, climate and social impact analyst at NCAR, observed that physical pressures on a region, resulting from highly mechanized technologies to force high food yields, when intensified by drought, lead to desertification, which is always damaging to humans and the environment. He quotes the 1977 United Nations Nairobi Conference on Desertification:

> The deterioration of productive ecosystems is an obvious and serious threat to human progress. In general, the quest for ever greater productivity has intensified exploitation and has carried disturbance by man into

> less productive and more fragile lands. Overexploitation gives rise to degradation of vegetation, soil and water, the three elements which serve as the natural foundation for human existence. In exceptionally fragile ecosystems, such as those on the desert margins, the loss of biological productivity through the degradation of plant, animal, soil and water resources can easily become irreversible, and permanently reduce their capacity to support human life.[71]

The difference between the High Plains and Ethiopia's Sahel is only a difference of degree.

One response, surely in desperation and panic, is abandonment of the region, as had been recommended and legislated in the 1930s. There are no dramatic new solutions such as the irrigation technologies that transformed the 1930s Dust Bowl into a 1960s garden. The high-efficiency water management practices, promoted today to stretch out supplies for longer-term farming, will inevitably run into the wall of depletion. Up to 60 percent of Ogallala water is currently consumed in flood irrigation not by plants, but lost to evapotranspiration and soil seepage. A spring 1989 Texas water resources study reported a 5 to 25 percent increase in groundwater consumption to raise identical crops.[72] Water importation is unlikely, despite the recommendations of the 1982 Department of Commerce and Corps of Engineers reports. Transportation of water from the Canadian Rockies according to the NAWAPA plan was estimated in the mid-1970s at $300 billion. Talk has again started about importing Canadian water.[73] Canada has, it is noted, 9 percent of the world's fresh water with only twenty-five million people to use it. Already Canada's opposition Liberal Party has publicly wondered, during the U.S.–Canada free-trade talks, "Why are certain commodities like beer spelled out as excluded from the agreement, but not water?" Importation from the Great Lakes faces high energy costs (water weighs eight pounds a gallon and would have to be lifted more than 2,500 feet from Lake Michigan to southwest Kansas because of altitude differences). Great Lakes governors have already gone on record opposing sale of their resource.

Stephen Schneider of NCAR warns, "What is new [in the greenhouse threat compared to the Dust Bowl] is the potential irreversibility of the changes that are now taking place."[74] Swedish scientist Bert Bolin stressed the severity and universality of the subject: "Climatic problems will be part of people's lives over the next century."[75] Americans have great difficulty thinking and acting in a long-term framework larger than four-year presidential terms and annual reports of corporations. Climate thinking in this case must span more than

18,000 years (since the last ice age), or 25,000 years (the supposed age of the Ogallala water). As University of Chicago atmospheric scientist V. Ramanathan noted, "By the time we know our theory is correct, it will be too late to stop the heating that has already occurred," and is only temporarily trapped in today's oceans.[76] Ramanathan also labeled the current warming as a test of the greenhouse effect by "an inadvertent global experiment" that has now reached "the crucial stage of verification."[77]

Uncertainty makes all planners, policy makers, scientists, and politicians uneasy. The problem with the potential impact of drought upon the Ogallala region is that this uncertainty is multiplied by inherent murkiness and contingency of both the hydrological cycle and climate.[78] The *New York Times* editorialized on 27 January 1989:

> Climatologists will argue for many years whether the greenhouse warming has started. But there's every reason to take action immediately, and not wait until that debate is concluded. Once warming begins, its momentum will continue—even if gas emissions could be stopped immediately—for the three decades or so that it takes to heat the oceans. At that point the planet will again be in equilibrium, but at a much higher temperature than that of the initial warming signal.

Schneider, writing in *Science*, concluded, "Whether the uncertainties are large enough to suggest delaying policy responses is not a scientific question per se, but a value judgment."[79] It is a matter of decision making with imperfect information. Decisions are not in the hands of scientists but in politicians' hands.

At one time, as recently as the 1930s, there was no significant groundwater irrigation of crops on the High Plains. As irrigation became practicable in the 1940s and 1950s, farmers first used it as "extra water" that supplemented the marginal twenty inches of regional rainfall. But when wells and pumps and unmuffled engines became commonplace, with floods coursing through aluminum pipes and the center-pivots making their circles, irrigation became a necessity. It was the means—water on demand—to provide high crop yields (intensive farming) over ever-increasing acreages (extensive farming). What was once extra water became necessary water.

If the greenhouse effect turned southwestern Kansas and the Oklahoma–Texas panhandle into an arid desert climate—as is predicted—then Ogallala water consumption would be doubled or tripled, not only to compensate for no rain, but also because hotter winds and the blinding sun would speed evaporation on irrigation-flooded fields and the growing plants would transpire their moisture more rapidly. In the summer of 1988 Ogallala pumping deficits sometimes went from the usual two feet a year, serious as that was, up to five feet a year.[80]

One problem is that today's "successful" farming, as a subcategory of economics, is deeply committed to flawed practices that appear to guarantee future failure.[81] Today the Plains farmer uses water-needy crops that depend upon soil and water mining, costly equipment and liberal applications of chemicals. It is clear that vast food surpluses are being created at great economic cost not passed on to the consumer, grain trader, or foreign buyer. Surpluses have in turn promoted an expanding beef industry on the High Plains, in which six pounds of grain produce one pound of beef. Cattle feedlots, that run tens of thousands of cattle through their pens every year, demand a minimum of eight to ten gallons of water per head per day. Historically, meat consumption grows as living standards and social expectations rise; this is true globally. Today, the Plains are locked into high water consumption to grow the wheat and water the beef.

Farmers are held in the grip of market forces and government price supports that tend to accelerate production. These encourage the narrow economic opposite of sustainable development—"transitional unsustainability"—by inducing farmers to use excessive amounts of pesticides and fertilizers and to waste underground and surface waters in irrigation.[82] The result is environmentally blind. The hidden ecological cost of lost water, spent soil, and bankrupt farmers is not easy to measure in dollars. Farming skills are not in unlimited supply. A skipped generation, if farmers fail, means the loss of accumulated knowledge. Almost universally, the farmers I interviewed spoke of the inability of a new generation of farmers to accumulate the capital to continue irrigation as water levels decline. This is attenuated by the reality of aging wells, pumps, motors, and sprinklers that will have to be replaced in the next decade.

Most consumers of High Plains groundwater still treat it as a "free good," available to the first taker at no cost for the water itself.[83] It takes only $15 to pump an acre-foot using natural gas and $30 using electricity. Hence, this free water has been generously consumed on profligate levels. As long as no price is attached to the Ogallala resource, water for irrigation appears to offer an extraordinary double benefit that is a dangerous masking of reality: (1) water raises yields dramatically, and (2) water can be ignored as a real farming cost. But free goods tend to be squandered. Under current marketplace pressures, resources are conserved only when they become scarce and expensive.

Waters laws, such as prior appropriation—"use it or lose it"—also counter sustainable development. Pierre R. Crosson and Norman J. Rosenberg of the think-tank, Resources for the Future, write that "markets are not well equipped to protect resources such as water . . . in which it is difficult to establish property rights."[84] An imperfect first step is not to try to change agriculture, but to change economic analysis to include environmental costs. Clearly the old distinctions are contradictory and unworkable.

In this context farmers recognize that irreplaceable water is as singular as they are. They often assign it a uniquely separate task that distinguishes it from other agricultural needs like equipment or pesticides. Policy analysis Arthur Maass and Raymond L. Anderson write that

> farmers typically refuse to treat water as a regular economic good, like fertilizer, for example. It is, they say, a special product and should be removed from ordinary market transactions so that farmers can control conflict, maintain popular influence and control, and realize equity and social justice.[85]

If water can be valued, and used, outside the economic marketplace or environmental reserve, the debate over the ethical duty of water can take place more in the world of law and public interest.

Such ethical imperatives are at the heart of public policies that shape the agricultural paradigm. Ethical questions—doing good things—go far beyond prices, markets, and efficiency. Doing the good, and measuring its level of goodness, includes the cultural, social, and political indicators that reflect the broader agricultural paradigm. Philosopher and policy analyst Mark Sagoff writes of "important shared values" to which the public will sacrifice prices and efficiency.[86] As long as the public remains sympathetic to the needs and services of the family farm, it will subsidize it. Social regulation expresses not simply individual self-interest but public values we choose collectively. The role of government is not merely to correct market errors, but to reflect a sense of national well-being. In American history, so-called "benevolent" goals, such as the anti-slavery movement, women's rights, the Marshall Plan, open immigration, urban welfare, and environmental protection accurately reflect, it is claimed, not primarily self-interest and market efficiency, but widely held national values. Americans are willing to support policies not tied to the profit motive, nor to recent expansionist growth patterns. Profits and the good society may not be the same.

Thoughtful farmers say that copious amounts of water are still wastefully pumped from the Ogallala. On this basis, Americans are still at the *pioneer stage* of groundwater use. Historically, the pioneer farmers who settled on the virgin plains in the last century were not efficient stewards of the land, and sadly the habit persists. They practiced a diminishing kind of farming in which the land was mined for a single purpose: crop after crop of corn or wheat. It was uneconomic and not self-sufficient. Land exploitation was accepted as necessary for their survival. Today it is land *and* water exploitation. (Energy use deepens the problem.)

The pioneer stage is also called *extensive farming*, or the planting of all possible acreage. It is still widely practiced on the High Plains. When the pioneers were replaced by permanent settlers, commercial markets institutionalized extensive farming. Agricultural economist Willard W. Cochrane observed that "this behavior was not irrational" but it was done "with reckless abandon."[89] American farmers needed to turn a profit; the way to keep production costs down was to expand on the cheapest resource, which was land. For most of American farm history, labor was scarce, capital was expensive, and land was cheap.

Over the last thirty years, groundwater became another disposable resource to stretch out extensive farming practices. Groundwater depletion is the underground equivalent of topsoil erosion. Once the pumping and sprinkling machinery put it in easy reach, water allowed the Plains farmer to continue to prosper even as land became more costly, labor became scarce, and capital became the heaviest burden of all. All this as grain prices stayed too low for survival without federal supports. Hence, High Plains irrigation, despite its technological sophistication, is a version of extensive farming on a pioneer level. This is rarely the path toward sustainable development.

The pioneer analogy does break down. Under today's high-production scenario, yield per acre is not low, as it was on the early wheat fields. Instead, High Plains irrigation is also *intensive farming*. It is both extensive—very large farms—and intensive—highest possible yields per acre. To serve this practice, the vast supply of groundwater from the Ogallala plays the role of the free good, doubly exploited for both extensive and intensive reasons.[88] True replacement cost is not figured in. This will end when the "lagging cost line" on Ogallala groundwater rises to the levels already reached by other agricultural resources, such as land, machinery, energy, and labor.[89] What is surprising is that groundwater has yet to be added to the Plains agricultural balance sheet, both short-term and long-term. The last thirty years of consumption means there is a long-term cost due in the future. It involves the likely failure of both local farmers and high-yield grain production.[90]

Kenneth A. Cook, of the Center for Resource Economics, calls for "a new social contract between farmers and society. For its part, society will have to recognize the enormous responsibility farmers already bear to conserve natural resources and protect the environment. Taxpayers will have to be willing to share more of that burden—probably a great deal more—as external costs of agricultural production becomes internalized."[91] But instead of cross-compliance across conflicting economic and environmental lines reflected in the recent farm bills, it is argued that Congress should "decouple" support payments from the usual commodity programs and "recouple" them to environmental recovery.[92] It this way, agriculture's environmental externalities, such as groundwater, could be internalized. "Congress must restructure the nation's farm policy, placing conservation at the core—not just at the periphery."[93]

The 1990s are likely to become a time of radical reconfiguration for Plains farming in the old Dust Bowl region. At present, "Maximum economic yield" (MEY), urging highest fencerow-to-fencerow production, is still the controlling viewpoint. The pace at which Ogallala water is consumed could even rise. A potential shift from MEY to MSY ("maximum sustainable yield")[94] would conserve Ogallala water for another generation, but it might not make enough net profit to keep today's farmers on their land. The polarities remain unresolved.

Another approach is to ask what is the advantage to sustaining the Plains? Why bother when the economic burden appears too great for farmers to survive, when their fields produce surplus food, and if they are going to run out of water anyhow? Once more, a larger point of view that includes social and cultural factors clears the air.

Costs of new dams, reservoirs, canals, and distribution systems have been rising in Asia, Africa, and Latin America. Irrigation on the High Plains does not face these large capital costs. High Plains irrigation is intensely localized and small-scale, using free-standing in-field pumps and sprinklers owned by the individual farmer. In most of the world, including California and Arizona, irrigation involves large, publicly owned and debt ridden systems of dams and diversion projects covering thousands of square miles, with tunnels and aqueducts moving water hundreds of miles from source to farmer, and creating high levels of waste through evaporation, besides threatening major environmental degradation. In contrast, the High Plains irrigator is free from distant technological breakdowns, independent of meddlesome collective decision making and complex water regulations. Large-scale projects must attach a high price to water and depend upon heavy public subsidies to keep costs at a manageable level for intensive-use farming.

In its individualized framework, Ogallala water is directly translated into improved crop yields. On the Plains, it is the farmer's point of view that prevails, and not the engineer's infrastructure.[95] Large systems tend to deliver water on a fixed schedule; the local farmer can match water with the needs of crops. Results are far superior on the Plains compared to many other irrigation systems elsewhere in the world, notably in India. This is the result of independent management decisions that can be directly responsive to local crop needs. This freedom is an unexpected benefit from the precarious limitation of the central High Plains: it has no major rivers or lakes to dam or tap for irrigation. There is no equivalent to the Sacramento, Columbia, or Colorado Rivers on the High Plains. The flatness of the land, lacking deep valleys, prevented large-scale dams by the Reclamation Service.

Another unusual advantage is the absence of major cities on the central and southern Plains. Garden City in Kansas, Guymon in Oklahoma, and even Lubbock in Texas will never become a Denver or Los Angeles or Phoenix. Farmers who cannot break even when their water costs scale up to $70 an acre-foot cannot compete with cities who can afford from $2,000 to $6,000 an acre-foot, as is the case with Arizona. Farming is water intensive. The water a typical Plains farmer needs for a years' wheat crop on his section of 1,280 acres could serve twenty-four typical American families annually. The metropolitan pressures that are driving farmers out in California or Arizona are unlikely to appear on the High Plains. Finally, although industrial use of water is six times more efficient than farming, the Plains have the advantage of little heavy industry. Except for the threatening rise of the use of Ogallala water for oilfield recovery, the Plains are remarkably free of heavy water pollution.

The central High Plains, for many reasons, are strategically attractive for continued irrigated farming. As cities spread in California and Arizona, and consume water once allocated to farmers, the groundwater of the Ogallala aquifer will take on increased importance in keeping the High Plains an internationally important food production region. When high prices reflect water scarcity in urban regions worldwide, the advantages of the nonurban, nonindustrial Ogallala region will make its agricultural sustainability even more attractive. This in turn should encourage more intensive water conservation to extend the lifespan of irrigation on the High Plains.

Irrigation on the Plains is still in a self-destruct mode and the Ogallala aquifer is still a nonrenewable resource. As such, the Ogallala today is a representative microcosm of the difficult global search for sustainable development. The Ogallala belongs to the world because humanity today is a globally dominant species whose needs spin a web of mastery across the earth. When food from a radius of thousands of miles enters a single shopping cart, or when bags of Kansas grain stamped USAID avert total starvation in Africa's Sahel, the whole world depends upon the Ogallala. As a result, the clear fresh waters of the Ogallala are being unnaturally gulped up at ten times their trickling pace of replacement. Over the next fifty years, when the world's food needs multiply five or ten times, the Ogallala waters, fulfilling Adam Smith's eighteenth-century prediction, will be as precious as diamonds.

How is it that water, which is so very useful that life is impossible without it, has such a low price—while diamonds, which are quite unnecessary have such a high price?

NOTES

1. W.E. Riebsame, S.A. Changnon, and T.R. Karl, *Drought and Natural Resources Management in the United States: Impacts and Implications of the 1987–1989 Drought* (Boulder, CO: Westview Press, 1990).
2. Edwin Kessler et al., "Duststorms from the U.S. High Plains in Late Winter 1977: Search for Cause and Implications," *Proceedings of the Oklahoma Academy of Science* 58 (1978): 116–28.
3. John Borchert, "The Dust Bowl in the 1970s," *Annals of the Association of American Geographers* 61 (March 1971): 13.
4. Dennis Farney and Bruce Ingersoll, "Drought Damages Bush's Chances in Farm Belt; Rain Now Would Be Too Late for Many Victims," *Wall Street Journal*, 27 June 1988.
5. Kevin E. Trenberth, Grant W. Branstator, Phillip A. Arkin, "Origins of the 1988 North American Drought," *Science* 242 (23 December 1988): 1640–45.
6. Quoted in "The Heat is On," *Time*, 19 October 1987, p. 63.
7. Bruce Ingersoll, "Extensive Erosion in Great Plains Tied to Dust Storm Is at Worst Level Since 1955," *Wall Street Journal*, 27 June 1988.
8. B. Drummond Ayres, Jr., "Vast Parched Stretches of U.S. Await Hot Summer," *New York Times*, 15 May 1988.
9. "Mobilizing to Help Farmers Through the Drought," *New York Times*, 27 June 1988.
10. Letter from Leland B. Taylor to *Journal of Soil and Water Conservation* 45 (May–June 1990), p. 357.
11. Jonathan R. Laing, "Greenhouse Effect: The Drought Sets the Grain Pits on Fire," *Barrons*, 27 June 1988.
12. Quoted in Laing, "Greenhouse Effect," *Barrons*, 27 June 1988.
13. Julia Flynn Siler, "Losses Bring Gains For Farmer in Futures," *New York Times*, 4 August 1988; see also Siler, "Drought Means Deluge in Grain Pit," and Keith Schneider, "World Grain Supplies Are Dropping," *New York Times*, 4 August 1988.
14. Quoted in Laing, "Greenhouse Effect," *Barrons*, 27 June 1988.
15. Scott Kilman and Richard Gibson. "Killing Drought Raises Food Prices, Portends Worsening of Inflation," *Wall Street Journal*, 14 June 1988; Barbara Rudolph, "The Drought's Food-Chain Reaction," *Time*, 11 July 1988.
16. "Worldwide Effects of the Drought," *(Newark) Star-Ledger*, 27 June 1988.
17. John F. Burns, "Drought Also Lays Waste to Canada's Farm Belt," and Edward A. Gargan, "Flash Floods and Drought Ravage China," *New York Times*, 3 August 1988.
18. Keith Schneider, "Drought Stirs Debate on Wheat Export Subsidies," *New York Times*, 29 June 1988.
19. Keith Schneider, "The Green Revolution: How Much Farther Can It Go?" *New York Times*, 21 August 1988.

20. Lester Brown et al., *State of the World 1989* (New York: W.W. Norton, 1989), 3–58.

21. But it made front-page news almost daily in national newspapers. See, for example, Keith Schneider, "1988 Drought Evokes Ghost of Dust Bowl," *New York Times*, 7 July 1988.

22. William Robbins, "On the Farm, A Disaster That Wasn't," *New York Times*, 16 October 1988.

23. Bruce Ingersoll, "Extensive Erosion in Great Plains Tied To Dust Storm Is at Worst Level Since '55," *Wall Street Journal*, 27 June 1988.

24. William Robbins, "Dry Soil Blows Away, Carrying Hope With It," *New York Times*, 7 August 1988.

25. The efforts received front-page attention nationally; see Keith Schneider, "Scientists Trying to Give Crops An Edge Over Nature's Forces," *New York Times*, 1 August 1988.

26. John Perkins, Evergreen State College, interview by author on his forthcoming NSF-funded study of the politics of agricultural research, November 1990.

27. Quoted in Schneider, "1988 Drought Evokes Ghost of Dust Bowl."

28. Letter by John T. Bird, Hays, Kansas, "Kansan on Kansas," *Wall Street Journal*, 17 April 1989.

29. "Bitter Harvest: A Seasoned Farm-Belt Watcher Assesses the Damage of the Drought," *Barrons*, 27 June 1988.

30. Kilman and Gibson, "Killing Drought Raises Food Prices," *Wall Street Journal*, 14 June 1988.

31. Despite its brevity (116 pp.), the Ogallala study, *You Never Miss The Water Till . . . (The Ogallala Story)*, by Morton W. Bittinger and Elizabeth B. Green, is exceptionally detailed, accurate, and useful (Littleton, CO: Water Resources Publications, 1980).

32. Quoted in Andrew Cassel, "As Water Level Sinks, Concerns Rise," *Philadelphia Enquirer*, 29 May 1989, and Gary Baker, phone interview, 27 June 1989; see also "Water Tables Drop in Parts of Kansas," *Southwest Daily Times*, 8 June 1989.

33. Interview by author, 27 June 1989.

34. Although Donald E. Green's book, *Land of the Underground Rain* (Austin: University of Texas Press, 1973), is about irrigation on the Texas High Plains alone, it is still the best source on the technological innovations that made large-scale agricultural irrigation possible.

35. William Robbins, "On the Farm, A Disaster That Wasn't," *New York Times*, 16 October 1988; Philip Shabecoff, "Draft Report on Global Warming Foresees Environmental Havoc in U.S.," *New York Times*, 20 October 1988; "Worldwide Effects of the Drought," *(Newark) Star Ledger*, 27 June 1988.

36. *Another Revolution in U.S. Farming?* (Washington: Agricultural Economic Report No. 441, ESCS, USDA, 1979), 42–75.

37. Donald Worster, *Dust Bowl: The Southern Plains in the 1930s* (New York: Oxford University Press, 1979), 223–24.

38. Mary Fund, *Water in Kansas: A Primer* (Whiting, KS: Kansas Rural Center, 1984), 7.

39. Lloyd E. Dunlap, Edwin D. Gutentag, and James G. Thomas, "Use of Ground Water During Drought Conditions in West-Central Kansas," report prepared for USGS 1979 Spring Meeting, Garden City, Kansas.

40. Data received in May 1986 and May 1987 from the Garden City, Kansas, and Liberal, Kansas, offices of the Soil Conservation Service, USDA.

41. Data based on interviews, 5 May 1988, with Andy Erhart, agricultural advisor, and Al Rauhut, sales manager, Henkle Drilling and Supply Company, Garden City, Kansas, and confirmed by data from Kenny Ochs, Gigot Irrigation Company, Garden City, and from the USDA Soil Conservation Service office in Garden City.

42. The interviews are an essential element in the author's forthcoming book-length study, *Ogallala: Water for a Dry Land*, scheduled for publication in 1992 by the University of Nebraska Press.

43. Interview with Wayne Wyatt, Manager of the District, in Lubbock, Texas, May 1987. See also *Rules of High Plains Underground Water Conservation District No. 1*, 1954; Frank A. Rayner, *Government and Groundwater Management* (Lubbock TX: High Plains Underground Water Conservation District No. 1, 1975), 1–4, 10; "The Case for Local Regulation," *The Cross Section* 28 (December 1982): 1–4; and Frank L. Baird, *District Groundwater Planning and Management Policies on the Texas High Plains: The Views of the People* (Lubbock: High Plains Underground Water Conservation District No. 1, July 1976), 4–5.

44. See Opie, Ogallala: Water for a Dry land, esp. chapter 5.

45. Data provided by the Liberal, Kansas, office of the Soil Conservation Service, USDA, May 1987. See also "Groundwater Supply Problems," in *Revised Management Program III: Rules and Regulations, and Policies and Standards* (Garden City, KS: Southwest Kansas Groundwater Management District No. 3, 1986), 6–7; Kansas Water Office, *Agricultural Water Conservation: Irrigation Plan Guidelines* (enclosed in letter to "Fellow Kansans" from the Kansas Water Office, Topeka, dated 18 December 1986); Kansas State Board of Agriculture, Division of Water Resources, "Administrative Policy No. 88–3" (attached to letter dated 12 December 1988, addressed to UAII County Conservation Districts); Earl B. Shurtz, *Kansas Water Law* (Wichita: Kansas Water Resources Board, 1967); and see Mary Fund and Elise Watkins Clement, *Distribution of Land and Water Ownership in Southwest Kansas* (Whiting KS: Kansas Rural Center, 1982).

46. Quoted in Kip Lowe, "Groundwater Future a Continuing Concern," in the *Colby (Kansas) Free Press*, 15 June 1990; see also "Groundwater District Halts Water Rights," *Atwood (Kansas) Citizen-Patriot*, 22 February 1990, and reports and publications by Northwest Kansas Groundwater Management District Four, and the author's interviews of the District Manager, Wayne Bossert, in April 1991.

47. Quoted in Sue Shellenbarger, "U.S. Farmers Face an Easier Row to Hoe," *Wall Street Journal*, 25 October 1988.

48. See the sequence of articles in the *Wall Street Journal*: Bruce Ingersoll, "Drought Likely to Bring Down Acreage of Harvest to Record Low This Century," 27 September 1988; Carlee R. Scott, "Drought Lingers as Threat Winter Wheat Crop," 12 December 1988; Sue Shellenbarger, "Unforgiving Climate of Kansas Is Punishing Winter Wheat Again," 14 March 1989; Bruce Ingersoll, "U.S. Sees Wheat Stocks at 17-Year Low Unless Rains Temper Drought's Effect," 7 April 1989, and Sue Shellenbarger and Bruce Ingersoll, "Second Drought in a Row Is Threatening 40% of Farm Belt, Some Western States," and "Wheat

Futures in Kansas City Expected to Climb Following U.S. Prediction of 8% Drop in Harvest," both in 12 May 1989. The *New York Times* missed badly in its front-page story by Keith Schneider, "Serious Drought Seen as Unlikely in U.S. This Year," 20 February 1989; but it turned around by November: William Robbins, "Wheat Crop Faces Threat of Drought," 29 November 1989.

49. Telephone interview with Jeff Schmidt, District Conservationist, Soil Conservation Service, Liberal, Kansas, 21 June 1989.

50. William Robbins, "Winter Wheat Farmers Fear Second Year's Crop Failure," *New York Times*, 2 January 1990.

51. Irvin Molotsky, "Drought Has Eased, U.S. Reports," *New York Times*, 13 September 1988.

52. *Scientific Assessment of Climate Change* (Geneva: Intergovernmental Panel on Climate Change, WMO, UNEP, 1990); Paul E. Waggoner, "U.S. Water Resources Versus an Announced But Uncertain Climate Change," *Science* 251 (1 March 1991): 1002.

53. "Using Forests to Counter the 'Greenhouse Effect,'" *Science*, 26 February 1988, 973.

54. "Is the Greenhouse Here?" *Science* (5 February 1988): 559–61; Philip Shabecoff, "Global Warmth in '88 Is Found To Set a Record," *New York Times*, 4 February 1989; "Do We Know Enough to Act," National Research Council, *Current Issues in Atmospheric Change* (Washington, D.C.: National Academy Press, 1987), 23–27.

55. Philip Shabecoff, "The Heat Is On," *New York Times*, 26 June 1988; Philip Shabecoff, "Global Warming: Experts Ponder Bewildering Feedback Effects," *New York Times*, 17 January 1989.

56. Jesse H. Ausubel, "Does Climate Still Matter?" *Nature* 350 (25 April 1991): 650. See also Emmanuel Le Roy Ladurie, *Times of Feast, Times of Famine: A History of Climate Since the Year 1000*, trans. Barbara Bray (Garden City, NY: Doubleday, 1971).

57. Stephen H. Schneider, "The Greenhouse Effect: Science and Policy," *Science* 243 (10 February 1989): 771.

58. Author's visit to NCAR, May 1986.

59. Edward Lorenz, "The Predictability of Hydrodynamic Flow," *Transactions of the New York Academy of Sciences II* 25/4 (1963): 409–32, quoted in Gleick, *Chaos: Making of a New Science*, 25.

60. Reported in Richard A. Kerr, "Report Urges Greenhouse Action Now," *Science* (1 July 1988): 23–24; see also "Climatologists See Sharp Temperature Swings Over the Next Decade," (Newark) *Star-Ledger*, 29 July 1988.

61. Michael Oppenheimer, "How to Cool Our Warming Planet" (Op-Ed), *New York Times*, 23 July 1988.

62. "Is a Climate Jump in Store for Earth?" *Science* (15 January 1988): 259.

63. John Noble Wilford, "His Bold Statement Transforms the Debate On Greenhouse Effect," *New York Times*, 23 August 1988; but also see Philip Shabecoff, "U.S. Data Since 1895 Fail To Show Warming Trend," *New York Times*, 26 January 1989.

64. Stephen H. Schneider, "Climate Modeling," *Scientific American* 256/3 (May 1987), pp. 77, 80; and see also Michael H. Glantz and Jesse H. Ausubel, "The Ogallala Aquifer and

Carbon Dioxide: Comparison and Convergence," *Environmental Conservation* 11 (summer 1984): 123–31; Thomas R. Karl, Richard R. Heim, Jr., and Robert G. Quayle, "The Greenhouse Effect in Central North America: If Not Now, When?" *Science* 251 (1 March 1991): 1058–61.

65. Schneider, "The Greenhouse Effect," 772; see also J.I. Hanchey et al., *Preparing for Climate Change: Proceedings, Washington, D.C., 27–29 October 1988* (Rockville, MD: Government Institutes, 1988), 394–405.

66. Robert Lewis, "Global Warming: The Cold Facts," *(Newark) Star-Ledger*, 5 February 1989; William K. Stevens, "With Cloudy Crystal Balls, Scientists Race to Assess Global Warming," *New York Times*, 7 February 1989.

67. See the critique by Raymond J. Supalla, Robert R. Lansford, and Noel R. Gollehon, "Is the Ogallala Going Dry?" in *Journal of Soil and Water Conservation* (November-December 1982): 310–14; and see Michael H. Glantz, Barbara G. Brown, and Maria E. Krenz, *Societal Responses to Regional Climate Change: Forecasting by Analogy* (Boulder: ESIG/EPA Study, Environmental and Societal Impacts Group, NCAR, 1988), 13.

68. Ric Jensen, "Are Things Warming Up? How Climate Changes Could Affect Texas," *Texas Water Resources* 15 (Spring 1989); see also Judith Clarkson and Robert King, *Global Warming and the Future of Texas Agriculture: Impacts and Policy* (Austin: Texas Department of Agriculture, 1989); Daniel Dudek, "Economic Implications of Climate Change Impacts on Southern Agriculture," *Proceedings of the Symposium on Climate Change in the Southern U.S.: Future Impacts and Present Policy Issues* (Washington, D.C.: USEPA, 1987); Norman Rosenberg, "Drought and Climate Change: For Better or Worse?" in *Planning for Drought: Toward a Reduction in Societal Vulnerability* (Boulder, CO: Westview Press, 1987).

69. R.F. Diffendal, Jr., "Plate Tectonics, Space, Geologic Time, and the Great Plains," *Great Plains Quarterly* 11 (spring 1991): 93–94.

70. Michael H. Glantz and Nicolai Orlovsky, "Desertification: A Review of the Concept," *Desertification Control Bulletin* 9 (December 1983): 15–21; see also Donald A. Wilhite and Michael H. Glantz, "Understanding the Drought Phenomenon: The Role of Definitions," *Water International* 10 (1985): 111–20; Michael H. Glantz, "Politics, Forecasts and Forecasting: Forecasts are the Answer, but What was the Question?" in *Policy Aspects of Climate Forecasting*, ed. Richard Krasnow (Washington, D.C.: Resources for the Future, 1994), 81–95; Michael H. Glantz, "Drought Follows the Plow," *The World & I* (April 1988): 208–13; and Jonathan G. Taylor, Thomas R. Stewart, and Mary Downton, "Perceptions of Drought in the Ogallala Aquifer Region," *Environment and Behavior* 20 (March 1988): 150–75.

71. Glantz and Orlovsky, "Desertification," 15.

72. Ric Jensen, "Are Things Warming Up?"; see also Judith Clarkson, "Global Climate Change and Its Implications for Agricultural Productivity in Texas," *Grassroots* (fall 1988): 3, 21–24.

73. Howard Witt, "Canadians Fear U.S. Water Grab," *Chicago Tribune*, 11 July 1988.

74. Quoted in "The Heat is On" (cover story), *Time*, 19 October 1987, p. 59.

75. "The Greenhouse Effect," *(Newark) Star-Ledger*, 7 June 1988.

76. Quoted in "The Heat is On," *Time*, 19 October 1987, p. 63.

77. V. Ramanathan, "The Greenhouse Theory of Climate Change: A Test by an Inadvertent Global Experiment," *Science* 15 (April 1988): 293–99.

78. See Paul E. Waggoner, ed., *Climate Change and U.S. Water Resources* (New York: Wiley-Interscience, 1990).

79. Schneider, "The Greenhouse Effect," 771–81.

80. "Just Enough to Fight Over," *Time*, 4 July 1988, p. 14.

81. See Thomas S. Kuhn, *The Structure of Scientific Revolutions*, 2nd ed. (Chicago: University of Chicago Press, 1962, 1970).

82. Jim MacNeill, "Strategies for Sustainable Economic Development," *Scientific American* 158–59, 163–64.

83. Wayne Bossert of Kansas District No. 4 estimates that pumpage in northwest Kansas ranges from 515 an acre-foot for natural gas to 530 an acre-foot for electricity, alongside capital costs on top of energy costs.

84. Pierre R. Crosson and Norman J. Rosenberg, "Strategies for Agriculture," *Scientific American* 261 (September 1989): 128.

85. A. Maass and R.L. Anderson, *And the Desert Shall Rejoice: Conflict, Growth and Justice in Arid Environments* (Cambridge MA: MIT Press, 1978), 5; see also F. Lee Brown et al., "Water Reallocation, Market Proficiency, and Conflicting Social Values," in *Western Water Institutions in a Changing Environment* (Napa, CA: John Muir Institute, 1980); and Kenneth Boulding, *Western Water Resources: Coming Problems and the Policy Alternatives* (Boulder, CO: Westview Press, 1980).

86. See the discussions in Sagoff, *The Economy of the Earth: Philosophy, Law, and the Environment* (New York: Cambridge University Press, 1988); see also Willard D. Cochrane, *The Development of American Agriculture* (Minneapolis: University of Minnesota Press, 1981), 137, 183–86, 320.

87. Cochrane, *Development of American Agriculture*, 137.

88. Ibid., 321–24.

89. But see Earl O. Heady, "National and International Commodity Price Impacts," in *Water Scarcity: Impacts on Western Agriculture*, ed. Ernest A. Engelbert (Berkeley: University of California Press, 1984), esp. 280–91.

90. See William Lockeretz, ed., *Environmentally Sound Agriculture* (New York: Praeger, 1983); Gordon J. Douglass, *Agriculture Sustainability in a Changing World Order* (Boulder: Westview Press, 1984); Joel Sokloff, *The Politics of Food* (San Francisco: Sierra Club Books, 1985); Clyde Kiker, "Comments," *Agriculture and Human Values* 313 (summer 1986): 71–74; and John Opie, "The Precarious Balance: Matching Market Dollars and Human Values in American Agriculture," *Environmental Professional* 10 (spring 1988): 36–45.

91. Cook, "The Environmental Era of U.S. Agricultural Policy," 366.

92. Boschwitz, "Building a Conservation-Centered Farm Policy," 451.

93. Ibid.

94. MEY and MSY are reviewed in another context by Arthur F. McEvoy, "Toward an

Interactive Theory of Nature and Culture: Ecology, Production, and Cognition in the California Fishing Industry," in *The Ends of the Earth: Perspectives on Modern Environmental History*, ed. Donald Worster (Cambridge: Cambridge University Press, 1988), 219–29.

95. Sandra Postel, *Water for Agriculture: Facing the Limits* (Washington, D.C.: The World-Watch Institute, World Watch Paper 93, 1989), 12, 40.

PERMISSIONS

Dan Flores, "Spirit of Place and the Value of Nature in the American West," reprinted from *Yellowstone Science* (Spring 1993), 6–10.
Robert MacCameron, "Environmental Change in Colonial New Mexico," reprinted from *Environmental History Review* 18 (Summer 1994), 17–39.
Dan Flores, "Bison Ecology and Bison Diplomacy: The Southern Plains from 1800 to 1850," reprinted from *Journal of American History* 78 (September 1992), 465–85. Copyright © 1992 by Organization of American Historians.
James E. Sherow, "Workings of the Geodialictic: High Plains Indians and Their Horses in the Region of the Arkansas River Valley, 1800–1870," reprinted from *Environmental History Review* 16 (Summer 1992), 261–84.
E. Gregory McPherson and Rene A. Haip, "Emerging Desert Landscape in Tucson," reprinted from *The Geographical Review* 79 (October 1989), 435–49.
Dorothy Zeisler-Vralsted, "Reclaiming the Arid West: The Role of the Northern Pacific Railway in Irrigating Kennewick, Washington," reprinted from *Pacific Northwest Quarterly* 84 (1993), 130–39.
Thomas R. Dunlap, "Wildlife, Science, and the National Parks, 1920–1940," reprinted from *Pacific Historical Review* 59 (2), 187–202. Copyright © 1990 by the American Historical Association, Pacific Coast Branch.
Richard West Sellars, "Manipulating Nature's Paradise: National Park Management Under Stephen T. Mather, 1916–1929," reprinted from *Montana: The Magazine of History* 43 (Spring 1993), 2–13.
Mark Harvey, "Battle for Dinosaur: Echo Park Dam and the Birth of the Modern Wilderness Movement," reprinted from *Montana: The Magazine of Western History* 45 (Winter 1995), 32–45.
Daniel Pope, "'We Can Wait. We Should Wait.' Eugene's Nuclear Power Controversy," from *Pacific Historical Review* 59 (3), 349–73. Copyright © 1990 by the American Historical Association, Pacific Coast Branch.
F. Lee Brown and Helen M. Ingram, "The Community Value of Water: Implications for the Rural Poor in the Southwest," reprinted from *Journal of the Southwest* 29 (Summer 1987), 179–202.

Mary Pardo, "Mexican American Women Grassroots Community Activists: 'Mothers of East Los Angeles,'" reprinted from *Frontiers: A Journal of Women Studies* 11 (1990), 1–7.

John Opie, "The Drought of 1988, the Global Warming Experiment, and Its Challenge to Irrigation in the Old Dust Bowl Region," reprinted from *Agricultural History* 66 (Spring 1992), 279–306; Copyright © 1992 by *Agricultural History*.

Index

Abert, James W., 80; on Indian horses, 100, 105
Acequia system, 51, 223, 228; local management of, 231
Ackermann, Diane: on evolution/topography, 33
Activism: Catholic Church and, 257n5; women and, 251, 255; working-class, 250–51
Adams, Ansel, 189
Adams, C. C., 148, 154
Adams, Charles, 169, 171
Agriculture: capitalistic, 39; commercial, 43; rural poor and, 226. *See also* Subsistence agriculture
Ainsworth, establishment of, 129
Albright, Horace M., 150, 154, 156n2; on coyotes, 153; Lane Letter and, 174n14; on park development, 163; winter olympics and, 164, 173n12
Albuquerque, 44; climate in, 55; environmental changes in, 57; livestock in, 45
Albuquerque Journal, on community value of water, 234–35
Alcalde mayor, land grants and, 52
Alfred A. Knopf, Brower and, 189, 190
Allen, Keith, 266
All the Pretty Horses (McCarthy), quote from, 32

Alltucker, John, 214
American Alpine Club, 188
American Association for the Advancement of Science, 169
American Civil Liberties Union, EWEB and, 206
American Society for Environmental History, 8, 179
American Society of Mammalogists, 148, 166
Anaya v. Public Service Company of New Mexico (1990), 177
Anderson, Clinton P.: Echo Park and, 191
Anderson, Raymond L.: on water allocation/equity, 230; on water as resource, 227; on water/economics, 279
Animal Communities of North America (Shelford), 151
Animal Ecology (Elton), 151
Anthony, Harold, 148
Anthony, Scott J., 91
Anza, Juan Bautista de: Comanche policy and, 65
Apaches, 67; bison-hunting lifestyle of, 68; ecological changes and, 58; High Plains and, 94; horses and, 97 (table); population of, 76; threats from, 44. *See also* Kiowa-Apaches; Plains Apaches
Appalachian Mountain Club, 188

293

Arapahos, 70; in Arkansas Valley, 94; at Big Timbers, 102; bison and, 73, 79; Great Plains and, 72; horse/robe trade and, 73; horses and, 95, 97 (table), 100, 101, 107; massacre of, 91, 92–93; population of, 76; Southern Plains and, 66. *See also* Southern Arapahos
Arbor Day, 119, 120
Arizona Daily Star: Arbor Day and, 120; tree-planting campaign by, 119
Arizona Native Plant Society, water conservation and, 123
Arkansas Agency, 106
Arkansas Valley, Indian peoples in, 94
Arroyos, formation of, 55, 56
Astor, John Jacob: robes for, 89n49
Atomic Energy Commission, Echo Park and, 186
Atomic Industrial Forum, EWEB and, 204
Attneave, Chris, 178, 206, 215
Audubon Society: management and, 148; predators and, 153
Ausman, La Verne G., 262–63
Austin, Mary, 19

Babbitt, Bruce: on water policy changes, 223
Bacon, Francis, 14
Bailey, Liberty Hyde, 14
Bailey, Vernon, 149
Baker, Alton, Jr., 209
Baker, Gary, 268; on soil moisture levels, 275
Baker, T. Lindsay: robe estimates by, 89n49
Ballantine Books, Bryerton and, 209
Bears, management of, 167
Beauty, Health, and Permanence (Hays), 178
Beckwith, E. G., 106; on Big Timbers, 103
Bedichek, Roy, 19
Beneficial use, 234, 271
Bennett, John: on adaptation theory, 94
Bent, Charles: robe trade and, 89n48; trading post of, 73

Bent, George, 103; trading post of, 96; watering sites and, 99
Bent, William, 79; on Comanches/smallpox, 80; trading post of, 96
Bent's Fort, 80; trade at, 79, 89n48; winter forage at, 101
Berry, Wendell: place as pause and, 34
Berthoud, Edward L., 106
Beyond the Hundredth Meridian (Stegner), 189
Big Timbers, 99; degradation of, 102–3, 106; winter camping at, 103
Biomes, 10, 32
Bioregionalism, 33, 35
Biosystems, 9, 10; controlling/dominating, 14
Bison: carrying capacity for, 78; climate cycle and, 78; cultural uses of, 75; decline of, 77, 88n40; evolution of, 69; horse and, 77; hunting, 72; mortality rate for, 74, 77; Plains Indians and, 74, 75; proliferation of, 74; wolf and, 74, 77, 86n32, 88n44; at Yellowstone, 167
Black blizzards, 262
Black Kettle, 91
Blyth & Co., EWEB and, 205
Bolin, Bert: on climatic problems, 276
Bollaert, William: on Texas Comanches, 88n39
Bonneville Dam, 202
Bonneville Power Administration (BPA), 202, 217; electrical system planning by, 205; EWEB and, 205
Boone and Crockett Club, 166
Borchert, Jay: on droughts/dust bowl, 261
Bossert, Wayne: on pumpage costs, 288n83
Botkin, Daniel, 9; on life/environment, 10
Boulding, Kenneth: on water as resource, 227
Boyle, R. V.: on horses/short grasses, 99
Boyle Heights, prison for, 245, 248, 249, 250
BPA. *See* Bonneville Power Administration

Bradley, Harold: Echo Park and, 192; film by, 187–88
Bradley, Richard: Echo Park and, 187
Brandon, William: on hyper-Indians, 68
Brant, Irving: Echo Park and, 192
Bretherton, Francis: greenhouse effect and, 262
Broeker, Wallace, 273
Brooks, Charles, 208
Brower, David, 8, 190; Echo Park and, 181, 187, 189, 193; film by, 193; float trips and, 188
Brown, Bill: on bison cultural uses, 75
Brown, F. Lee, ix; on water/money, 178
Bryant, Harold, 150
Bryerton, Gene: investigative series by, 209, 219n53
Buffalo Ranch, herd at, 166–67
Bureau of Biological Survey, 148, 158n15; management by, 165; predator control and, 149, 166; refuges by, 154
Bureau of Entomology, 165, 168
Bureau of Fisheries, 165, 168
Bureau of Land Management, Echo Park and, 182
Bureau of Plant Industry, management and, 165
Bureau of Reclamation, 184; CRSP and, 183; dams and, 185; Echo Park and, 181, 182, 191–92, 193, 194
Burlingame, E. C., 134

Caddoans, 71, 87n37
Cahalane, Victor, 150
Calvin, William: on consciousness/culture, 12
Canjilon, water for, 231
Caprock Canyonlands (Flores), 2
Carless Atom, The (Novick), 206
Carnegie Desert Laboratory, 119
Carrying capacity, 74; changes in, 77; of Great Plains, 70, 84n14; rainfall and, 78
Carson, Rachel, 8, 14

Castas, 44–45, 46, 52
Castillo, Aurora: on coalition building, 251–52; on MELA, 249; on motherhood, 250
Cebolla, water for, 231
Central Arizona Project, 233
Central Utah Project, Echo Park and, 183
Chacon, Fernando de, 46, 47
Changes in the Land: Indians, Colonists and the Ecology of New England (Cronon), 41
Chaos, 27n36; place and, 17
Chapman, Oscar: Echo Park and, 186
Cheyennes, 70, 105, 106; in Arkansas Valley, 94; at Big Timbers, 102; bison and, 73, 79; Great Plains and, 72; horse/robe trade and, 73; horses and, 95, 97 (table), 100, 107; population of, 76; Southern Plains and, 66. *See also* Southern Cheyennes
Chicago Board of Trade, grain trading on, 263
Chino, Wendell: on Indian water, 232
Chivington, John M., 105; massacre by, 92–93
Chouteau, August Pierre: horse/robe trade and, 72
Ciboleros, 78
Citizens for the Orderly Development of Electricity (CODE), 210
Clean Air Act (1963), 8
Clements, Frederick: climax-ecosystem theories of, 17
Climate, changes in, 55–56, 63n47, 84n16
Coalition Against the Prison, 250, 257n6, 258n26
Cochrane, Willard W., 280
CODE. *See* Citizens for the Orderly Development of Electricity
Colonias, water for, 231
Colorado River, development of, 222
Colorado River Aqueduct, 182
Colorado River Compact (1922), 182, 194, 222–23

Colorado River Storage Project (CRSP), 191, 198n47; Echo Park and, 185, 187, 194–95; support for, 183
Columbian Exchange, impact of, 41, 42–43, 53
Columbia River, hydropower from, 202
Comanches, 44, 65, 106; in Arkansas Valley, 94; at Big Timbers, 102; bison and, 73, 78, 94; divisions of, 67–68; ecological changes and, 58; epidemics/pandemics and, 89n50; ethnobotany of, 84n17; horses and, 88n39, 94, 95, 96, 97 (table), 99–101; hunting by, 68, 70; land and, 79; population of, 75–76, 89n50; smallpox epidemic and, 79–80; Southern Plains and, 68–69, 71, 72; starvation for, 76; trade with, 79, 96
Commodity value, water and, 227, 234–35, 238
Community value, water and, 227, 228–33, 234–35, 236
Community voice, 253
Consciousness: biological evolution and, 12; culture and, 12; current theories of, 12–13
Conservation, 113, 281; water, 122–23, 125, 278–79
Conservation and the Gospel of Efficiency (Hays), 8
Convention on Trade in Endangered Species (1973), 155
Cook, Kenneth A.: on social contact, 280
Cooper, Thomas, 114, 133, 139, 144n24; irrigation and, 135, 136; Kennewick canal and, 134–35, 141; NPI and, 137; reclamation projects and, 132
Corn: drop in, 263; water for, 270
Corps of Engineers, 184; water importation and, 276
Coyotes: elimination of, 153; elk and, 152
Crane, Stephen: quote of, vii
Cronon, William, viii, 59n3; culture/landscape and, 109n8; on nature/human society, 41
Crop insurance, 263, 265
Crosson, Pierre R.: on water conservation, 278–79
CRSP. *See* Colorado River Storage Project
Cuerno Verde, 65
Culture, 20, 246; consciousness and, 12; environment and, 2, 40; landscape and, 109n8; place and, 2; water and, 221
Curtis, Richard, 213

Darwin, Charles: Spencer and, 6
Da Vinci, Leonardo: Nature and, 11
Davis, W.W.H., 48
Dawkins, Richard, 11; on consciousness, 12
Dawn Lake, 229, 233
Deforestation, 48–49, 53, 57, 58; problems with, 51
Dell Haven Irrigation District, 131, 133, 135
De Mun, Jules: horse/robe trade and, 72
Dennert, Richard: on consciousness/culture, 12; on workings of mind, 12–13
Denver Post, Echo Park and, 183
Department of Agriculture: food prices and, 264; on grain, 263, 265
Department of Commerce, water importation and, 276
Department of Corrections: Boyle Heights meeting and, 245; strategy by, 257n10
Department of the Interior, Echo Park and, 182
Descartes, René, 14, 36
Deseret News, Echo Park and, 183
Desert landscaping, 117, 123–25
Deukmejian, George: ELA prison and, 244–45
DeVoto, Bernard, 19; Echo Park and, 186, 187
Dice, Lee, 148, 150, 152
Dinosaur National Monument, 185, 186; Echo Park and, 181, 183, 187; films about, 187, 188; protecting, 187, 191, 194
Discordant Harmonies (Botkin), 9

Diseases, 19–20, 77, 89n50; impact of, 79–80, 81
Dixon, Joseph, 148, 150, 158n15
Drought, 57, 228; anticipating, 261–62; greenhouse effect and, 264; groundwater and, 269; horse forage and, 98 (table); horses and, 101; impact of, 55, 79; irrigation and, 268; overcoming, 263, 266. *See also* Great Drought of 1988
Drury, Newton, 186, 197n18; Echo Park and, 185
Dunlap, Thomas, ix; on national parks/animal populations, 114–15
Dust Bowl, 261, 262, 263, 266, 268, 269, 270; greenhouse effect and, 276; irrigation and, 275

East Los Angeles: environmental struggles in, 255; have nots in, 259n46
Echo Park: battle over, 182, 183, 187, 188, 189, 190, 192, 94–96; CRSP and, 185, 194–95; Hetch Hetchy and, 194; preservation of, 181, 191; publicity for, 187–88, 190–91
Ecological Society of America, 169, 175n31
Ecology, 1, 55; animal, 147; monumentalism and, 37; Native American, 67; tourism and, 171
Ecology of the Coyote in the Yellowstone (Murie), 152
Ecosystems, 10, 32; culture and, 40; deterioration of, 17, 275–76; diversity/complexity of, 43
EFPC. *See* Eugene Future Power Committee
Eggert, Charles: film by, 188
Eisenhower, Dwight D.: Echo Park and, 191, 193
Elliott, Howard, 138, 139; canal repair and, 145n38; irrigation and, 140, 141
Elton, Charles, 151
Emory, W. H.: Big Timbers and, 103; on Indian horses, 105
Encinias, Judge: quote of, 177
Endangered Species Act (1966), 8, 155

Endangered Species Act (1969), 155
Endangered Species Act (1973), predator reintroduction and, 155
Engels, Friedrich, 15
Enlarged Homestead Act (1909), 69
Environment, 1, 54, 60n4, 94; culture and, 2, 40, 177, 178–79; demography and, 43–44; economic activity and, 47–49, 56; factors of, 3, 23n11, 55–59; life and, 10; linear model for, 43; nature and, 10; people in, 9–16; place and, 14; race/gender/culture and, 177, 178–79
Environmental Defense Fund, 8
Environmental determinism, 7, 34
Environmental history/historians, viii, 14, 65; approaches for, 1, 5, 6; comeback for, 8; conservation movement and, 113; criticism of, 16; methodologies/ideologies of, 18; Native American, 66; place and, 16; West and, 3–9, 18–21
Environmentalism, 5; middle-class, 177, 178; women and, 255
Environmental Review, publication of, 8
Epidemics, 77, 79–80, 81, 89n50; climate and, 19–20
Erosion: damage from, 266; increase in, 55, 56
Eugene Future Power Committee (EFPC): effectiveness of, 214, 215, 216; EWEB and, 207, 208, 217; fund raising by, 219n43; health/safety dangers and, 213; petition campaign and, 210, 212; university community and, 216
Eugene Register-Guard, 214; Bryerton in, 209, 219n53; EFPC and, 219n43; EWEB and, 204; petition campaign and, 212
Eugene Water and Electric Board (EWEB): BPA and, 205; EFPC and, 217; Governor' Task Force and, 211; growth strategy of, 205; nuclear power and, 202–7, 209, 213, 214–15; opposition to, 206, 207, 208, 209–11, 213; petition campaign and, 212; rebuttal by, 208; Trojan plant and, 209, 216

Everglades National Park, wildlife watching at, 154
Evernden, Neil: on human-Nature dualism, 13, 15; Nature and, 11
Evolution, 26n33; topography and, 33
EWEB. *See* Eugene Water and Electric Board
Ewers, John C.: on Comanches/smallpox, 80
Export Enhancement Program, spending by, 264
Extensive farming, 277, 280

Fading Trails (Interior), 154
Farming, extensive/intensive, 277, 280, 281
"Fauna of the National Parks" (Wright), 150
Federal Power Commission, power withdrawals and, 191
Filthy Fifties, drought during, 261
First nature, second nature and, 15
Fitzpatrick, Thomas, 76
Flader, Susan, 8
Flint Hills, environmental history of, 3, 5
Flores, Dan, viii–ix, 2, 40; on environment/region/place, 1
Food costs, changes in, 263, 264
"Food Habits of the Coyotes of Jackson Hole, Wyoming" (Murie), 152
Forage, 101, 266; horse, 98 (table)
Ford, Richard: on Tewas/land, 53
Fort Belknap Indians, water-rights lawsuit by, 232
Fort Lyon, 91, 92, 106; environmental impact of, 103
Fowler, Jacob, 72, 76, 95, 106; Big Timbers and, 99, 103
Fradkin, Philip L.: on West/water distribution, 221
Freeways, communities disrupted by, 252–53
Friends of the Earth, 8
Frontier thesis. *See* Turner thesis

Gabrielson, Ira: Echo Park and, 181
Gadsden Purchase (1854), Tucson and, 118
Gaia, 108n3; biological systems and, 9; geodialectic and, 93
Gallegos, Hernan, 45
Game Management (Leopold), 151
Ganado mayor/menor, 45
Geodialectic, 93, 108, 108n3; horses and, 96, 97–98; keeping pace with, 94
Giustina, Ehrman: EWEB and, 203
Glacier National Park, 161, 163, 184; dam proposal for, 184, 185; hatchery at, 168; wolves in, 155
Glacken, Clarence, 35
Glantz, Michael H.: on ecosystem deterioration, 275–76
Glen Canyon Dam, 191; Echo Park and, 183, 187, 195
Global climate modeling, 274–75
Global warming, 262, 274
Gofman, John, 212
Gongora, Cruzat y: grant nullification by, 52
Goodnight, Charles: on bison, 86n31
Gould, Stephen Jay: on contingency, 27n36; on evolution, 17; quote of, 1
Governor's Task Force, EWEB and, 211
Grains, drop in, 263
Grand Canyon: dam proposal for, 185, 192, 193; Echo Park and, 195
Grand Canyon National Monument, 185
Grand Canyon National Park, 164, 185; headquarters building at, 163
Grand Coulee Dam, 202
Grant, U.S., III: Echo Park and, 187
Grassroots politics, 252; in Mexican-American communities, 257n5; requirements of, 250; women and, 243–44, 245–46, 255
Grazing, impact of, 56, 98–99
Great Drought of 1988, 261, 267; heat wave and, 274; impact of, 265; relief from, 265; surviving, 271–72
Great Plains: bison on, 69, 84n16, 86n31;

carrying capacity of, 70, 84n14; culture and, 20; drought on, 262; human population peripheral to, 75; place and, 19
Great Plains, The (Webb), 7
Green History of the World, A (Ponting), 35
Greenhouse effect, 270; drought and, 264; Dust Bowl and, 276; impact of, 273–74, 275, 277–78
Greenpeace, 8
Green Revolution, 264, 265
Gregg, Josiah: on Rio Grande Valley, 48
Gribbin, John, 13
Grinnell, Joseph, 150, 153, 157n9; ecology and, 158n15; predator control and, 149
Groundwater: depletion of, 122, 275, 280; as disposable resource, 280; drought and, 268, 269; irrigation with, 277; pioneer-stage use of, 279; policies on, 271
Groundwater Management Act, 124
Gutierrez, Juana, 258n26; on family/activism, 251; on freeways/relocation, 252–53; on MELA involvement, 246, 248
Gutierrez, Martin, 258n26
Gutierrez, Ricardo, 258n26
Gutierrez, Veronica, 248, 258n26

Haip, Renee A., ix, 114
Hansen, James E.: on global warming, 274, 275
Harvey, Mark W. T., ix, 178
Hatch, Bus, 188
Hays, Samuel, 8; on middle-class environmentalism, 178
Hecht, Alan, 273
Hermiston Project, 139
Hetch Hetchy, 192; controversy about, 189; Echo Park and, 194
Hides, hunting for, 66, 76
High Plains: drought on, 267–68, 272–73; irrigation on, 267, 269–70, 277, 278, 280, 281, 282, 284n34
High Plains Indians: bison and, 95; environmental adaptation by, 107–8;

horses and, 40, 94, 95, 97, 102, 104, 107; hunting by, 104
Hill, James J., 131
Hispanics: community value of water and, 234; land and, 39; water and, 223–29, 236, 237, 238
History: mental/physical laws of, 23n11; society and, 13; spirit of place and, 37; writing, 12–13
Hobbes, Thomas: on human cultures, 16; Nature and, 11
Hoffman, John S., 273
Hogan, Elizabeth, 213
Holaday, Joseph, 178, 206, 212; EWEB and, 207–8
Hollis, Larry, 267
Hoover Dam, 182, 186, 190, 223
Horses: carrying capacity and, 77, 78; distribution of, 72, 98; environmental change and, 40; forage needs for, 98–99, 98 (table); geodialectic and, 96, 97–98; herding, 100–1, 102, 107; impact of, 45, 67, 95; maintaining, 98, 99–100; wintering problems for, 99–100, 101
Hughes, Donald, 8
Human beings: biological commonalities of, 11–12; environments and, 12; natural world and, 41
Human-Nature dualism, 11, 13, 16; first/second nature and, 15
Hunt, Robert and Eva: on water/social organization, 232
Hunting, 40, 47, 68, 70, 72, 104; for hides/robes, 66, 76, 80; sustainable, 66
Huver, Charles, 212
Hydro-Thermal Power Plan, 202, 205

Ickes, Harold, 186
Idea of Wilderness, The (Oelschlaeger), 35
Indian peoples: bison-hunting practices of, 40; as ecologists, 66; horses for, 40; land and, 39, 40; water for, 224–26, 227, 229, 232, 237, 238

Ingram, Helen M., ix; on water/money, 178
In Search of Schrödinger's Cat (Gribbin), 13
Intensive farming, 277, 280, 281
Irrigation, 53, 59–60n4, 119, 123, 134, 222, 277, 284n34; community involvement in, 231; drought and, 268; Dust Bowl and, 275; fairness in, 230; golden age of, 268; municipal/industrial water needs and, 236; railroads and, 129, 130; Spanish, 50–51
Isle Royale National Park: preservation and, 155; wildlife watching at, 154; wolves at, 159n35
Izaak Walton League, 188; Echo Park and, 187, 192

Jackson, Wes: sustainable agriculture and, 39
Jackson Hole National Monument, 184
Jacobs, Wilbur, 8
Johnson, Lyndon B., 8
Joint Power Planning Council (JPPC), 202, 205
Jumanos, 71, 83n9

Kaibab National Forest, 156n5; deer die offs in, 151
Kansa Indians, 5; Konza and, 10, 12, 15
Kansas Groundwater Management District No. 3, 271
Kelso, Maurice: on water as resource, 227
Kennewick: irrigation in, 129, 132, 133, 137, 139; Northern Pacific at, 130
Kennewick canal, 131, 140, 141, 145n41
Kennewick Commercial Club, 134
Kennewick *Courier*, on NP/Kennewick canal, 141
Kennewick Irrigation District, 141, 142, 145n41
Kent Decree (1910), 232
Keppel, Fred, 210
Kern, Larry: drought and, 271–72

Kings Canyon National Park, 184; dam proposal for, 185
Kiona canal, 131, 133, 139, 140, 141
Kiowa-Apaches: bison and, 79; population of, 76, 87n37
Kiowas, 78; at Big Timbers, 102; bison and, 68, 73, 90n52; Great Plains and, 72; High Plains and, 94; horses and, 94, 96, 97 (table); origin myths of, 83n9; population of, 76, 87n37; Southern Plains and, 66, 71; trade with, 79, 96
Kiowa Calendar, 76, 79
Konza, 9, 17; environmental history and, 21; Kansa Indians and, 10, 12, 15
Konza Prairie Research Natural Area, 5
Kramer, James: drought and, 271
Kuhn, Thomas, 38

LaFarge, Oliver, 230
Lake Mead, 186, 190
Land: analysis of, 54–55; conserving, 52; deterioration of, 51; frontier societies and, 60n5
Land grants, 52; water use and, 225
Land Institute, cattle forage and, 266
Land institutions, Spanish, 43, 50–51
Landscape: changes in, 20; culture and, 109n8; samples of original, 196n11; water-efficient, 122–23. *See also* Desert landscaping
Landscaping in the Desert, 124
Land Use, Environment and Social Change: The Shaping of Island County, Washington (White), 41
Lane, Franklin K., 162, 174n14
Lane County Building Trades Council, 210
Lane County Democratic Party, EWEB and, 206
Lane County Labor Council, nuclear power and, 211
Lane Letter (1918), 165, 171, 174n14
Las Madres de la Plaza de Mayo, 249, 259n29

Lawrence, D. H.: on spirit of place, 32
League of Women Voters: Novick and, 206; petition drive and, 210
Left Hand, 91
Leopold, Aldo, 14, 148, 151
Levy, Jerold: on human population/Southern Plains, 75
Lewis, Richard S.: on nuclear protestors, 207
Life force, 9
Life-systems/communities, 10
Life zones, in New Mexico, 42
Little Ice Age, climate change and, 63n47
Litton, Martin: Echo Park and, 188
Livestock, impact of, 45, 46
Living Wilderness, Echo Park and, 187
Llano Estacado, 80; bison on, 81
Locke, John, 14
Lodore Canyon, 181, 185
Long, Stephen: horse/robe trade and, 72
Lorenz, Edward, 274
Los Angeles Times, Echo Park and, 188
Louses, problems with, 102
Lovelock, James: on geodialectic, 93; life force and, 9
Lower Yakima canal, 132, 133, 139, 140
Lower Yellowstone Project, 139
Luxan, Diego Perez de, 45; on Rio Grande Valley, 48
Lyon, Thomas: on anti-dualistic notions, 14

McAtee, W. L.: Murie and, 152
McCall, Tom, 211
MacCameron, Robert, viii; on subsistence agriculture, 39–40
McCarthy, Cormac: on spirit of place, 32
McCorkle, J. S.: on horses/short grasses, 99
MacDonald, A. R., 119
McKay, Douglas: Echo Park and, 190
McKibben, Bill, 10
McLean, Dewey M., 273
McPherson, E. Gregory, ix, 114
Maass, Arthur: on water allocation/equity, 230; on water as resource, 227; on water/economics, 279
Malin, James, 7, 8, 9
Mammoth Cave National Park, dam proposal for, 185
Mammoth Hot Springs, zoo at, 167
Management era: opportunity/challenge in, 223–25; transition to, 224
Man and Nature (Marsh), quote from, 113
Man Who Killed the Deer, The (Waters), quote from, 229
Marcy, R. B., 96; on Comanches/herding, 100–1; on horse wintering problems, 99, 100; on Plains Indian horses, 104
Margulis, Lynn, 9–10; on human continuity, 16
Marine Mammal Protection Act (1972), 155
Market culture: as second nature, 16; sustainable agriculture and, 27n40
Marsh, George Perkins, 9; environmental movement and, 113; quote of, 113
Martin, M. P., 138, 145n33; irrigation and, 140, 141; NPI and, 140
Marx, Karl, 14; on human cultures, 16; on people/natural history, 15; on second nature, 16
Material culture, 20; ecological impacts of, 49; Spanish, 43
Mather, Stephen T., 115, 168; appointment of, 173n3; on bison, 167; national parks and, 163; National Park Service and, 171, 172; park development and, 162, 169; predator control and, 166; preservation and, 170, 171, 174n14; tourism and, 163, 164, 165
Matheson, Scott: on water, 231
Maximum economic yield (MEY), 281
Maximum sustainable yield (MSY), 281
Maxon, Andy: lawsuit by, 217
Mayordomo (ditch boss), 228
Means, Florence Cranell: on Hopis/drought, 229
MELA. *See* Mothers of East Los Angeles

Mellen, C. S., 135; irrigation and, 132, 136, 138; NPI and, 137
Melosi, Martin, ix, 114
Merchant, Carolyn: culture/landscape and, 109n8
Metes and bounds, using, 50
MEY. *See* Maximum economic yield
Mills, Harlow: on predator reintroduction, 155
Milwaukee Journal, on Echo Park, 187
Mining, impact of, 47–48
Minorities, environmental degradation and, 254
Mitchell, Robert: Sioux/Cheyennes and, 105
Molina, Gloria, 245, 248, 257n10, 258n12
Molina, Martha, 248; MELA and, 257–58n12
Monuments, formation/preservation of, 28n43, 37
Moody, Ralph: on water conflicts, 229
Moretta, John: Boyle Heights prison and, 248–49; MELA and, 250, 253, 254
Morris, Desmond: on territoriality, 33
Mother Ditch, La Acequia Madre, The (LaFarge), 230
Motherhood, defining, 250, 255
Mothers of East Los Angeles (MELA): class/ethnic identity and, 251–52; formation of, 244, 245–46; motherhood and, 250, 255; political involvement by, 243, 245, 246, 255–56, 258n13; school activities and, 247; self-criticism by, 252
Mount Rainier National Park, 163, 184; hatchery at, 168
MSY. *See* Maximum sustainable yield
Muir, John, 14, 188, 189; Hetch Hetchy and, 192
Murie, Adolph, 150, 154, 158n23; coyotes and, 151; preservation and, 152; on wolves/Dall sheep, 152
Murie, Olaus, 149, 158n23; coyotes and, 151, 152
Museum of Vertebrate Zoology, 149, 158n15

Nash, Gerald, ix
Nash, Roderick, 8, 181
National parks: animal populations and, 114–15; construction in, 163, 164; establishing, 37, 148; management of, 169–70, 172; purpose of, 147, 161; tourism and, 163, 164–65, 170, 172, 173n2; water rights for, 226
National Parks Association (NPA), 183, 184, 196n1; Drury and, 186; Echo Park and, 181, 187, 192, 195
National Park Service: bison and, 167; cannonball parks/primeval parks and, 184; conservation and, 170; controversies and, 161, 168–69; coyotes and, 153; creation of, 161; Echo Park and, 191; forest policy statement by, 175n27; management and, 148, 167–68, 172; mission of, 147, 161; predator control and, 149, 155, 165, 166; preservation and, 155, 162, 171, 173n2; tourism and, 164–65, 170
National Park Service Act (1916), 185, 190, 191
National Parks magazine, 184, 187
National Resources Inventory, Great Plains irrigation and, 269
National Wild and Scenic Rivers Act (1968), 8
National Wildlife Federation, Echo Park and, 192
Natural History of the Senses, A (Ackermann), 33
Nature: considering, 3; controlling/dominating, 14, 15; environment and, 10; Europeans and, 11; Greeks and, 11, 24n18; human societies and, 11, 41; non-human, 5; people and, 20; religion and, 54; Romans and, 24n18; social constructions of, 26n35; as storehouse, 35. *See also* First nature; Human-Nature dualism; Second nature
Nature Conservancy, Konza Prairie and, 5
Navajo Indian Irrigation Project (NIIP), 223, 238n7

Navajos, 45; threats from, 44
NAWAPA plan, 276
Nematodes, problems with, 102
Newell, Frederick H., 139
New Mexico Game Protective Association, Leopold and, 148
Newton, Sir Isaac, 14; Nature and, 11
New York Times: on crop insurance, 263; on Echo Park, 187; on greenhouse warming, 277
New York Zoological Society, 166
NIIP. *See* Navajo Indian Irrigation Project
North American Buffalo, The (Roe), debate about, 67
North Cascades National Park, 193
Northern Pacific Irrigation Company (NPI), 138–39, 140, 142; Kennewick and, 137–38
Northern Pacific Railroad, 114; irrigation and, 129, 130, 139; Kennewick canal and, 138, 141, 142; in Yakima Valley, 129, 130, 136–37, 139; Yellowstone National Park and, 170
Northern Rocky Mountain Wolf Recovery Team, 155
Northwestern Improvement Company (NWI), 137, 142; Kennewick and, 133, 134; land applications to, 136
Northwest Kansas Groundwater Management District Four, 271
Northwest Utilities, 217
Novick, Aaron: EWEB and, 212
Novick, Jane, 178, 212, 213; EWEB and, 207, 215; on nuclear power, 210; opposition by, 206, 217
Novick, Sheldon, 206, 213
NPA. *See* National Parks Association
NPI. *See* Northern Pacific Irrigation Company
"Nuclear Dilemma, The" (Bryerton), 209
Nuclear power: benefits of, 201–2; support for, 210–11

NWI. *See* Northwestern Improvement Company

Oakes, John: Echo Park and, 192
Odum, Eugene: climax-ecosystem theories of, 17
Oelschlaeger, Max, 31, 35
Official Story, The (film), 249
Ogalalla aquifer, 276, 277; beneficial use of, 271; conserving, 281; crop yields and, 281–82; depletion of, 270–71, 275; drought and, 269, 270; future of, 282; irrigation water from, 262, 268, 278, 279, 280
Oklahoma State Water Board, 271
Olson, Sigurd: Echo Park and, 181
Olympic National Park, 184
Oñate, Juan de, 53
One-Eye, 91
Opie, John, ix, 9, 179; *Environmental Review* and, 8
Oppenheimer, Michael: on global warming, 274
Oregon Environmental Council, nuclear power and, 210
Oregon Environmental Foundation, 212
Oregon State University Extension Service, warm-water irrigation and, 208
Organic Act (1916), 162, 168, 170, 171, 172nn2, 3; National Park Service and, 161
Osborn, Alan J.: on horse distribution/High Plains, 98
Otis, Raymond: on Hispanics/water, 228
Overgrazing, 53, 58; problems with, 46, 51, 57
Owl Woman, 96

Packard, Fred: Echo Park and, 181, 185
Paez Hurtado, Juan, 52
Palmer Drought Index, using, 272
Pardo, Mary, ix, 179
Parker Dam, 182

Parkman, Francis: on Arapahos/horses, 101
Parks, Keith, 217; on EWEB, 215; on Novick, 212; on nuclear power opposition, 218n20
Pasco: irrigation at, 132; Northern Pacific at, 129–30, 131
Pawnees: bison and, 77; horse herding by, 109n9
People-Nature dualism, 20, 21; climax-ecosystem theories and, 17–18
Perils of the Peaceful Atom, The (Curtis and Hogan), 213
Perkins, John, 8
Perlman, David: float trip and, 187
Phillips, John, 153
Pike, Zebulon, 95, 99
Pinchot, Gifford, 195; natural resource use and, 170
Pine Cone, Leopold and, 148
Pinkett, Harold, 8
Pioneer stage, 279, 280
Place, 1, 3; chaos/complexity and, 17; culture and, 2; environment and, 14, 16; individualism/capitalism and, 2; as pause, 34; region and, 32; space and, 32, 34; West as, 18–21. *See also* Sense of place; Spirit of place
Plains Apaches: horses and, 94, 96; trade with, 96
Plazas, 44
Point Reyes National Seashore, 193
Ponting, Clive, 35
Pope, Daniel, ix, 178
Porter, Charles O., 204, 209
Portland General Electric Company, 202
Powell, John Wesley, 189
Predator control, 149, 154, 155, 165–66
Preservation, 152, 155, 162, 171, 173n2; of monuments/parks, 28n43; progress and, 149; scientists and, 156n1
Price, Byron: EWEB and, 203, 204; on nuclear power opposition, 218n20
Primeval parks, 184, 186

Prior appropriation, 7, 226, 278
Province of New Mexico: census of, 61n10; as defensive/missionary outpost, 60n5
Pueblo Indians, 35, 45; acculturation of, 52–53; cultural borrowing from, 53; environmental impacts and, 46, 53; irrigation by, 59–60n4; land and, 42, 53, 60n4; population of, 61n10; religion/environment and, 54; subsistence agriculture and, 53–54; trade with, 47; water development and, 223
Pueblo Revolt (1680), 44, 49, 61n10, 68

Quality-of-life issues, women and, 254, 255

Railroads: development of, 129; irrigation and, 130; national parks and, 163
Rainbow Bridge, protecting, 195
Rainfall, 19–20, 69; carrying capacity and, 78
Ramanathan, V.: on warming, 277
Ramos, Lucy: on community voice, 253
Ranchos, 44
Rancho Seco plant, 215
Reagan, Ronald: disaster-relief measure and, 263
Reclamation Service, 138, 282; Yakima Valley and, 141
Red Rock Lakes Migratory Waterfowl Refuge, 154
Redwoods National Park, 193
Region, 1; place and, 32
Regionalism, dominance of, 32, 33
Resources: caring for, 233–34; control of, 177; exploitation of, 177
Reynolds, Steve, 231
Rio Abajo, 44, 45, 46; environmental impacts at, 51
Rio Arriba, 44, 46; environmental impacts at, 51; Spanish population at, 49–50
Rio Grande Compact, 237, 241n56
Rio Grande Valley: climate in, 55; colonial

population of, 43, 44–45, 56; deforestation in, 48–49; environmental change in, 41–42, 43, 45, 47–49, 51, 56–57; hunting in, 47; poor minorities in, 225; storage facilities/delivery systems on, 223
Rio Puerco, 46
Robes, hunting for, 76, 80
Robles, Erlinda, 248; on husband/activism, 250–51; on motherhood, 250; on school/mothers, 247
Roe, Frank: debate about, 67
Roosevelt, Franklin D., 192, 195
Roosevelt, Theodore, 138
Rosa, Richard, 212
Rosenberg, Norman J.: on water conservation, 278–79
Rothman, Hal: on environmental/cultural change, 61n8
Rousseau, Jean Jacques, 14
Rural poor: agriculture and, 226; in southwest, 225; water and, 222, 224, 225–27, 235–36, 237
Russell, Carl P., 150

Sacramento Municipal Utility District, Rancho Seco and, 215
Sagebrush Rebellion, 195
Sagoff, Mark: on shared values, 279
St. Vrain, Ceran, 73; robe trade and, 89n48
Salt Lake Tribune, Echo Park and, 183
Salt River Indians, water rights and, 232
Salt River Project Water Users Association, 232
Sampson, W. C.: irrigation and, 137
Sand Creek, 106; horses at, 107; massacre at, 91–92, 93, 105, 107; winter camping at, 103
Sandoz, Mari, 19
San Francisco Chronicle, on Echo Park, 187
San Juan-Chama diversion project, 223, 238n7
Santa Barbara, oil spill at, 8
Santa Cruz de la Cañada, 44; environmental changes in, 57; livestock at, 45

Santa Cruz River, 120
Santa Fe, 44; climate in, 55; environmental changes in, 57; livestock at, 45
Santa Fe Trail, 103, 106
Satanta, quote of, 39
Saturday Evening Post, DeVoto and, 186
SAWARA. *See* Southern Arizona Water Resources Association
Schaffer, William M.: chaos theory and, 17
Schmidt, Jeff: on sand/static electricity damage, 268
Schneider, Stephen, 274; on greenhouse effect, 273, 276, 277; studies by, 275
Scientific Monthly, The: Adams in, 169
Scientists and Citizens for Atomic Power, 210, 212
Second nature: first nature and, 15; market culture as, 16
Sectionalism, dominance of, 32
Sekul, Joseph D.: on COPS, 257n5
Selfish Gene, The (Dawkins), 11
Sellars, Richard, ix, 115
Semple, Ellen: on biomes/ecoregions, 32
Sense of place, 19; developing, 31, 34; individualism/capitalism and, 2
Sequoia National Park, 161, 164; headquarters building at, 163
"Shall We Let Them Ruin Our National Parks?" (DeVoto), 186
Shelford, Victor, 151
Shepard, Paul: 25n25; quote of, 31
Sherwood, Morgan, 8
Shields, Wayne: CODE and, 210
Shifferd, Kent: *Environmental History* and, 8
Shoshones, 67, 68; ethnobotany of, 84n17
Sierra Club, 183; Bradley film and, 188; Echo Park and, 181, 187, 192, 193, 195; Eggert film and, 188; environmental racism and, 179; float trip by, 187, 188; Hetch Hetchy and, 189; nuclear power and, 210
Sierra Club Bulletin, Echo Park and, 187

Significance of Sections in American History, The (Turner), 32
Silent Spring (Carson), 8
Simmons, Marc, 46
Sioux, 105; hunting by, 68
Skinner's Butte, war memorial at, 214, 216
Smith, Adam, 36, 282
Snake River Pumping Project, 132
Snyder, Gary: place as pause and, 34
Social change, 6; women and, 255
Social regulation, public values and, 279
Soil Conservation Service (USDA), 262, 265; ghost crops and, 272
Sonora Desert, Tucson and, 117
Southern Arapahos, horses and, 94, 96, 97
Southern Arizona Water Resources Association (SAWARA), water conservation/desert landscaping and, 123
Southern Cheyennes: horses and, 94, 96, 97; land and, 20; massacre of, 91, 92–93
Southern Plains: adaptation strategies on, 109n9; agricultural census for, 69–70; bison on, 69, 70, 74, 76, 79, 80, 81, 84n16; carrying capacity of, 70; ecological changes on, 76; game on, 73; human population on, 75, 87n33; hunting on, 76; Indians on, 75–76; rainfall on, 69
Space, place and, 32, 34
Spencer, Herbert, 6
Spirit of place, 33, 34–35; defining, 31, 32, 37; Euro-American culture and, 36; investigating, 32; western, 37, 38
Split Mountain Canyon, 185
Sporhase decision, 237
Staked Plain, 80; bison on, 81
Stegner, Wallace, 19; Echo Park and, 195–96; on Powell, 189; "Wilderness Letter" and, 190
Sterling, Keir, 8
Stevenson, Adlai: Echo Park and, 192
Stone and Webster (engineering company), 211; nuclear steam supply system by, 205
Straus, Michael: Echo Park and, 185

Strongyles, problems with, 102
Subsistence agriculture, 39–40, 53–54; environmental diversity and, 43
Sunnyside Project, 139
Sunshine Climate Club, 121
Sustainable development, 39, 114, 280, 282; market culture and, 27n40; water laws and, 278
Swadesh, Frances: on social relations, 54

Tarr, Joel, 114
Territorial Imperative, The (Morris), 33
Tewas, land and, 53
Texas Groundwater Management District No. 1, 270–71
Tharp, William, 101–2
Third World: environmental degradation and, 254; labor migration from, 254
"This Atomic World" (EWEB), 204
This Is Dinosaur (Stegner), 189, 190, 193
Thompson, Ben H., 150
Thoreau, Henry David: on anti-dualistic notions, 14
Three Mile Island, accident at, 215
Tieton Project, 139
Tiffany, John: EWEB and, 203, 208
Tohono O'odham: community value of water and, 234; harvest ceremony of, 229; water for, 232
Topography, 35; evolution and, 33
Topophilia, 33–34
Topophilia (Tuan), 2
Tourism: ecology and, 171; national parks and, 163, 164–65, 170, 172, 173n2
Traces on the Rhodian Shore (Glacken), 35
Treaty of Fort Holmes (1835), 78
Treaty of Guadalupe Hidalgo (1848), 66, 226
Trojan plant, 202, 203, 209, 216; support for, 215
Truman, Harry: Echo Park and, 192
Tuan, Yi-Fu: on place as pause, 34; on place/environment/culture, 2; on topophilia, 33–34

Tucson: desert landscape in, 125; environmental transformation for, 114, 117, 118–22; tourism in, 121, 125
Tucson Water, 119
Turner, Frederick Jackson, 7, 32; environmental history and, 6; Webb and, 18
Turner thesis, 6, 32

Undesirable projects, opposition to, 244, 245
United Nations Food and Agricultural Organization, drought and, 264–65
United Nations Nairobi Conference on Desertification, 275
U.S. Army: horses of, 104, 105–6; predator control and, 165
United States Forest Service: Adams on, 171; Echo Park and, 182; management and, 165; natural resource use and, 170
United States Geological Survey, EWEB and, 204
Upper Arkansas Agency, per capita horse ownership/tribe in, 97 (table), 107 (table)
Urban settlement, geography of, 18–19
Utah Water and Power Board, CRSP and, 183
Utes, 52, 102; horses and, 68; threats from, 44

Vecsey, Christopher: on religion/nature, 54
Vigas, 48, 49
Vigil, James, Jr., 245, 257n10
Villalobos, Frank, 257n7; Boyle Heights prison and, 248, 249
Villas, 44, 46, 53; environmental changes at, 57–58

Wall Street Journal: on EFPC, 215; on Great Drought, 267
War is Kind (Crane), quote from, vii
Washington Post, on Echo Park, 187

Washington Public Power Supply System, 205
Waste incinerators, locating, 244, 245
Water: battles over, 224–25, 230; commodity value and, 234–35, 238; community and, 228–30, 232; community value and, 227, 230–36; conserving, 52; control of, 221–22, 225–27, 236, 241n57; cultural values and, 221; economics and, 232, 278–79; emotional relationship to, 234; money and, 221, 232; rural poor and, 222; social organization and, 232; as special resource, 227–28, 229; symbolism of, 227, 228
Water allocation, 226; equity and, 230; fairness of, 230; Spanish law on, 229
Water conservation, 278–79; promoting, 122–23; urban vegetative resources and, 125
"Water is Different Syndrome" (Kelso), 227
Water laws, sustainable development and, 278
Water management, 224, 239n9, 276; irrigation farmers and, 270–71
Water politics, disreputable image for, 237–38
Water rights, strata of, 226
Waters, Frank: on water/community survival, 232; on water/myth, 229 Watkins, Arthur: CRSP and, 198n47; Echo Park and, 191
Webb, Walter Prescott: on environment, 6, 7; Great Frontier and, 36; Turner and, 7, 18
Wedel, Waldo R., 99
West: environmental history of, 20; as place, 18–21; shaping of, 20; studying, 9
Western history, environmental history and, 3–9, 18
West Umatilla Project, 139
Wheat: drop in, 263, 266; drought and, 272; prices for, 264; water for, 270

Wheelock, T. B., 96, 106; on Indian horses, 105
White, Richard, viii, 59n3, 66, 109n9; culture/landscape and, 109n8; environmental history and, 6; on nature/human society, 41
White Thunder, 96
Whitfield, John, 79
Whitfield, W. D., 96
Wilderness Act (1964), 8, 182, 190
Wilderness and the American Mind (Nash), 8
Wilderness Condition, The (Oelschlaeger), quote from, 31
"Wilderness Letter" (1960), 190
Wilderness River Trail (film), 188
Wilderness Society, 183; American West and, 184; Drury and, 186; Echo Park and, 181, 187, 192, 193, 195
Wild Life Division, 147; coyotes and, 153; preservation and, 155
Wilkinson, L. E., 211
Williams, Raymond: on Man/Nature, 25n20
Winters doctrine, 232
Wissler, Clark: on biomes/ecoregions, 32
Woeikof, Alexander Ivanovich, 9
Wolf: bison and, 74, 77, 86n32, 88n44; Dall sheep and, 152; eradication of, 166; population of, 77; restoration of, 155, 161
Wonderful Life (Gould), 27n36; quote from, 1
Wood, King, Dawson, Love and Sabatine (law firm), EWEB and, 205
World Watch Institute, agricultural resiliency and, 265

Worster, Donald, viii, 8, 26n31, 39, 269; culture/landscape and, 109n8; geodialectic and, 93; on market culture/sustainable agriculture, 27n40; on subsistence agriculture, 43
Wright, George, 150, 151, 154

Xeriscape, 114, 123

Yakima Commercial Club, 134
Yakima Irrigating and Improvement Company (YI&I), 130, 133, 137, 143n4; ditches by, 131; Kennewick and, 132, 134, 135
Yakima Valley, 131; irrigation in, 137; Kennewick project and, 134; Northern Pacific in, 129, 130, 136–37, 139; Reclamation Service and, 141
Yampa Canyon, 181
Yellowstone National Park, 37; bison at, 167; dam proposals for, 173n13; establishment of, 161; hatchery at, 168; hotel at, 163; Northern Pacific and, 170; preservation and, 155; tourism and, 167; wolf reintroduction in, 161
Yellow Wolf, 80, 91; horse raid by, 104
YI&I. *See* Yakima Irrigating and Improvement Company
Yosemite National Park, 37, 161; golf course at, 173n11; hatchery at, 168; Hetch Hetchy and, 189; Indian Field Days at, 164
Yosemite Valley, 164; development of, 161
Young, Stanley Paul, 149

Zahniser, Howard: Echo Park and, 181, 185, 193
Zeisler-Vralsted, Dorothy, ix, 114

Contributors

F. LEE BROWN, Professor. Department of Economics, University of New Mexico.

THOMAS DUNLAP, Professor. Department of History, Texas A&M University.

DAN FLORES, John Hammond Professor of History. Department of History, University of Montana.

RENEE A. HAIP was a graduate student in architecture at the University of Arizona.

MARK HARVEY, Associate Professor. Department of History, North Dakota State University.

ROBERT MACCAMERON, Professor. SUNY, Empire State College.

E. GREGORY MCPHERSON, Professor. Environmental Horticulture, University of California, Davis.

JOHN OPIE, Distinguished Professor. Department of History, New Jersey Institute of Technology.

MARY PARDO, Professor. Department of Chicano Studies, California State University, Northridge.

DANIEL A. POPE, Professor. Department of History, University of Oregon.

DICK SELLARS, Historian. National Forest Service.

JAMES E. SHEROW, Associate Professor. Department of History, Kansas State University.

DOROTHY ZEISLER-VRALSTED, Assistant Professor. Department of History, University of Wisconsin at La Crosse.